SWINE FEEDING AND NUTRITION

ANIMAL FEEDING AND NUTRITION

A Series of Monographs and Treatises

Tony J. Cunha, Editor

Distinguished Service Professor Emeritus
University of Florida
Gainesville, Florida

and

Dean, School of Agriculture
Professor of Animal Science
California State Polytechnic University
Pomona, California

Tony J. Cunha, SWINE FEEDING AND NUTRITION, 1977

W. J. Miller, DAIRY CATTLE FEEDING AND NUTRITION, 1979

Tilden Wayne Perry, BEEF CATTLE FEEDING AND NUTRITION, 1980

in preparation

Tony J. Cunha, HORSE FEEDING AND NUTRITION, 1980

SWINE FEEDING AND NUTRITION

Tony J. Cunha
Department of Animal Science
School of Agriculture
California State Polytechnic University
Pomona, California

ACADEMIC PRESS New York San Francisco London 1977
A Subsidiary of Harcourt Brace Jovanovich, Publishers

ACADEMIC PRESS, INC.
111 Fifth Avenue, New York, New York 10003

United Kingdom Edition published by
ACADEMIC PRESS, INC. (LONDON) LTD.
24/28 Oval Road, London NW1

Library of Congress Cataloging in Publication Data

Cunha, Tony J
Swine Feeding and nutrition.

(Animal feeding and nutrition)
Includes bibliographical references and index.
1. Swine–Feeding and feeds. I. Title.
SF396.5.C86 636.4′08′4 77-5694
ISBN 0–12–196550–3

PRINTED IN THE UNITED STATES OF AMERICA

80 81 82 9 8 7 6 5 4 3 2

To the late Mr. and Mrs. F. T. Roseberry
Los Banos, California

and to my wife Gwen and family

this book is dedicated in appreciation

Contents

5 Protein Requirements of the Pig

6 Carbohydrates and Fiber

7 Fatty Acids, Fat, and Energy

8 Water

Foreword

This is the first volume in a series of books on animal feeding and nutrition. This book discusses swine feeding and nutrition in detail. Others are being written on beef cattle feeding and nutrition, dairy cattle feeding and nutrition, poultry feeding and nutrition, and horse feeding and nutrition. These will be followed by sheep and goat feeding and nutrition and other pertinent works.

Proper feeding and nutrition are very important since feed constitutes a major share of the cost of animal production. Research and new developments on feeding and nutrition have been quite numerous in recent years. This has resulted in new feed-processing methods, changes in diets, and more supplementation with vitamins, minerals, and amino acids. New developments in the use of antibiotics and feed additives have made them even more essential as intensification of animal production occurs. All of these and other changes as well necessitate that top authorities in the field collate all available information in one volume for each species of farm animal. The volume of scientific literature is so large and its interpretation so complex that there will be a continuing need for summarizing and interpreting these new developments in up-to-date books.

There is a great deal of emphasis now on increasing the world's food supply. There are at least 500 million people who are seriously lacking in protein and calorie intake. There are additional hundreds of millions who suffer from malnutrition. Animals provide a very important share of the world's food intake. In the United States, for example, 44% of the food consumed comes from animal products. The developing countries have 60% of the world's livestock and poultry but produce only 22% of the world's meat, milk, and eggs. Better feeding, breeding, and management of their animal production enterprises would greatly increase their food supply. It is hoped that this series of books on animal feeding and nutrition will be of some assistance in the United States and in world food production.

To introduce this new series of books is a challenge and a privilege. It is my pleasure to write this first volume, to advise on this project, to thank the other authors working on books, and to especially thank the staff of Academic Press for their concerted effort in producing this series.

Tony J. Cunha

Preface

In 1957, I wrote a book bearing the same title. Because of the many advances made in the past twenty years, there was a need for a completely revised work. This book is the result. It provides information helpful to those interested in swine feeding and nutrition. It was designed to be especially valuable to beginners in swine production, to established swine raisers, and to those who are concerned, directly or indirectly, with swine feeding or nutrition. This book will be very helpful to feed manufacturers and dealers and others concerned with producing the many different nutrients, supplements, feeds, antibiotics, and other feed additives, as well as other ingredients used in swine diets. It will also be useful to county agents, farm advisors and consultants, veterinarians, and to teachers of vocational agriculture. The book was also designed to be of value to college and university students and teachers in courses on feeds and feeding, swine production, swine nutrition, and animal nutrition. The text contains basic information for students in these courses. Moreover, it contains many key references for those interested in obtaining further information on a particular subject.

In the first chapter the past, present, and what might occur in the next 50 years in the swine industry are discussed. Chapter 2 reviews the many factors that can affect nutrient requirements and needs. It helps explain why nutrient requirement standards are only a guide in formulating diets. Chapters 3 to 10 contain concise, up-to-date summaries of minerals, vitamins, protein, amino acids, carbohydrates, fiber, fatty acids, fat, energy, water, enzymes, and antibiotics and other antimicrobial compounds. The nutrient requirements of the pig are discussed and compared with National Research Council recommendations. Deficiency symptoms for all nutrients are discussed and most nutritional deficiencies are illustrated with photographs. The needs of the pig for various nutrients are presented to give the reader a basis for determining what good, well-balanced diets should contain. The practical application of this basic information is in each of these eight chapters.

The relative value of feeds for use in swine diets is discussed in Chapter 11. Feeds used in all areas of the world are reviewed. The last three chapters deal with feeding the baby pig, growing-finishing pigs, and the breeding herd. The advantages and disadvantages of early weaning and prestarter, starter, grower, finishing, gestation, and lactation diets are discussed. Sample diets are given from a number of universities which can be used as a guide in formulating diets. The use of silage and other high fiber feeds are discussed. These last three chapters make use of the basic nutrition information discussed in previous chapters.

It is hoped this book will be helpful throughout the world. My first book enjoyed worldwide distribution and was translated into Spanish. There is not much difference in the nutrient requirements of swine in different countries. The minor differences which may occur can be taken care of by the safety factor which is usually applied by the feed industry and swine producers making use of nutrient requirement standards. Thus, this book should be useful in all swine-producing areas of the world.

In preparing this book I have had the benefit of suggestions from many eminent scientists in the United States and abroad. I wish to express my sincere appreciation to them and to those who supplied photographs and other material used. I especially want to thank the following: G. E. Combs, H. D. Wallace, and J. H. Conrad (Florida); J. L. Krider and W. M. Beeson (Purdue); R. H. Grummer and G. Bohstedt (Wisconsin); J. E. Burnside (Georgia); V. W. Hays (Kentucky); D. E. Ullrey and E. R. Miller (Michigan); D. H. Baker (Illinois); E. T. Kornegay (VPI); V. C. Speer (Iowa); W. G. Pond (Cornell); L. E. Hanson and R. J. Meade (Minnesota); R. C. Wahlstrom (South Dakota); A. J. Clawson (North Carolina); H. S. Teague (USDA); E. R. Peo (Nebraska); H. H. Heitman (California); Homer Fausch (Cal Poly–Pomona); Russ Anderson (Cal Poly–San Luis Obispo); Jesse Bell (Fresno State); R. Braude (England); J. P. Bowland (Canada); A. S. Jones (Scotland); G. A. Lodge (England); I. A. M. Lucas (Canada); Dudley Smith (Australia); J. H. Maner (Brazil); and H. Clausen (Denmark).

Tony J. Cunha

1

Past, Present, and Future in the Swine Industry

The swine industry has progressed a great deal in the past 50 years. It has changed from producing a lard-type pig to a more meaty and heavier muscled animal. It has changed from an industry with little expertise to one in which many diets are now computer programmed, in which some sophisticated housing with temperature control is being used, and in which breeding practices and management have undergone many undreamed of changes since 1925. Some highlights of these changes follow.

I. FEEDING AND NUTRITION

Fifty years ago, the use of pasture was a must in producing pigs. It supplied vitamins, minerals, amino acids, and other factors neither the farmer nor the scientist had yet heard about. It covered up many of the deficiencies of corn which was the main grain used. At that time, many experiment stations recommended corn, good quality pasture, and water. In dairy regions skim milk was used. Tankage was also used where available. Slop feeding of pigs was practiced widely. Some used charcoal and wood ashes. This type of feeding program was inefficient, and required at least 500 lb of feed/100 lb of gain under the best of conditions. A review of Professor W. W. Smith's book on ''Pork Production'' published in 1922 showed that under experimental conditions most universities achieved a 100-lb gain with about 450–500 lb of feed. However, the farmer probably achieved the same results with about 550 or more pounds of feed. It was not uncommon for some pigs to be from one to two years of age before they were marketed. Today, records at Swine Evaluation Centers show pigs producing 100 lb of gain with an average of about 300 lb of feed. The few top-achieving pigs in these tests are putting on 100 lb of gain with less than 250 lb of feed or about half what it took 50 years ago.

Pasture is no longer needed to finish pigs for market provided a well balanced diet is used. More economical gains can now be produced in complete confinement. The confinement diet is now supplemented with calcium, phosphorus, salt and the trace minerals. It is also fortified with vitamins A,D,E,K, riboflavin, niacin, pantothenic acid, B_{12} and choline. In some cases biotin and pyridoxine are added. The feed is also fortified with an antiobiotic or some other antimicrobial compound. In some cases, amino acid fortification is being used. Sophisticated operators are having these diets balanced by computer which provides them with the most profitable diets for the feeds and prices available at the time.

Unfortunately, there is still more to learn about handling sows in complete confinement to get as good results as can be obtained on pasture. But each year more progress is made in solving the problems of reproduction under complete confinement with no access to dirt and/or pasture.

Rate of gain has continually increased during the past 50 years. The best evidence for this continual increase comes from a summary of 30 years of research data kept by Dr. R. Braude of England at his research laboratory. From 1940 to 1970, the rate of gain of his pigs continually increased. Rate of gain in 1940, 1950, 1960 and 1970 was 1.28, 1.48, 1.59 and 1.75 lb daily respectively. Rate of gain has increased in the United States to the point where pigs now go to market at an average age of 6 months. The top-achieving pigs reach market weight at about 4 months of age. During the next 50 years, pigs should be going to market at an average age of about 4.5 months in the United States. Rate of gain should increase still more.

II. CROSSBREEDING

Fifty years ago straightbreeding was the main method of producing pigs. Now the use of rotational crossbreeding systems predominates. The sows are kept as crossbreds, and purebred boars of two to four different breeds are used in rotation. Crossbreeding is being used because the resulting hybrid vigor results in better sows that will farrow and wean larger numbers of pigs. In 1957, the Iowa Station reported that 3 or 4 breed crosses used in rotational crossbreeding produced about 40% more pork than straightbreeding to one breed of sire.

III. MULTIPLE FARROWING

Fifty years ago, farmers farrowed once a year with a few farrowing twice per year. They farrowed in the fall or spring, with the largest number of sows being farrowed in the spring. Now, the top producers are farrowing on a year-round basis. Some farrow every 2 months, and the large, sophisticated operations farrow every month. Farrowing more frequently provides for more efficient use of labor, equipment and facilities. It also provides pigs to be marketed throughout the year. Multiple farrowing has changed the market price for hogs so there is a more uniform price throughout the year as compared to seasonal highs and lows which used to be the case with one or two farrowing periods during the year. Fifty years from now, farrowing should occur throughout the year.

IV. MECHANIZATION AND HOUSING

In the early 1920's, very simple housing and little, if any, mechanization was used. Pigs were hand-fed, and slop barrel feeding was practiced widely. Now the better swine producers have housing with air ventilation and a few have temperature control during the winter and summer. Slotted floors and waste disposal systems are becoming a part of swine facilities. This increase in the use of mechanization and sophisticated housing provides for more efficient use of scarce and high priced labor, ease of handling and managing the animals, efficiency of operation, as well as maximum comfort for the pigs. Fifty years from now, most pigs will be exposed to less changes in environmental conditions than the humans taking care of them. Temperature will be controlled to provide optimum results during growth and reproduction.

V. EARLY WEANING

Fifty years ago, pigs were weaned at 8–12 weeks of age and sometimes even later. The weaning age started to decrease, and now some producers are weaning at 3 weeks of age. The majority who wean early, however, are weaning at 4–5 weeks of age. However, the trend is downward. Fifty years from now, pigs will be weaned very shortly after birth and the sow will serve primarily as an incubator for them. This will occur as litter size is increased considerably and as the technology becomes available to make it more profitable to wean pigs shortly after birth. A litter size of 14–16 pigs should be a common occurrence in the year 2025. Baby pigs will be raised in cages and fed milk replacers shortly after birth. This will decrease baby pig losses to just a few percent as compared to the 20–25% death loss which occurs between birth and weaning now.

VI. SIZE OF PRODUCTION UNITS

Fifty years ago, swine operations were small and run mostly in combination with other farm enterprises. Since then, swine operations have gradually increased in size. However, it has been only in the last 10–15 years that really large swine operations have developed extensively. There are now many 500–1000 sow units in the United States. Western Europe has a number of 1000 sow operations. In some of the state farms in Eastern Euorpean countries, as many as 6000 sows in one operation exist. Australia has a 4400 sow unit under one manager. However, the largest concentration of swine operations in the United States will continue to be 50–150 sow units for some time to come. This is the size unit which a farmer and his family can take care of themselves. These units are as efficient as the very large sow operations. This is because of the extra "tender-loving-care" the farm family can give their animals. This can save an extra pig or two at farrowing. Hired labor is scarce and good farm labor is even more so. Therefore, the family swine farm, which is operated efficiently, is not in danger of being squeezed out by the large units for many years to come.

VII. AGE OF SWINE PRODUCERS AND EFFECT ON SIZE OF PRODUCTION UNITS

In the early 1920's, young men were not reluctant to enter the swine business alone or in partnership with their parents. In the last 20 years, however, there has been less interest by young men toward entering the swine industry. They are more interested in other areas, especially in jobs with a 40-hour week. As a result, the average age of swine producers has been increasing. In a recent survey, Dr. James H. Bailey of South Dakota University found that only 12.5% of the swine producers were in the 20–30-year-old category. Over 63% of them were over 40 years of age. If this trend continues, the alternative is fewer swine farms and increasingly larger operations. Therefore, by the end of the next 20 years, a greater increase in large swine units will occur. By the year 2025, the small swine farm units will be relatively few in number.

VIII. DISEASES AND PARASITES

Diseases and parasites took a heavy toll 50 years ago. They are still a problem today, but much less so. Improper diesease and parasite control can still cut returns by about 20%. Some pigs are now being raised as SPF (specific pathogen-free) animals. If done properly, this eliminates atrophic rhinitis, virus pneumonia and other diseases. Parasite problems are being reduced by confinement feeding because there is less chance for completion of the parasite's life cycle. This will especially be the case as slotted floors or other methods of handling animals are developed to prevent contact with their feces. As confinement feeding increases, an increase in certain infectious diseases occurs. Eventually, however, preventive measures will be developed for combating them. Real progress will occur when diseases and parasites can be prevented rather than merely treated after they occur. Vaccination against viral, bacterial and parasitic diseases will increase. By the year 2025, disease and parasite problems will be minimal in the sophisticated and well-managed swine operations.

IX. PROGRESS IN PRODUCTION EFFICIENCY

A glance at Tables 1.1 and 1.2 shows the progress that has occurred when some of the old type diets were compared to modern day diets at Minnesota and Purdue Universities. These old type diets gave better results when tested now than if they had been fed under the conditions which existed 50 years ago. The results show, however, that pigs grow faster with less feed and with less cost per pound of gain with modern-day diets. The results indicate the tremendous value of modern feeding programs in swine operations.

Recently, researchers at Pennsylvania State University compared a diet the university recommended in 1921 to a 1971 diet. The results obtained are shown in the following tabulation:

Diet	Gain/day (lb)	Feed/100 lb gain (lb)
1921	0.84	361
1971	1.65	283

The results show a very large difference in performance. New technology developed in the next 50 years will likewise increase efficiency of production still further.

X. PROGRESS IN CARCASS QUALITY

Table 1.3 shows the changes which occurred with the winners at the National Barrow Show from 1951 to 1970.

TABLE 1.1
Growth and Feed Efficiency of 51-lb Pigs Fed Three Different Diets to 200 lb

Diet fed	Avg. daily gain	Feed/100 lb gain
1910	0.45[a]	6.98[a]
1930	1.20	3.85
1953	1.81	3.44

[a]These pigs were not doing well and so were changed to the 1953 diet when they weighed 135 lb. Data obtained from Dr. L. E. Hanson, University of Minnesota.

TABLE 1.2
Results of a 1909 versus a 1959 Diet in Dry Lot[a]

	1909 Diet	1959 Diet
Initial weight (lb)	44	44
Final weight (lb)	199	199
Number of days to reach 200 lb	135	91
Average daily gain (lb)	1.15	1.71
Average daily feed intake (lb)	4.90	5.47
Feed per pound of gain (lb)	4.25	3.20
Feed cost per pound of gain (cents)	11.4	9.0

[a]Data obtained from J. H. Conrad and W. M. Beeson, Purdue University.

TABLE 1.3
Average Measurements of National Barrow Show Winners

Period	No. animals	No. breeds	Length (in.)	Thickness back fat (in.)	% Ham	Loin eye (in.^{2})	Ham-loin index[a]
1951–1955	199	10	29.9	1.59	13.54	4.44	79
1956–1960	294	10	30.8	1.42	14.06	4.28	83
1961–1965	276	9	31.0	1.36	15.40	4.66	100
1966–1970	669	9	30.7	1.18	16.88	5.07	118
1970 Champion		—	30.8	0.70	18.64	8.05	167
Change from 1951–1970		—	+2.7%	−25.9%	+24.6%	+14.2%	+149%

[a]Index = 10 points/in.^{2} loin eye + 10 points/% ham over 10% of adjusted liveweight.

Especially noticeable in Table 1.3 is the increase in ham and in the ham-loin index plus the decrease in the thickness of back fat which has occurred. This is due to the changeover to a meatier, more heavily muscled pig in the ham and loin area. Also very interesting is how the 1970 champion barrow compares with the average of the other pigs. The 8.05 in^2 of loin eye area which it had is close to being twice as much as the average pig has in the United States now.

These data might be compared to those reported from the Indiana Swine Evaluation Station in Table 1.4. The data from the Indiana Swine Evaluation Station compare very favorably to those from the National Barrow Show during the 1965 period.

The data from the Florida Station in Table 1.5 show the big difference in various criteria between the averages obtained at the Center versus the top pen and the top pig. This example shows the opportunity available to increase rate of gain, feed efficiency, and carcass quality by swine producers.

Since carcass quality traits show a high rate of heritability, swine producers can make considerable progress in the future by evaluating the carcass quality of their animals and selecting for these traits. It is obvious they can make much more progress as one looks at the figures shown in Tables 1.3–1.6. The late Dr. W. A. Craft of Iowa State University reported the following percentage heritability estimates on carcass attributes: body length, 59; loin eye area, 48; thickness of back fat, 49; percent ham, 58; and carcass score, 46. These values indicate one can make rapid progress by selection for carcass quality. Proof of this is that Dr. H. Clausen of Denmark has, by breeding and selection, been able to produce a few pigs which have maximum carcass quality even though they are

TABLE 1.4
Indiana Swine Evaluation Station Data (1965)

	Number of test pigs	Average of all pigs	Range	
			Low	High
Feed per pound of gain (lb)	—	296	260	368
Chilled carcass weight (lb)	226	144	132	163
Ham weight (lb)	230	30.5	27.0	34.8
Loins (lb)	230	24.0	21.4	29.0
Percent of carcass as hams, loins	226	38.0	33.6	42.7
Back fat (in.)	231	1.38	0.96	1.73
Loin eye area (in.^{2})	231	4.47[a]	3.07	6.40[b]
Carcass length (in.)	231	30.8	28.0	31.8

[a]In 1959, the average loin eye area was 3.79 in.^{2}.
[b]Since these data were published they have had one pig with a 9.09 in.^{2} loin eye area.

TABLE 1.5
Records at Swine Evaluation Center at Live Oak, Florida from September 1, 1965, to August 31, 1970

	Average of all pens during 5 years	Top pen record	Top individual pig
Age at 200 lb (days)	162	124	111
Average daily gain (lb[a])	1.64	2.11	2.42
Feed per 100 lb gain (lb)	333	276	—
Back fat thickness (in.)	1.36	0.99	0.73
Carcass length (in.)	30.03	31.81	32.50
Loin eye area (in.^{2})	4.13	5.66	6.45
Percent lean cuts of carcass weight	53.61	60.31	62.60
Percent lean cuts of liveweight	—	43.93	45.56
Percent ham	—	17.74	18.91
Percent loin	—	13.54	14.27
Ham-loin percent	38.02	42.66	44.82
Dressing percent	—	75.35	77.67
Index score	106.1	159.0	—

[a]Average starting weight was 71 lb.

self-fed and consume all the feed they want. Previously, he had to limit the feed intake of pigs to obtain maximum carcass leanness and quality. So. U.S. producers need to pay more attention to evaluating their pigs and to selecting and culling according to carcass quality attributes.

The key to the future is in producing leaner pigs with less waste fat. But there is also a need for studies on the "eating qualities" of pork. These include determining the factors which affect flavor, juiciness, palatability, tenderness, aroma and other criteria which go into making pork cuts more appealing to the consumer. The United States is lacking this kind of information.

In the September 15, 1971 issue of *Hormel Farmer* the results obtained at the Swine Testing Stations of Iowa, Nebraska, Minnesota and South Dakota for the pigs fed there during the test period were summarized. The top results obtained are shown in Table 1.6 (all of these were for one pig except the percent ham which was for 4 pigs). The information in this table indicates how a top pig will perform. The data are important since what is attained by one pig can be achieved by others as programs are developed to accomplish these goals.

Table 1.7 shows an estimate of the average swine carcass characteristics in the United States from 1925 to those predicted for 2025. These figures were obtained by contacting swine authorities, meat packing company specialists and others. The data are an average or consensus of their replies. It is interesting to note that

TABLE 1.6
Results Obtained with Top Achieving Pigs in Midwest Testing Stations

Rate of gain (lb)	2.40
Feed per 100 lb gain (lb)	224
Carcass length (in.)	32.5
Back fat (in.)	0.85
Loin eye area (in.^{2})	8.10
Percent ham	17.8
Age at 220 lb (days)	129

the average figures are quite different from what the top pigs will do (see Tables 1.3–1.6). This shows room for carcass improvement which exists in the United States. It also shows the large difference which exists between the top swine carcass and the average carcass produced in the United States. It is interesting to look at the estimates which the various specialists gave as the best results to be expected by the year 2025. They were as follows: average back-fat thickness, 0.5 in; percent ham and loin, 55%; and loin eye area, 12 in^2. Many swine specialists think there is a limit to how much the ham, loin and rib-eye area can be increased. The pig must still have enough heart, lung and digestive tract capacity. The changeover to a heavy muscled pig also means that certain nutritional requirements may be altered. Therefore, in order to produce a high quality carcass and a high rate of reproduction, nutritional requirement studies need to keep pace with the new type pig being produced.

XI. EXCESS FAT IN SWINE

The waste fat problem in swine is very important. Feed efficiency will not be improved until excess fat is eliminated. The average pig in the United States requires 3.3–3.5 lb of feed per pound of gain from weaning to market weight. Yet the top pig in a 1971 Swine Evaluation Center in the United States required

TABLE 1.7
Estimates of Average Carcass Characteristics in United States

	1925	1975	2025
Average back fat thickness (in.)	2.5	1.6	1.2
Percent ham and loin	32	39	45
Loin eye area (in.^{2})	3	4.1	5.5

TABLE 1.8
Percent Fat, Lean and Bone in Market, Barrow and Gilt Carcasses[a]

	USDA grade number			
Separable components in carcass	1	2	3	4
Percent fat	38.2	45.5	49.5	54.8
Percent lean	50.9	43.3	40.7	36.9
Percent bone	10.9	11.2	9.8	8.3

[a]Florida Agricultural Experiment Station. Mimeo No. 70-3, 1969 by A. Z. Palmer, H. R. Cross and J. W. Carpenter. These data might vary some with pigs of different breeding.

only 2.24 lb of feed per pound of gain or about two-thirds as much feed. This is a good example of what producers of meat-type pigs can do toward decreasing feed needs. It should be stressed that about 90% of the barrows and gilts marketed in the United States have more separable fat than separable lean in their carcass. This is because only about 10% of the barrows and gilts marketed in the United States are USDA grade number 1 (see Table 1.8). USDA Marketing Bulletin No. 51 states that 8, 42, 36, 12 and 2% of the barrows and gilts slaughtered in the United States in 1968 were in USDA grades 1,2,3,4, and utility, respectively. It is estimated that instead of 8%, the figure is now 10%, or slightly over, for the barrows and gilts making USDA grade 1 in the United States. Therefore, greater emphasis must be placed on producing meat-type pigs by a larger number of the swine producers in the U.S.

The data in Table 1.8 show that only the pork carcasses in USDA grade number 1 have more separable lean than separable fat in their carcass. But, even in this grade, the carcass has 38.2% separable fat, which is still too high and can be decreased in the future.

In the grade numbers 2,3 and 4, all the carcasses have more separable fat than separable lean. In grade number 4, pork carcasses have 48% more separable fat than separable lean. This is a very wasteful and expensive carcass to produce and one the consumer objects to. In fact, the consumer objects to the excess fat in grades 2,3 and 4. Until carcass fat can be decreased to more acceptable levels, the U.S. consumer will not increase its level of pork consumption to an appreciable extent, if at all.

XII. REPRODUCTION IN MEAT-TYPE ANIMALS

A word of caution is needed on programs initiated to develop more of a meat-type pig. Some have been bred to the point where they are too heavy in the ham and loin area and too light in the front end. Some of these pigs do not have enough lung, heart, and digestive tract capacity and are encountering problems during the latter part of the growth period and especially during reproduction. Dr. H. Clausen of Denmark, who has done such a good job of developing the meat-type Danish Landrace, admits this has happened to some extent in Denmark. Now, they are trying to breed away from it. Therefore, as breeding programs are developed for producing a meat-type pig, special attention must be given to assuring that the animals can function properly throughout growth and also during reproduction.

Selection of a meat-type animal should not be done at the expense of meat quality. Care must be taken to insure that acceptable quality levels are maintained. Past experience with extremely heavy muscled, trim animals indicates that problems with carcass quality become more frequent. In the future, the economical animal will be one which has the correct combination of economic factors, i.e., gainability, cutability, and palatability. The animal must also be able to reproduce and not be susceptible to abnormal conditions which have been encountered. Pale soft exudative pork (PSE) and pork stress syndrome (PSS) may have resulted from selection for meat-type animals.

XIII. COST OF PORK PRODUCTION WILL DECREASE

The cost of feed represents about 60–80% of the cost of producing pork. It will vary depending on the size of operation, litter size, rate of gain, feed efficiency,

TABLE 1.9
Estimated Average for United States

	1925	1975	2025
Average daily gain from birth to market (lb)	0.70	1.2	1.8
Feed per 100 lb gain from birth to market (lb)	550	350	250
Number of pigs weaned per litter	5.12 (1924)	7.4	12.0[b]
Average weaning age (weeks)	10	6[a]	2[d]
Average age to market pigs (months)	11	6[a]	4.5
Pigs finished in confinement (%)	Very few	46[a]	90[c]
Sows fed in complete confinement away from pasture and dirt (%)	None	4[a]	75[c]

[a]Information obtained from a nationwide survey by Dr. James H. Bailey, South Dakota State University in 1971.

[b]This is assuming super-ovulation problems with sows will have been solved and many swine producers will be farrowing 14–16 pigs/litter and raising them in cages with milk substitutes.

[c]There will be areas, as in Europe now, where sows and pigs will still be fed on pasture even though land is scarce and population pressure great.

[d]The sophisticated swine producers will be weaning at 1 day of age whereas others will take as long as 2–3 weeks of age, and a few will wean at an older age.

leanness of pig, and many other factors. Cost per pound of gain has decreased during the past 50 years. It should decrease still further during the next 50 years as litter size is increased, death loss is curtailed, rate of gain and feed efficiency are increased, a leaner pig is produced, and as the efficiency of other production practices are increased (see Table 1.9).

Feed required per pound of gain will decrease as the amount of lean increases and as carcass fat content decreases in the future. High amino acid grains, such as opaque 2 corn, can cut the protein supplementation needed almost in half. Of course, much still remains to be learned in developing these grains so they will consistently reduce protein needs. The greatest increase in feed efficiency will come if litter size can be appreciably increased and especially if 14–16 pigs can be weaned per litter. If the present death losses of 20–25% between birth and weaning can be cut to just a few percent, this will likewise increase pork production efficiency considerably.

XIV. SUMMARY

Much progress has been made during the past 50 years in the swine industry in the United States. As great as this has been, even more can be accomplished in the next 50 years (Tables 1.7 and 1.9). The greatest area for improvement is in carcass quality (including eating quality as well). The average swine producer needs to do a better job of keeping abreast of changes occurring in the industry. The producer must also more quickly adapt new production and management techniques as they are shown to be practical, economical and of value to producing a better quality product to meet consumer needs. There is too much of a gap between the quality and performance of the top-achieving pig and the average pig in the United States. This area provides the greatest challenge, and, if accomplished, the goals set for the year 2025 could be increased even more. Almost all the goals for the year 2025 are already being met by a few pigs in the United States. A few years ago, the USDA estimated that the average farmer took 10 years to adapt a new research finding which was shown to be practical and economical. The United States has approached the time where this time lag cannot be afforded. A visit to the National Barrow Show and then to an average market, where pigs are being brought in for sale and slaughter, shows that there are still too many pigs being marketed that are not too different from the overfat pig of 20 or more years ago. Therein lies the challenge. Change must occur more quickly and with more producers, especially the average swine producer.

2

Problems in Supplying Feed Nutrients for the Pig

I. INTRODUCTION

Feed represents about 60–80% of the cost of producing hogs. These costs are lower when one is producing feeder pigs, and are a higher percent of total costs when one is finishing pigs. Thus, successful swine production requires a carefully planned and efficient feeding program. Diets that would suffice a few years ago are not adequate now (Fig 2.1). Swine production is becoming more intensified and specialized.

Through proper breeding and selection, rapid growing strains of swine have been developed. Many swine producers now plan to market their hogs at close to 5 months of age instead of the 7–8 months that was accepted practice a few years ago. Moreover, they are selecting and breeding for larger litters and are breeding their sows to farrow twice a year. In some cases, where producers practice early weaning, a sow produces about 5 litters in 2 years. Gilts are being bred to farrow their first litters at an earlier age than in the past, also.

II. NEED WELL-BALANCED DIETS

To meet this intensified and stepped-up hog production program, properly balanced, high quality diets must be fed. This means that nutritional needs of the pig for protein, minerals, vitamins, fats and carbohydrates must be met fully for profitable and efficient production. No matter how careful one is in his breeding, management, and disease control program (and these are very important), one cannot make a profit unless the hogs are fed properly. For these reasons, those concerned with swine feeding need to know the nutrient requirements of the pig, the characteristics of a good diet, and the nutritive value of the feeds used in swine diets. They also need to learn how to put all this information together into a

Fig. 2.1. This shows the progress made since 1910 in the nutrition of the pig. New developments in swine feeding should make it possible to obtain even more pork with less feed in the future. (Courtesy of L. E. Hanson, Minnesota Agr. Exp. Sta.)

well-balanced and economical feeding program. Finally, they need to recognize the critical periods in swine feeding and learn to meet their demands.

III. PREVENTING SMALL PIG LOSSES

Extensive surveys show that between 20 and 25% of all pigs farrowed die before they reach market (Fig. 2.2). This results in a tremendous loss to swine producers each year. Not all swine losses can be traced to faulty nutrition, but nutritional deficiencies account for a good part of them. Certainly, one can visualize what havoc would occur if a feed or car manufacturer lost 20–25% of his product before it was sold. Thus, the problem of small pig deaths needs considerable study to eliminate as much as possible the handicap of ''small pig losses'' which the swine producer has to contend with year after year. Producers can eliminate these losses only by practicing a better job of feeding.

The problem of adequate nutrition for the sow and the very young pig is a challenging one. It is not always given needed attention on the average farm. The figures in the following tabulation show how much feed is lost by the death of young pigs (1).

Age at death	Amount of feed lost (lb/pig)
Birth	140
10 weeks	260
18 weeks	360
26 weeks	602
34 weeks	990

Fig. 2.2. This is a weak litter of pigs due to a lack of vitamins in the diet. Most, if not all, pigs born in this weakened condition die. Proper feeding will prevent this heavy toll of small pig losses. (Courtesy of T. J. Cunha and Washington State University.)

By using current feed prices, one can determine readily how much the loss of each pig means in dollars and cents. These figures will also show why it is so important to do a good job of feeding in order to eliminate as many small pig losses as possible.

IV. WHAT MAKES A GOOD DIET

In balancing swine diets, most nutritionists think in terms of nutrients and not just feeds, as used to be the case. A "nutrient" is any food constituent which aids in the support of animal life. Today, these known nutrients consist of 10 essential amino acids, 17 vitamins, 13 or more essential mineral elements, essential fatty acids, carbohydrates, and unidentified factors which need to be taken into consideration in compounding a diet (Fig. 2.3).

These nutrients need to be furnished in such proportion, level, and form as will nourish properly the age pig to which they are fed. Not only is the amount of a nutrient important, but it needs to be fed in the correct proportion with other nutrients for maximum utilization. This amount and proportion will usually vary with the age of the pig and the stage of its life cycle. For example, young, growing pigs have different requirements from those of older animals. The nutritional requirements of the sow are different from those of the growing pig. Requirements during lactation are greater than during gestation.

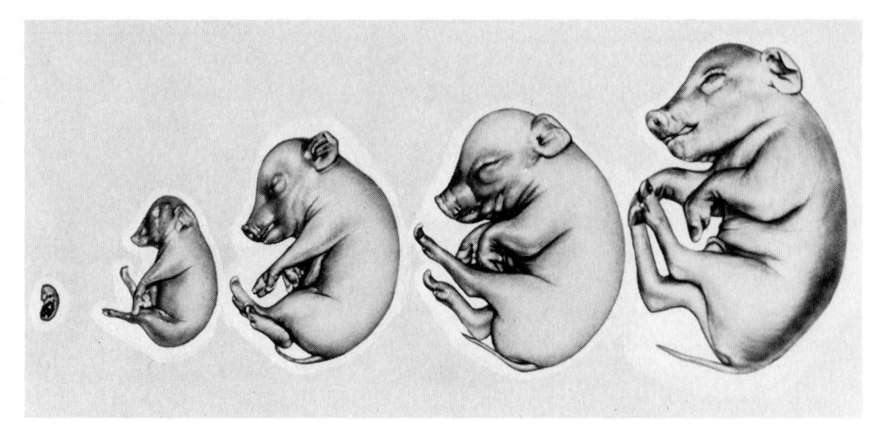

Fig. 2.3. Preserved pig embryos at 30, 60, 80, 90 and 106 days of development during gestation. Note that the pig is still very small at 30 days. (Courtesy of R. M. Bethke, Ralston Purina Company, St. Louis, Missouri.)

A good diet also must be palatable. Unless the pig consumes the diet readily, it is not a good one, regardless of how well-balanced it may be or what the chemical analyses may indicate. The pig is the final judge of how palatable a feed is. The fineness of grind affects palatability. The amount of mineral and high fiber feeds in the diet also affects palatability. Certain feeds and nutrients added to the diet increase palatability and others decrease it. All this must be kept in mind in arriving at a combination of feeds and supplements to make a well-balanced diet. This means that a knowledge of feeds and supplements is very necessary with regard to their overall effect on palatability or desirability of the diet for the pig.

Well-balanced diets will usually contain a variety of feeds and supplements. A variety of feeds tends to improve the balance of the diet and thus prevent certain nutritional deficiencies. In addition, a variety of feeds will usually make the diet more palatable. However, with the advancing knowledge of nutrition, it has become easier to supplement simple diets with the nutrients which they may lack.

A diet must be economical. It takes skill and know-how to balance a diet properly and make it an economical one at the same time. Thus, it is important to know the relative values of feeds, so that one may take advantage of price differences and changes. The alert feeder or feed manufacturer will always be on the watch for price differences and changes which present an opportunity for increased profits. Many times a good knowledge on this score will mean the difference between profit and loss.

Toxic substances must be avoided in compounding diets. This means that one needs a knowledge of injurious substances which feed ingredients may contain. Fluorine in raw rock phosphate, selenium in grains from high selenium areas, gossypol in cottonseed meal, ergot in rye, scab in barley, and limonin in citrus seed meal are examples of some of these substances. Fluorine can be removed from raw rock phosphate to make it safe for swine feeding. Cottonseed meal is now being processed so that the free gossypol is reduced to very low and safe levels. This meal can be safely fed, with good results, as the sole protein supplement for swine, although most feeders use it as part of the protein supplement fed. Ergot in rye will cause abortions and should not be fed to pregnant sows.

Discussion on these toxic substances and other limitations of various feeds will be included in later chapters in this book. This short discussion will serve to illustrate the point that a feeding test is the final criterion of the value of a feed.

Many feeds will show promise from a chemical analysis standpoint. The analyses serve as a good guide in determining the value of a feed. But the final test will be how the pig performs when the feed is given to him in a well-balanced diet. A routine chemical analysis may not show that the feed contains a toxic substance, nor will it be able to judge how palatable the feed may be. Many feeds will not give good results if used in high levels in the diet, but will give excellent results if included at a low level. This means that new feeds should be studied at various levels in the diet to obtain a good evaluation of them.

Some feeds may produce soft pork if fed long enough and at high enough levels. Soybeans, peanuts, chufas, hominy feed, rice bran, rice polish, sesame seed, sunflower seed, flaxseed, mast, buckwheat, garbage, and other feeds produce soft pork. Since pigs with soft pork are discounted heavily on the market, this problem requires serious consideration. When these feeds are used, they should be fed at low levels, and the pigs should be fed feeds which harden the fat after they weigh 75–125 lb. Thus, the effect of feeds on the quality of the carcass is another important aspect of diet formulation.

Feeds which are apt to become rancid must be guarded against. Rancid diets are low in palatability. Moreover, rancidity destroys certain nutrients in the intestinal tract. For example, rancidity destroys certain vitamins. Proper transportation and storage of feeds will lessen this rancidity problem.

Feeds and supplements need to be adapted to the areas in which they will be used. In the Corn Belt, diets are built around corn. In the West, the Southwest, and other neighboring areas, swine diets include more barley, wheat, oats and grain sorghums. In the Southeast, corn is used to a great extent in swine feeding. In the Northeast and near large cities, garbage feeding operations occur. Thus, it is important that feed manufacturers make supplements which will balance the home-grown feeds available in the various areas of the country.

V. UNDERFEEDING AND OVERFEEDING

The first problem to be met from a nutritional standpoint is providing the pig enough to eat. This is still the biggest and most serious problem in swine feeding.

If a pig does not get enough feed, it is apt to lack energy, as well as protein, minerals, and vitamins.

This is a very important point. Many swine producers will buy the most carefully balanced supplement or complete feed they can find; but they fail to feed enough of it to their pigs. As a result the animals become deficient in certain nutrients. This problem should be given more consideration by feed manufacturers. They can suggest on the feed tag that a certain amount of the feed or supplement be fed daily to supply the nutrient requirements of pigs of various

Fig. 2.4. Note difference in condition of two sows during lactation. The sow on the bottom is being fed an adequate diet whereas the one on the top is not. (Courtesy of T. J. Cunha and University of Florida.)

weights, or the manufacturers can provide literature and education programs for their customers giving this information in more detail. No matter how the information is provided, it is very important that swine producers be impressed with the fact that a well-balanced feed will not do the job unless it is fed in adequate amounts.

Using unbalanced diets is also a form of underfeeding. Many swine producers still make this mistake. They are reluctant to buy the necessary ingredients to balance their diets. But doing this is being "penny-wise and pound-foolish." It is profitable to buy whatever nutrients, supplements, or feeds are necessary to provide the pig with its daily requirements of protein, minerals, and vitamins. Rapid growth requires less feed per pound of weight increase than slow growth. This means that, in most cases, pigs should be fed for market in as short a period as possible. There are some exceptions to this; a diet having limited energy, but still supplying the required protein, minerals, and vitamins, may fill a need. This is especially true when regulating the quantity of feed may control the degree of fatness of the animal. Breeding animals and prospective herd replacement animals are usually not fed in excess. Animals that are too fat will not reproduce as well as those kept in moderate condition (Fig. 2.4).

VI. REGULARITY AND CHANGES IN FEEDING

Regularity of feeding will repay the feeder. Animals respond to regularity, the same as humans do, and it should be part of a good feeding program. Sudden changes in the diet may throw an animal off-feed. Any major changes in the diet should be made gradually. Several days should be taken to change from one important feed to another. Swine producers never notice much of the effect of abrupt changing from one feed to another. It occurs, however, especially with very young pigs. Sometimes it is slight, but other times it may cause diarrhea, and the animal may go off-feed for several days. Avoiding any such loss—if it can be avoided—is to the benefit of the feeder.

VII. NUTRIENT REQUIREMENTS OF THE PIG

Information is not available on all the nutrients required by the pig. Tables 2.1 and 2.2 supply facts on the nutrients for which there is fairly good information available. Each one of these nutrients will be discussed in greater detail in Chapters 3, 4 and 5, which deal with minerals, proteins, and vitamins. Other nutrients, not included in Tables 2.1 and 2.2, will also be discussed in Chapters 3, 4 and 5.

TABLE 2.1
Nutrient Requirements of Growing Swine Fed *ad libitum:* Percentage or Amount per Pound of Diet

Nutrients		Requirements				
	Liveweight (lb):	11–22	22–44	44–77	77–132	132–220
	Daily gain (lb):	0.66	1.10	1.32	1.65	1.98
Energy and protein						
Digestible energy[a] (kcal)		1590	1590	1500	1500	1500
Metabolizable energy[a] (kcal)		1527	1527	1441	1441	1441
Crude protein[b] (%)		22	18	16	14	13
Inorganic nutrients (%)						
Calcium		0.80	0.65	0.65	0.50	0.50
Phosphorus		0.60	0.50	0.50	0.40	0.40
Sodium		—	0.10	0.10	—	—
Chlorine		—	0.13	0.13	—	—
Vitamins						
β-Carotene (mg)		2	1.6	1.2	1.2	1.2
Vitamin A (IU)		1000	800	600	600	600
Vitamin D (IU)		100	90	90	60	60
Vitamin E (mg)		5	5	5	5.0	5
Thiamine (mg)		0.6	0.5	0.5	0.5	0.5
Riboflavin (mg)		1.4	1.4	1.2	1.0	1.0

Niacin[c] (mg)	10	8.0	6.4	4.5	4.5
Pantothenic acid (mg)	6	5.0	5.0	5.0	5.0
Vitamin B_6 (mg)	0.7	0.7	0.5	—	—
Choline (mg)	500	410	—	—	—
Vitamin B_{12} (μg)	10	7	5.0	5.0	5.0
Amino acids (%)					
Arginine	0.28	0.23	0.20	0.18	0.16
Histidine	0.25	0.20	0.18	0.18	0.15
Isoleucine	0.69	0.56	0.50	0.44	0.41
Leucine	0.83	0.68	0.60	0.52	0.48
Lysine	0.96	0.79	0.70	0.61	0.57
Methionine + cystine[d]	0.69	0.56	0.50	0.44	0.41
Phenylalanine + tyrosine[e]	0.69	0.56	0.50	0.44	0.41
Threonine	0.62	0.51	0.45	0.39	0.37
Tryptophan	0.18	0.15	0.13	0.11	0.11
Valine	0.69	0.56	0.50	0.44	0.41

[a]These suggested energy levels are derived from corn-based diets. When barley or medium- or low-energy grains are fed, these energy levels will not be met. Formulations based on barley or similar grains are satisfactory for pigs weighing 44–220 lb but feed conversion will normally be reduced with the lower-energy diets.

[b]Approximate protein levels required to meet the essential amino acid needs. If cereal grains other than corn are used, an increase of 1–2% of protein may be required.

[c]It is assumed that all the niacin in the cereal grains and their byproducts is in a bound form and thus is largely unavailable.

[d]Methionine can fulfill the total requirement; cystine can meet at least 50% of the total requirement.

[e]Phenylalanine can fulfill the total requirement; tyrosine can fulfill 30% of the total requirement.

The suggested requirements given in Tables 2.1 and 2.2 are based on experimental evidence and were worked out by the Committee on Nutrient Requirements of Swine of the National Research Council (NRC) (3). The values in Tables 2.1 and 2.2 represent an approximate, but not necessarily an exact, average of the experimental results. Further discussion of each of these requirements will be given in Chapters 3, 4 and 5.

The recommended nutrient levels given in Tables 2.1 and 2.2 are to be used as a guide. In most cases, they do not contain a margin of safety. The recommended levels are slightly lower in some cases than levels shown to be needed by certain investigators. This is discussed in detail for each nutrient in Chapters 3, 4 and 5. Thus, feed manufacturers and swine feeders will find it desirable, in some cases, to increase the level of certain nutrients whenever transportation and storage conditions may cause some deterioration of that nutrient or when they feel that a higher level might be desirable for some other reason.

A. Variation in Requirements of Animals

Most of the experimental work on the nutritive requirements of animals is based on the performance of a group of animals. The average of the group is used to determine rate of gain and feed efficiency. When deficiency symptoms are the criteria of nutrient adequacy, lack of these symptoms in a group of pigs is usually the basis for the investigators' conclusions. In any group of animals, there may be some pigs that are borderline in the nutrient and thus show no deficiency symptoms. They are deficient in the nutrient, nevertheless, and it may only result in some decrease in rate of gain or feed efficiency. Thus, although the nutrient may be adequate for a majority of the animals, it may be deficient for some animals who have a higher requirement or for some reason do not utilize it efficiently.

One such example is a study at Michigan State University (2). This showed that a level of 4.15 mg of pantothenic acid per pound of feed was adequate for only 5 of the 10 pigs in the experimental lot. These five pigs gained normally and at no time showed any signs of a deficiency. The other five pigs, however, showed typical pantothenic acid deficiency symptoms as evidenced by locomotor incoordination. Thus, it must be borne in mind that individual animals vary in their requirements of nutrients. This means that some margin of safety is desirable in compounding diets and especially with nutrients that are not very stable and may be slowly and gradually destroyed by long storage.

B. Variation in Availability of Nutrients in Feeds

Undoubtedly there are differences in the availability of nutrients, depending on their form. There is also a difference between the availability of these nutrients as

TABLE 2.2
Nutrient Requirements of Breeding Swine: Percentage or Amount per Pound of Diet

Nutrients	Liveweight (lb):	Requirements		
		Bred gilts and sows 242–550	Lactating gilts and sows 308–550	Young and adult boars 242–550
Energy and protein				
Digestible energy (kcal)		1500	1500	1500
Metabolizable energy (kcal)		1440	1440	1440
Crude protein (%)		14	15	14
Inorganic nutrients (%)				
Calcium		0.75	0.75	0.75
Phosphorus		0.50	0.50	0.50
NaCl (salt)		0.5	0.5	0.5
Vitamins				
β-Carotene (mg)		3.7	3.0	3.7
Vitamin A (IU)		1864	1500	1860
Vitamin D (IU)		125	100	125
Vitamin E (mg)		5	5.0	5
Thiamin (mg)		0.7	0.45	0.7
Riboflavin (mg)		1.8	1.5	1.8
Niacin (mg)		10	8	10
Pantothenic acid (mg)		7.5	6	7.5
Vitamin B_{12} (μg)		6.4	5	6.4
Amino Acids (%)				
Arginine		—	0.34[b]	[c]
Histidine		0.20[a]	0.26[b]	[c]
Isoleucine		0.37	0.67[b]	[c]
Leucine		0.66[a]	0.99[b]	[c]
Lysine		0.42	0.60[b]	[c]
Methionine + cystine		0.28	0.36[b]	[c]
Phenylalanine + tyrosine		0.52[a]	1.00[b]	[c]
Threonine		0.34	0.51[b]	[c]
Tryptophan		0.07	0.13[b]	[c]
Valine		0.46	0.68[b]	[c]

[a]This level is adequate; the minimum requirement has not been established.

[b]All suggested requirements for lactation are based on the requirement for maintenance + amino acids produced in milk by sows fed 11–12 lb of feed per day from which amino acids are 80% available.

[c]No data available; it is suggested that the requirement will not exceed that of bred gilts and sows.

determined by a chemical analysis or a microbiological assay and as determined by the use the pig will make of them. Their use by the pig may be different from the figures published on the nutrient analyses by various assay methods. This does not mean that analytical values are not valuable—they definitely are. It does mean, however, that some degree of reservation should be exercised and that provision should be made for taking care of this possible difference in availability. Just one example of this difference in availability is the Purdue finding by Dr. W. M. Beeson that zinc in soybean protein is less available than that in casein. This is due to the phytic acid in soybean protein forming a complex which makes the zinc less available. A swine feeding trial is the final criterion for determining whether a certain combination of feeds mixed by a feeder or feed manufacturer is adequate in certain nutrients. As certain diets vary in their makeup with different feeds, there undoubtedly is some difference in the availability of the nutrients contained therein. This problem requires careful consideration by those concerned with compounding diets.

C. Variation in Results with Natural versus Purified Diets

Most of the nutrient requirements worked out for the pig have been obtained by using purified diets. Yet this information is applied directly to natural or practical diets as fed on the farm. As compared to purified diets, there is some difference in the availability of nutrients in natural diets. Many of the nutrients, such as vitamins and minerals, are added to purified diets in relatively pure form. In natural diets, these vitamins and minerals elements are contained in the feeds in their natural state. In most cases, they are in different forms than when they are fed in purified or laboratory diets. Moreover, the two different diets will have a different effect on the intestinal synthesis of certain nutrients which subsequently will affect their need in the diet. Thus, there is some difference in requirements as worked out with purified and natural diets. Two examples of this difference in diets are the Purdue finding by Dr. W. M. Beeson that zinc in soybean protein is less available than that in casein (already stated above) and the recent Michigan study by Dr. E. R. Miller showing that vitamin D needs were higher with a soybean protein diet than with a casein protein diet. The zinc requirements were 18 ppm with the casein diet and 50 ppm with a soybean diet. The vitamin D requirements were 45 IU per pound of feed with a casein diet and 227 IU with a soybean protein diet.

There also is apt to be some difference in amino acid requirements as worked out with synthetic amino acids as compared to the need for them when they are supplied in natural or practical diets. The digestibility of protein varies, depending on the feed used. Thus, amino acid requirements worked out with one diet may not necessarily apply with another kind of diet. This means that some care must be used in applying results obtained with purified diets directly to natural diets. This does not impair the value of the use of purified diets in nutritional work; but rather indicates that these same nutrients also need to be studied with natural diets. The data obtained with purified diets serve as a valuable guide for studying problems encountered in natural diets.

D. Variation in Deficiency Symptoms

Symptoms for mineral, amino acid, and vitamin deficiencies will be described in Chapters 3, 4 and 5. Single nutrient deficiencies will seldom be encountered under farm conditions. In most cases, multiple nutrient deficiencies will occur. As a result, a complex deficiency will arise; this may be a combination of symptoms described for various single nutrients, or it may be something entirely different. Conditions such as reduced appetite and growth or unthriftiness are common to malnutrition in general. The nature of the deficiency may be detected only by careful review of the dietary history of the animal and by close observation of the symptoms. A fruitful field is open for future studies involving multiple deficiencies such as occur on the farm under varying conditions of feeding. Almost no experimental information is available on multiple deficiencies in the pig.

Nutritional deficiencies may exist without the appearance of definite deficiency symptoms. A nutritional deficiency may show itself only by slight tissue depletion which may have very little, if any, effect on the performance of the animal. As the deficiency becomes more severe, however, it will effect chemical processes in the body and will eventually result in symptoms which can be observed by looking at the animal.

E. Variation in Treatment of Deficiencies

Many nutritional deficiencies produced experimentally can be treated by supplying the missing nutrient. If treatment begins early enough, most, if not all, of the symptoms can usually be cured by supplying the missing substance. This will not always be the case, though, if the deficiency is of long standing and certain changes have occurred in the body which cannot be repaired by feeding the missing nutrient. This must be borne in mind when giving a group of deficient pigs a highly fortified diet to cure nutritional deficiencies. If the pigs have been deficient too long, it may be too late to cure them, although those which are the least deficient may respond to treatment and recover well. They should not be depended on, however, for herd replacement animals. Their reproductive systems and possibly other body organs may not be normal enough to produce and wean large and heavy litters. Severe deficiencies are detrimental to the animal and, in many cases, cause permanent harm. Thus, a good feeding program should eliminate starvation periods in which pigs are allowed very little feed and thus become deficient in certain nutrients.

F. Stress Conditions

It is difficult to list all the factors that create stress conditions, since some are not known. Changes (especially sudden and big changes) in temperature, moisture and humidity can cause a stress. Improper and irregular feeding and management are stress factors. Poor housing and muddy lots can be serious stress factors. All of these factors, plus other stress conditions, can alter the need for certain nutrients. This is one reason why many feed manufacturers like to use field trials for evaluating the adequacy of nutrient and feed additives needed in their diets. These field trials, if properly conducted, may be more applicable to average farm conditions than many university experiments. This is because university trials are usually conducted with good quality animals, good feeds, good management and with good sanitation. On the average farm, however, there will usually be average animals, average management and average sanitation. Moreover, it is very difficult, and sometimes impossible, to duplicate stress conditions which occur in the field at the university. This is the reason why certain stress feeds are used and why certain nutrients and additives are fed at higher levels to take care of stress conditions on the farm. Higher levels of nutrients and feed additives, however, should be fed with caution since some can cause harmful effects if used at too high a level. Thus, one needs to exercise good judgment when increasing nutrient levels above NRC requirements.

G. Subclinical Disease Level

This refers to a disease condition that occurs at a low level, even though nothing seems to be wrong with the pig's appearance. However, it causes the pig to perform below the level it should. A good example is the response to antibiotics. When hogs are placed in a new house, they respond very little, if any, to antibiotics. The longer they are kept in the house, however, the greater the response to antibiotics. For lack of a better explanation, this seems to be due to a buildup of microorganisms in the house which causes a subclinical disease level in the pigs. Muddy lots, changing climate and a lack of sanitation all have an effect on the subclinical disease level encountered on the farm. This can vary considerably and accounts for the variability in the response obtained in research laboratories and on farms to antibiotics and other antimicrobial compounds.

H. Antimetabolites

There are many antimetabolites in feeds. They can increase the need for the nutrients they are similar to in chemical structure. This is an area in which very little information is available on the pig.

I. Nutrient Interrelationships

Experimental information on nutrient interrelationships is only in its early stages. As more of these interrelationships are solved, they will answer some of the unexplained results obtained in nutrition studies. Just a few examples of these interrelationships are choline and methionine; methionine and cystine; phenylalanine and tyrosine; niacin and tryptophan; calcium, manganese, and copper; zinc, copper, and protein; copper, zinc, and iron; vitamin D, calcium, phosphorus, and magnesium; iron and phosphorus; molybdenum, copper, and sulfur; sodium and potassium; biotin and pantothenic acid; B_{12} and methionine; and vitamin E and selenium. Nutrition will not be thoroughly understood until all the many interrelationships of nutrients are identified. Neither can one accurately determine the requirement of certain nutrients until their interrelationships with other nutrients are known. These requirements will be modified by the level of other nutrients. This discussion explains why there is some variation in the nutrient requirements reported by different investigators.

J. Peroxidizing Fats in Diet

The presence in the diet of peroxidizing fats can destroy certain nutrients. This will especially occur in the presence of catalyzing minerals in the absence of suitable stabilization. Thus, one should minimize any effect peroxidizing fats may have since this could increase the need for certain nutrients in the diet. Making sure that diets do not become rancid is one means of preventing peroxidizing effects. Another is to add a suitable antioxidant to the diet.

K. Minerals Added to Diet

Certain minerals salts are more destructive to vitamins than others. In general, the carbonates and oxides are less destructive than the other forms. However, there are exceptions to the use of only carbonates and oxides. For example, iron oxide has very little availability, and iron carbonates are less available than iron sulfate in the pig. Thus, the form of mineral element used can affect its availability as well as the stability of other nutrients.

L. Criteria Used to Determine Nutrient Requirements

Quite a few nutrient requirements have been worked out without the use of a really complete study. Many of them involve data on only rate of gain and feed efficiency. An example of what effect this may have on nutrient requirements is a Michigan State study by Dr. E. R. Miller on pyridoxine. He showed that a level of 0.5 mg pyridoxine per kilogram of solids was nearly adequate for good growth and feed conversion with the baby pig. When the blood (hemoglobin, red blood cells, etc), and urinary xanthurenic acid constituents were taken into consideration, however, the pyridoxine requirements was shown to be higher. It was greater than 0.75 mg, but less than 1.0 mg per kilogram of solids. If histopathological data had been obtained, the requirement might have been even higher. This example indicates that some nutrient requirements of the pig may be higher when complete studies involving growth, reproduction, data on blood and other body constituents as well as histopathology are obtained. Only with such complete studies can accurate data on nutrient requirement be obtained.

M. Nutrients and Compounds in Water

This is an area which is virtually unexplored and on which considerable emphasis should be placed in the future. Water is a source of minerals and other compounds. Since the pig will consume 2–3 lb of water per pound of feed, the level of minerals in water should be taken into consideration when determining nutrient requirements. Nitrites, sulfites and other chemicals in the water can destroy certain nutrients. It is also possible that high levels of sulfates and other compounds in water can cause diarrhea and other disturbances in the pig.

N. Energy Content of Diet

The energy content of the diet will definitely affect nutrient needs. For example, amino acid requirements increase as the caloric density and protein level increases. Dr. R. F. Sewell of the University of Georgia showed that the addition of 10% fat to the diet increased the requirement for pantothenic acid by 50%. Many other examples could be cited. These, however, are sufficient to illustrate that the caloric density of the diet can alter nutrient requirements.

O. Intestinal Flora

The flora is affected by the type of diet and nutrients fed. Therefore, the intestinal synthesis of nutrients or the requirements of nutrients by the intestinal microflora can be changed. This in turn will affect the requirement for certain nutrients in the diet. This accounts for recent reports indicating a need for certain vitamins which normally one would expect the pig to synthesize in sufficient quantities for its needs. Unfavorable intestinal flora can also have detrimental effects on the pig.

P. Faster Growth

Pigs are being bred for a faster growth rate and increased litter size. This may increase certain nutrient needs. Certain deficiencies, for example, will not show up unless a good rate of growth occurs. The faster rate of growth and increased feed efficiency may account for the occurrence of certain deficiencies previously not encountered. Therefore, nutrient requirements determined quite a few years ago may not be adequate for the modern-day pig with faster growth rate, less feed per unit of gain and a lean carcass.

Q. Effect on Carcass Quality

More emphasis needs to be placed on carcass quality when nutrient levels are being studied. It may be possible that a certain level of a nutrient may not affect rate of gain or feed efficiency but it may affect carcass quality. For example, Danish data show that the kind of protein supplement used affects the fat content of the carcass although it may not affect rate and efficiency of gain. Recent studies also indicate that amino acid supplementation and protein level will affect carcass quality although they may not affect rate of gain. Indications are that higher levels of protein will increase carcass quality. This means that as a leaner pig is produced, certain nutrient requirements could be affected.

R. Will the Growing Pig Be Kept for Reproduction?

Most studies on nutrient needs during growth have not been followed up with gestation and lactation studies to determine if the nutrient level was adequate to develop a normal reproductive tract during the growing period. Until this is done, in many cases the level of a nutrient adequate for growth may not be adequate for pigs kept for breeding purposes later on. Some examples to illustrate this are the studies performed at Wisconsin, Washington and Purdue which showed that the diet fed during growth influenced the ability of the pigs to conceive, reproduce and lactate many months later. In many cases, the diets which gave the best results during growth gave the poorest results later on during gestation and lactation. This emphasizes the fact that studies are needed which involve the entire life cycle of the pig and which take into account the specific nutrient needs during each phase and how they affect requirements during the other phases. These are the type of studies needed to acquire accurate information on the need for nutrients and feed additives in swine diets.

S. Type of Floor Used

Slotted floors are increasing in use. They cause the pig to have less access to its feces. It is not known to what extent coprophagy (eating own feces) is practiced by the pig. The rat practices coprophagy, and its growth rate is decreased if it does not have access to its feces. Thus, it is possible the same may occur with the pig. At any rate, this possibility should be studied if the trend toward slotted floors proceeds to the point where the pig no longer has contact with its feces.

T. Economics of Nutrient Level

The cost and return of using various levels of nutrients and feed additives in the diet is very important. The returns must be such as to warrant the cost involved. Moreover, one must always keep from using nutrients at excessively high levels to prevent the harmful effect that might occur.

U. Other Possibilities

There are a number of other factors which may affect nutrient needs. Some of them are (a) environmental temperature, (b) destruction of nutrients by light or irradiation, (c) hormones, antibiotics, and other feed additives in the diet, (d) absorption of nutrients on colloids in the feed and in the intestinal tract, (e) toxins from aflatoxins or fungal infections, (f) enzymes, such as thiaminase, in feeds which destroy nutrients.

VIII. SUMMARY

Feed represents a large percentage of the cost involved in swine production. Therefore, every effort should be made to feed well-balanced diets. Baby pig losses still amount to 20–25% of all pigs farrowed. Better feeding and management practices can eliminate a good share of these costly losses.

Nutrient requirements of the pig vary, depending on many factors. This makes it difficult to give exact nutrient recommendations to take care of all the conditions encountered on farms. The National Research Council (NRC) nutrient requirement standards are only a guide to be used in formulating diets. Some margin of safety needs to be used with certain nutrients. The safety factor to use will vary with the different situations encountered.

The only sure way to know whether a diet lacks certain nutrients for the pig is to run a feeding trial and determine whether the addition of the nutrient is beneficial. This also means that feed formulators need to conduct trials to determine the nutrients and feed additives required with their type of diet and under the conditions it will be used. Only then will they definitely know whether their diets are adequate in nutrients and/or feed additives.

REFERENCES

1. Luecke, R. W., J. A. Hoefer, and F. Thorp, Jr., *J. Anim. Sci.* **12,** 606 (1953).
2. McMillen, W. N., *Mich. Agric. Ext. Serv., Bull.* **299** (1949).
3. Cunha, T. J., J. P. Bowland, J. H. Conrad, V. W. Hays, R. J. Meade, and H. S. Teague, N.A.S.—N.R.C. publication (1973).

3

Mineral Requirements of the Pig

Mineral deficiencies still cost swine producers large sums of money each year. This is an unnecessary loss. Minerals are low in price and can be fed easily to all classes of swine.

I. RELATION OF SOIL MINERALS TO SWINE FEEDS

Soils in the United States are declining in fertility. In a single human generation, the soil fertility on a farm can be depleted by failure to add crop residues, manure, and other forms of plant foods. Many animal ailments, which are on the increase, may be due in part to this decline in soil fertility. As this decline increases, farmers will have to supply not only minerals now deficient but also other minerals which may eventually become low in the soil. It is not enough to think of feeding minerals to livestock. One should also think of adding minerals to the soil, which is gradually becoming depleted. Minerals added to depleted soils bolster the mineral content of plants grown thereon. Added minerals also increase the tonnage yield of the forage and other nutrients in the plant.

II. INCREASED PRODUCTION AND CONFINED CONDITIONS INCREASE MINERAL NEEDS

Many farmers are obtaining higher production rates; in turn, this has increased the mineral needs of swine. For example, using antibiotics and adequate vitamin supplementation has increased the growth rate of swine. This more rapid growth undoubtedly has affected mineral requirements and suggests a need for re-evaluating presently recommended levels. Similarly, the increased feeding of swine in close confinement has made it even more essential to feed diets adequate in minerals. Increased production, close confinement, and declining soil fertility increase the need to supply adequate minerals in the diet.

III. VITAL FUNCTIONS OF MINERALS AND EFFECTS OF DEFICIENCY

Minerals perform important functions in the animal's body. In addition to being constituents of bone and teeth, mineral elements serve the body in many other ways. Nearly every process of the animal body depends on one or more of the mineral elements for proper functioning. Minerals are just as essential for growth, reproduction, and lactation as are proteins, fats, carbohydrates, and vitamins.

A lack of minerals in the diet may cause any of the following deficiency symptoms: reduced or poor appetite; expensive, poor gains; rickets; soft or brittle bones; beading of the ribs; stiffness or malformed joints; posterior paralysis (''going down in back'') (Fig. 3.1); goiter; animals which are unthrifty looking; pigs born hairless; failure to come in heat regularly; poor milk production; weak or dead young; and many other ailments (1–7). Thus, it is not too surprising that a pig fed a mineral-deficient diet will eventually die if the deficiency is severe enough and is not corrected.

IV. THE MINERAL CONTENT OF THE ANIMAL BODY

Calcium and phosphorus make up over 70% of the ash in the body. For the most part, in the body these two mineral elements occur combined with each

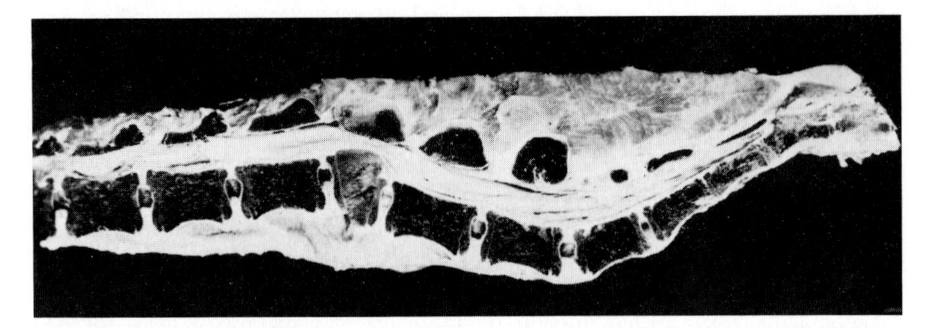

Fig. 3.1. A cause of posterior paralysis. Split section of backbone of pig, showing a fractured lumbar vertebra. When a vertebra collapses, it presses on the spinal cord and thus causes paralysis of the rear extremities. This is referred to as ''going down in the back.'' It can be caused by a calcium, phosphorus or a vitamin D deficiency. (Courtesy of G. Bohstedt, University of Wisconsin.)

other. An inadequate supply of either one in the diet will limit the utilization of the other. Approximately 80% of the phosphorus and 99% of the calcium in the body are present in the bones and teeth. These figures indicate the importance of calcium and phosphorus in the diet and the role they play in giving rigidity and strength to the skeletal structure. Although the other minerals are contained in smaller amounts in the animal's body, their presence is just as important as that of calcium and phosphorus (Fig. 3.2).

V. ESSENTIAL MINERAL ELEMENTS AND THOSE APT TO BE DEFICIENT

At present, the following fourteen mineral elements have been shown to perform essential functions in the body. They thus must be present in the diet: calcium, phosphorus, sodium, chlorine, copper, iron, cobalt, iodine, manganese, magnesium, sulfur, zinc, potassium and selenium. Fluorine might also

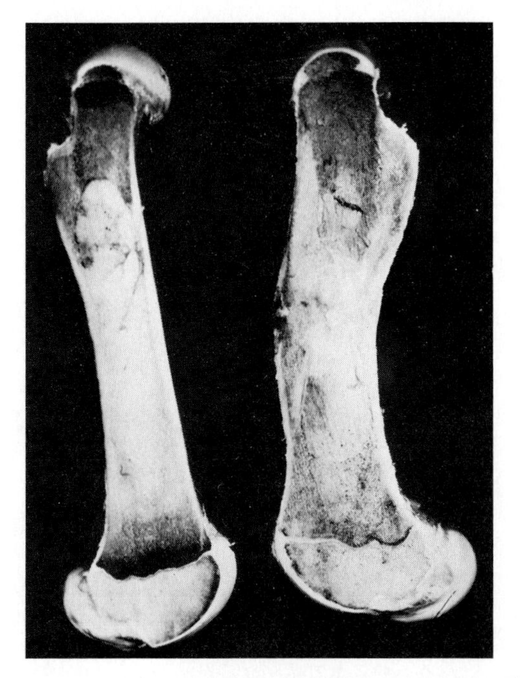

Fig. 3.2. Why hogs break their legs. Note abnormal thigh bone on right. The thickness of the bone is no indication of its strength. Nature has thickened the bone so as to help overcome its weakness, which is evident from the spontaneous fracture near its middle. A pig with such a thigh bone is a cripple, walking with difficulty. This is due to a lack of calcium, phosphorus or vitamin D. (Courtesy of G. Bohstedt, University of Wisconsin.)

be included in the list of essential mineral elements. At high levels it is harmful, but at low levels it has decreased tooth decay; this means it affects teeth beneficially in some manner. Molybdenum has recently been found to be an enzyme constituent. Thus, it might also be classified as an essential mineral element. It is also possible that other minerals may be essential in the body.

Of the essential mineral elements, ten are likely to be deficient in swine rations. These are calcium, phosphorus, sodium, chlorine, cobalt, iron, copper, zinc, iodine and selenium. Cobalt probably is not needed, however, if the diet contains sufficient vitamin B_{12}. So far as is known, deficiencies of the other four essential mineral elements are not encountered, since the feeds used probably contain them in sufficient quantities. There are some indications, however, that manganese and magnesium supplementation may be beneficial under certain situations. It may be that as soils decline in fertility other mineral elements may also become deficient in swine diets.

VI. CALCIUM AND PHOSPHORUS

These two minerals are discussed together because there is a close relationship between them. Calcium and phosphorus make up over one-half of the minerals in milk. They also comprise about three-fourths of the minerals in the entire body of an animal. Thus, it is important to supply these two minerals in swine diets. Swine suffer more from a lack of calcium than from the lack of any other mineral, with the exception of common salt.

A. Calcium Most Apt to Be Lacking

There is more likely to be a deficiency of calcium than of phosphorus in swine diets (Fig. 3.3). This is because grain and protein supplements make up a very

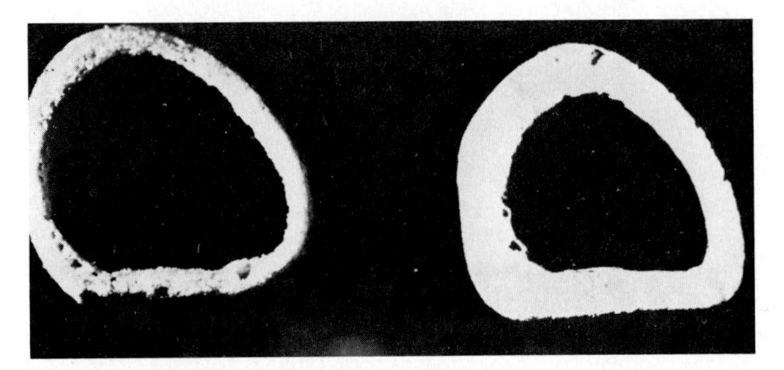

Fig. 3.3. Calcium deficiency. Cross section of leg bone. Note the thickness and strength of the bone from the pig fed an adequate calcium diet as compared to that of a calcium-deficient pig on the left. (Courtesy of the late A. G. Hogan, University of Missouri.)

large portion of the feed consumed by these animals. By comparing the content of feeds and the requirement of the pig for calcium and phosphorus, one can readily see why calcium is most apt to be lacking in swine diets. Cereal grains, which make up the bulk of swine diets, are quite low in calcium.

Grains are fair sources of phosphorus, and protein supplements are good sources; so a well-balanced diet usually contains almost enough, if not enough, phosphorus to supply the needs of the pig. Table 3.1 shows the calcium and phosphorus content of some feeds which are widely used (see Table 11.5 for the calcium and phosphorus level of feeds).

B. Minerals Needed When Animal Proteins Are Fed

If the protein supplement in swine diets is made up largely of fish meal, tankage, meat and bone scraps, or milk by-products, usually no minerals other

TABLE 3.1
Calcium and Phosphorus in Some Widely Used Feeds[a]

Feeding stuff	Calcium (%)	Phosphorus (%)
Protein concentrates		
Wheat bran, 15% protein	0.14	1.15
Cottonseed meal, 41% protein	0.20	1.05
Skim milk, dried, 34% protein	1.28	1.02
Wheat standard middlings, 16% protein	0.09	0.53
Linseed meal, solvent, 36% protein	0.40	0.90
Soybean oil meal, solvent, 44% protein	0.29	0.65
Corn gluten feed, 26% protein	0.40	0.80
Fish meal, menhaden, 60% protein	5.11	2.88
Meat and bone meal, 50% protein	10.76	5.33
Feather meal, 86% protein	0.33	0.55
Meat meal, 55% protein	8.27	4.10
Peanut meal, solvent, 60% protein	0.20	0.63
Grains		
Barley	0.24	0.36
Oats	0.06	0.27
Corn, yellow	0.02	0.28
Milo	0.03	0.28
Wheat, hard winter or spring	0.05	0.37
Brewer's dried grains, 25% protein	0.29	0.53
Roughage		
Alfalfa meal, 17% protein, dehydrated	1.44	0.22
Requirement of pig for calcium and phosphorus	0.50–0.80	0.40–0.60

[a]These data were obtained from 1977 National Research Council Publication on Nutrient Requirements of Poultry.

than a small amount of salt are needed. These protein supplements usually supply enough additional calcium and phosphorus (Fig. 3.4) to satisfy the needs of the pig. They may also supply enough additional copper and iron to sufficiently enrich swine diets. Unfortunately, however, animal protein feed sources are in short supply and only small amounts are used in swine diets. In many diets, they are left out entirely.

C. Minerals Needed When Plant Proteins Are Fed

When little or no animal protein supplement is fed, and the grain feeds are balanced with soybean oil meal, peanut meal, cottonseed meal, or other plant protein concentrates (all low in calcium, but good in phosphorus), it becomes necessary to feed a calcium supplement.

D. The Calcium-Phosphorus Ratio and Vitamin D

To obtain proper calcium and phosphorus utilization, three requirements must be met. First, a sufficient supply of calcium and phosphorus must be fed; second, a suitable ratio between them must be maintained; and third, a sufficient amount

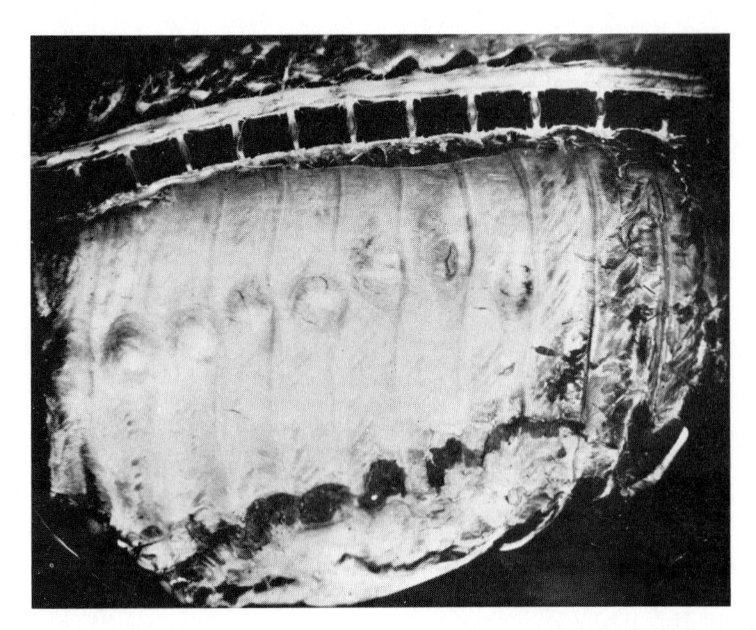

Fig. 3.4. Beading of the ribs. Large knobs develop at the juncture of the bony with the cartilaginous parts of the ribs. This is caused by a calcium or phosphorus deficiency. (Courtesy of G. Bohstedt, University of Wisconsin.)

of vitamin D must be available. The ratio of calcium to phosphorus will vary depending on the weight and age of the pig and the level of calcium and phosphorus in the diet. If the calcium intake is too high, for example, and the phosphorus level is just about adequate, then the excess calcium could cause a deficiency of phosphorus to occur. This is due to excess calcium tying up some phosphorus in the intestinal tract and causing it to not be absorbed and thus excreted in the feces.

The proper calcium to phosphorus ratio is about 1.1–1.5:1.0. Sow's milk contains a calcium:phosphorus ratio of about 1.3:1. More important than the ratio is making sure that an adequate level of both calcium and phosphorus is present. These two mineral elements are more efficiently utilized when they are present in a certain ratio to each other.

Vitamin D is needed for maximum calcium and phosphorus absorption and utilization and for normal bone and teeth development. Studies have shown that animals receiving no vitamin D excrete an excessive amount of calcium and phosphorus in the feces. Vitamin D is of more benefit as the ratio of calcium to phosphorus gets more out of balance. But, if the calcium:phosphorus ratio becomes too much out of line, then vitamin D alone will not be of much help in alleviating the situation.

E. Sources of Calcium and Phosphorus

Tables 3.2 and 3.3 list sources of calcium and phosphorus and their approximate percentage in various mineral supplements. The composition of these mineral supplements varies somewhat depending on the purity of the raw material and the method used in processing.

F. Effect of Excess Calcium and Phosphorus

A large excess of either calcium or phosphorus interferes with the absorption of the other. With an excess of either one, the other tends to become tied up as the insoluble tricalcium phosphate, which the pig cannot absorb. This explains why it is important to have a suitable ratio between calcium and phosphorus; if one mineral is too high in relation to the other, it accentuates the lack of the mineral which is low. Thus, if a swine producer using a diet low in phosphorus were to include a very high level of calcium in the diet, the phosphorus level available to the pig would become even lower (Fig. 3.5). In addition to excess calcium increasing the need for phosphorus, it may also increase the need for zinc, copper, manganese and possibly other nutrients. Therefore, it is wise to avoid an excess of calcium since it can increase the need for other nutritional factors.

TABLE 3.2
Typical Analyses of Phosphate Compounds[a]

Compound	Phosphorus content (%)	Calcium content (%)	Sodium content (%)	Nitrogen content (%)	Fluoride content (%)
Defluorinated phosphates manufactured from defluorinated phosphoric acid					
Monocalcium phosphate	21.0	16.0	—	—	0.16
Dicalcium phosphate	18.5	21.0	—	—	0.14
Defluorinated phosphate	18.0	32.0	—	—	0.16
Defluorinated wet-process phosphoric acid	23.7	0.2	—	—	0.18
Defluorinated phosphates manufactured from furnace phosphoric acid					
Monocalcium phosphate	23.0	22.0	—	—	0.03
Dicalcium phosphate	18.5	26.0	—	—	0.05
Tricalcium phosphate	19.5	38.0	—	—	0.05
Monosodium phosphate, anhy.	25.5	—	19.0	—	0.03
Disodium phosphate	21.5	—	32.0	—	0.03
Sodium tripolyphosphate	25.0	—	30.0	—	0.03
Feed-grade phosphoric acid	23.7	—	—	—	0.03
High-fluoride phosphates					
Soft rock phosphate	9.0	17.0	—	—	1.2
Ground rock phosphate	13.0	35.0	—	—	3.7
Ground low-fluorine rock phosphate	14.0	36.0	—	—	0.45
Triple superphosphate	21.0	16.0	—	—	2.0
Wet-process phosphoric acid (undefluorinated)	23.7	0.2	—	—	2.5

[a]From 1974 National Research Council publication on "Effect of Fluorides in Animals" (8).

TABLE 3.3
Calcium Analyses of Some Mineral Sources

Ingredient	Calcium (%)
Limestone	38.0
Oyster shell	38.0
Calcite, high grade	34.0
Dolomitic limestone	22.0
Gypsum	22.0
Wood ashes	21.0

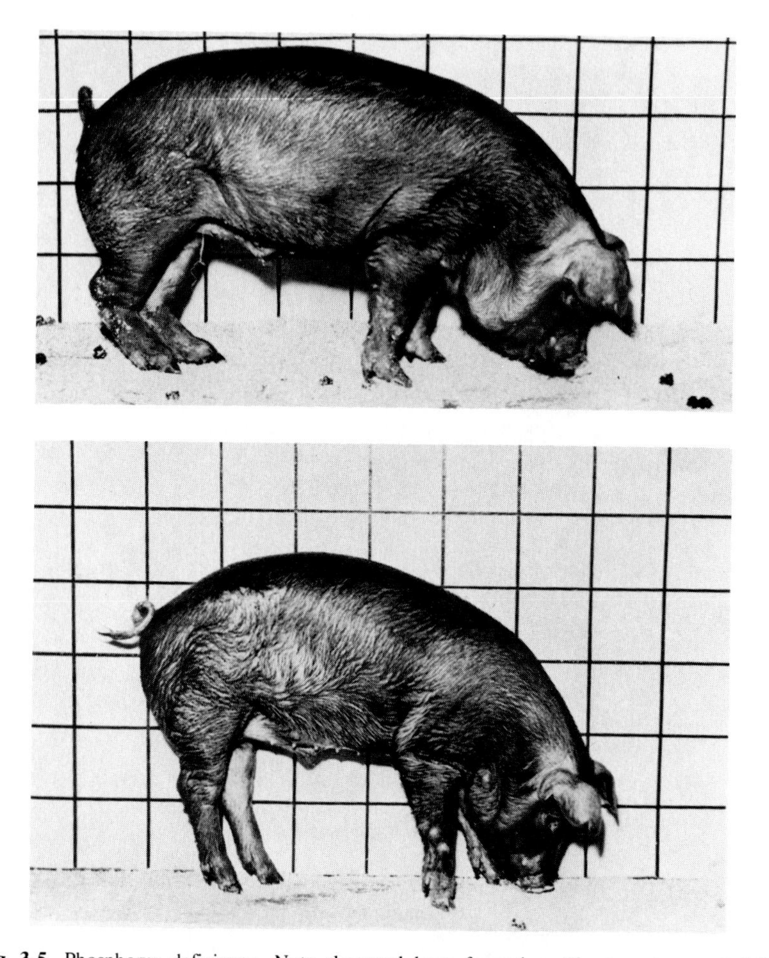

Fig. 3.5. Phosphorus deficiency. Note abnormal bone formation. The top pig was fed from weaning on low phosphorus diet containing 0.18% phosphorus. The bottom pig was fed same diet except it contained 0.55% phosphorus. (Courtesy of W. M. Beeson, Purdue University.)

Excess phosphorus can also have harmful effects. It also increases calcium needs by tying up some calcium in the intestinal tract so that it cannot be absorbed. There is also some indication that excess phosphorus increases zinc needs. Large amounts of iron, magnesium, and aluminum in the diet should be avoided, since they interfere with the absorption of phosphorus by forming insoluble phosphates. (See Fig. 3.6).

Fig. 3.6. Calcium, phosphorus and vitamin D deficiency. Note similarity of deficiency symptoms in these 5-week-old pigs. The photos show the following deficiencies: top, calcium; middle, phosphorus; bottom, vitamin D. (Courtesy of E. R. Miller and D. E. Ullrey, Michigan State University.)

G. Availability of Phytin Phosphorus

About half or more of the phosphorus in cereal grains is in the form of phytin. Phytin consists of the vitamin inositol combined with phosphorus and other minerals. About 20–50% of the phytin phosphorus is available to the pig (9). The availability of phytin phosphorus is influenced by the level of vitamin D, calcium, the calcium to phosphorus ratio, zinc, alimentary tract pH, and other factors. A good guide is to assume that no more than about 50% of the phosphorus in plant feeds is available to the pig. By contrast, the phosphorus in inorganic mineral supplements is usually available above 80%. Thus, the level of calcium and phosphorus needed in the diet will vary and will depend on the feeds and mineral supplements being fed. Further, it should be stressed that the calcium and phosphorus level determined by a chemical analysis of a feed gives only the total present. It does not indicate the amount available to the pig.

H. Factors Affecting Calcium and Phosphorus Needs

Different results are obtained under varying conditions on calcium and phosphorus needs. For example, the dietary calcium and phosphorus levels needed for maximum rate of gain and feed efficiency may not give maximum bone ash and bone strength. (Fig. 3.7). Scientists are not in agreement on how much bone ash and strength is actually needed to prevent feet and leg problems (Fig. 3.8). Some scientists feel that a level below that needed for maximum bone strength is satisfactory for feeding pigs under farm conditions. They recommend against a

Fig. 3.7. Calcium deficiency. Note abnormal bone development and rachitic condition in advanced stage of deficiency. Lack of calcium retards normal skeletal development. (Courtesy of the late N. R. Ellis, USDA.)

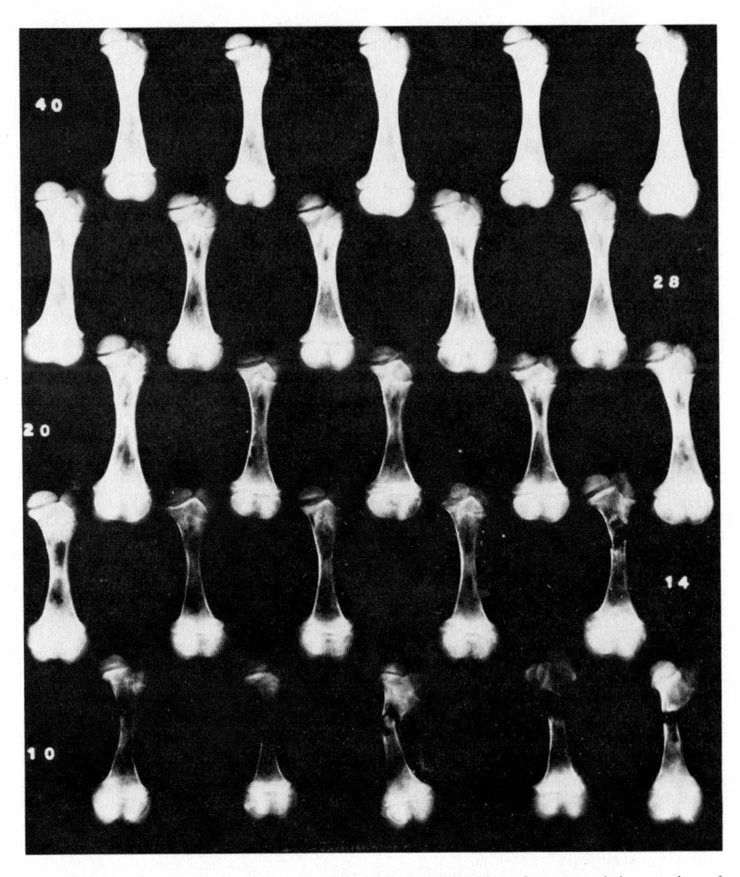

Fig. 3.8. X-ray reproductions of the left femur of pigs 35 days of age receiving various levels of phosphorus for a 28-day period. The number at the end of the row represents the level of phosphorus (0.10, 0.14, 0.20, 0.28 and 0.40%) in the diet. Note the increased density as the level of phosphorus in the diet increased. (Courtesy of G. E. Combs and Iowa State University.)

high level of calcium and phosphorus if it will decrease rate of gain and feed efficiency. They also wonder if the highest strength of the bone has any practical value or benefit for the growing-finishing pig. There are differences in the availability of calcium and phosphorus in various feeds and mineral supplements. This accounts for the differences reported on calcium and phosphorus requirements since they are reported as total calcium and phosphorus levels in the diet rather than on the amount which is actually available to the pig. Therefore, research reports from various Experiment Stations can vary from studies reported at other Stations. Consequently, the best guide to use is the 1973 National Research Council recommendations shown in Tables 3.4 and 3.5.

I. Requirements of the Pig for Calcium and Phosphorus

Table 3.4 shows the calcium and phosphorus levels recommended for the growing-finishing pig by the United States National Research Council (9).

There is still much to be learned on the exact calcium and phosphorus needs of the pig to fit most situations on the farm. Until these are all understood, the levels shown in Tables 3.4 and 3.5 should be used. These levels reflect what the majority of experiments throughout the world indicate should be used at the present time and are a good guide to follow.

If breeding stock is fed a lower level of feed intake daily than shown in Table 3.5, the level of calcium and phosphorus in the diet needs to be increased. This is to insure that the sow obtains enough total calcium and phosphorus daily to take care of the needs for reproduction. For example, some producers feed 3 lb of feed during the first 10–11 weeks of gestation. If this is done, then 1.1% calcium and 0.75% phosphorus should be used in the diet. In general, the requirements for calcium during reproduction and lactation is about 1.5 times that of phosphorus.

The calcium and phosphorus requirements can also be influenced by how long the sow suckles the pigs. For example, early weaning at 3–4 weeks of age is not as hard on the sow as when the pigs are weaned at 5–8 weeks of age. The larger the number of pigs suckled and the heavier the sow milks, the more stress is placed on calcium and phosphorus needs. A requirement for extra calcium and phosphorus can be taken care of by having minerals available for self-feeding by the pigs.

J. Self-Fed Minerals

It is a good idea to provide pigs free access to a good mineral mixture containing calcium, phosphorus, salt and the trace minerals. This is low cost insurance to protect against a possible need for minerals in addition to those included in the diet. Some pigs will differ in mineral requirements because of growth rate or

TABLE 3.4
Calcium and Phosphorus Requirements of the Growing-Finishing Pig (9)

Weight of pig		Percent of mineral element in total diet	
lb	kg	Calcium	Phosphorus
11–22	5–10	0.8	0.6
22–44	10–20	0.65	0.5
44–77	20–35	0.65	0.5
77–132	35–60	0.50	0.4
132–220	60–100	0.50	0.4

TABLE 3.5
Calcium and Phosphorus Requirements for Sows and Boars (9)

	Percent in the total diet	
	Calcium	Phosphorus
Gilts and sows—during gestation period		
Fed an average of 4.4 lb of feed daily	0.75	0.5
Gilts and sows—during lactation		
Gilts fed 11 lb and sows fed 12.5 lb of feed daily	0.75	0.5
Boars—young and adult		
Young boars fed 5.5 lb and adult boars fed 4.4 lb of feed daily	0.75	0.5

productivity level. Some will consume more feed than others. There will also be some variation in the mineral content of feeds as well as in the availability of the minerals they contain. This, plus other factors, indicate that having a mineral mixture available provides an opportunity for pigs needing extra minerals to obtain them from the mineral box. If the pigs do not need the minerals, they will usually leave them alone. Therefore, one should balance the diet by including the proper level of minerals in it and should also provide a mineral mixture accessible to the pig just in case an extra amount is needed.

VII. SALT

Salt contains both sodium and chlorine. The body contains about 0.2% sodium. Some of it is contained in the bones, but by far the greatest amount is in fluids outside the cells. Chlorine is found both inside and outside of body cells. Both sodium and chlorine play very important roles in the body. Milk contains a considerable amount of sodium and chlorine, and a lack of salt in the diet will ultimately decrease milk production. Sow's milk contains about 340 ppm of sodium and 1000 ppm of chlorine (10). Salt serves both as a condiment and a nutrient. As a condiment it stimulates salivary secretion. Saliva contains enzymes important in feed digestion. When salt intake is low, the body adjusts to conserve its supply. Urine output of sodium and chlorine nearly stops. On the other hand, high salt intake triggers greater excretion of sodium and chlorine by the kidney, and water needs increase. In this way, the body adjusts to a wide range of salt concentration so long as ample water is available. The commonly used diets do not contain enough salt to supply the needs of the pig, so diets need salt supplementation.

A. Value of Salt in Diet

Sodium, chlorine, and sodium chloride deficiencies in the pig cause a decrease in daily gain and efficiency of feed utilization (Fig. 3.9). Deficient pigs have been shown to keep licking their cages for salt. An example of the value of salt in the diet was shown in a Purdue experiment (11). Pigs receiving no salt in their diets required 174 lb more feed/100 lb of gain, and their rate of gain was only half as fast as that of pigs given adequate salt. One pound of salt, costing only one or two cents, saved 287 lb of feed. In another Purdue trial (12), the salt-fed pigs ate 23% more feed and gained at nearly double the rate on 33% less feed.

Fig. 3.9. Salt deficiency. Note difference in size of the pigs not supplemented with salt on top as compared to the pigs fed 0.5% salt in the diet. (Courtesy of G. E. Combs, University of Florida.)

One pound of salt saved 185 lb of feed. In a Florida trial (13), 13-pound pigs fed no salt gained 0.28 lb/day, consumed 0.98 lb of feed daily and required 3.5 lb of feed/lb of gain. The pigs fed 0.5% salt in the diet, gained 1.44 lb/day, consumed 3.95 lb of feed daily and required 2.74 lb of feed/lb of gain. These three studies indicate the importance of salt in the diet of the pig.

B. Salt Requirements of the Pig

The 1973 National Research Council publication on Swine Nutrient Requirements (9) recommends as follows: "Under practical swine feeding conditions, salt needs are met by including 0.25–0.5% salt in the diet or by allowing swine free access to salt alone or in a mineral mixture." This recommendation applies throughout the life cycle of the pig.

A 1974 study at Purdue (14) concluded that a level of 0.3% salt in the total diet was the overall optimum level to use for swine. A 1975 Florida study (13) showed that the addition of 0.2% salt to the diet was satisfactory for the growing-finishing pig. There are many factors that can alter the level of salt supplementation needed in swine diets. Therefore, most persons add 0.3–0.5% salt to all swine diets. This assures them of having enough salt in the diet. It is a good practice to self-feed salt in addition to that supplied in the diet, in case the pigs need more than is included in the diet. A mineral box with salt is low cost insurance against a deficiency occurring. If the salt level in the diet is adequate, the pigs will usually not eat it.

C. Effect of Excess Salt

The Wisconsin Station (15) produced toxic symptoms in pigs (Fig. 3.10), which were starved for salt for several months, and then given feed with 6.0–8.0% salt (on a dry matter basis) all at once with no drinking water. The most marked symptoms were nervousness, staggering, marked weakness, paralysis of the hind legs and general paralysis. Excessive salt in swine diets is a very rare occurrence. One should be aware of the danger, however, in case it might occur. To avoid toxicity, the following precautions should be followed: (1) avoid starving pigs for salt; (2) make sure that pigs always have access to plenty of clean, fresh water. This should always be the case regardless of whether excess salt is fed. Excess salt is not a problem if plenty of water is available.

D. Salt in Manure

Many have asked whether the salt in the feed has any effect on the land for crop production if the manure is added to the soil. In a 1974 study (16) at Purdue, 20, 40 and 60 tons of liquid swine waste was added per acre to corn land. Wastes

Fig. 3.10. Salt toxicity. In top photo, the pigs are beginning to show signs of salt poisoning when fed 8% salt in diet and no water. In bottom, the pigs are in advanced stages and exhibiting clonic spasms of legs. (Courtesy of G. Bohstedt and R. H. Grummer, University of Wisconsin.)

from pigs fed 0.2–0.5% salt in the diet were used. The exchangeable soil sodium concentration increased in the soil at both levels of salt feeding. However, the sodium accumulation was not high enough to cause any problem with soil productivity. They stated that the available nitrogen, phosphorus and potassium contents of animal wastes are still probably the most important factors determining the optimal rates for application of animal wastes for corn production with minimal adverse effects on soil productivity and environmental quality. More studies are needed on this problem, but this study and others indicate that the salt in the diet and in the manure has no adverse effect on the soil or crop production if the manure is added to the soil at reasonable levels.

E. Salt Is a Good Carrier for Trace Minerals

Since swine consistently need salt, it serves as a dependable and safe carrier for the trace minerals—copper, iron, zinc, manganese, iodine, cobalt and selenium. Selenium is now considered one of the trace elements to use in swine diets. The Food and Drug Administration approved its use for swine in February 1974. Selenium can be used in salt which is to be added to the diet. But, it is still not allowed in salt which is to be self-fed to swine. Eventually, however, this should also be approved since it is safe to do so.

Trace minerals are needed in only small amounts in diets. Most farmers do not have the facilities for weighing nor for properly mixing the trace minerals in their diets. They are needed in parts per million. One part per million corresponds to 0.0001% of the diet or to 0.002 lb per ton. So, these very small quantities require excellent weighing facilities, specialized mixing equipment and highly qualified personnel. Moreover, if potassium iodide is used as the source of iodine, it needs to be stabilized with an antioxidant to protect it from destruction. So, the technology needed indicates the farmer is wise to buy trace mineralized salt or a trace mineralized mineral mixture, rather than try to mix the trace minerals into the diet himself.

VIII. MINERAL SALTS VARY IN MINERAL ELEMENTS

Many persons are not aware that mineral salts will vary in their mineral element content. The writer has encountered many who think that one pound of copper sulfate, for example, supplies one pound of copper. Copper sulfate contains only 25.5% copper. Thus, it takes a little over 4 lb of copper sulfate to supply one pound of copper itself (elemental copper). A few examples of mineral salts and the percent of the mineral element they contain are given in Table 3.6 to illustrate this point. Table 3.6 can also be used as a guide in determining the percent of mineral element in some of the more commonly used mineral salts.

Converting Parts per Million to Other Terms

Many mistakes are made in converting parts per million to other terms. Unless one understands how to do this properly, it is best not to get involved in it. This is another reason why farmers should not try mixing their own trace minerals unless they are especially well qualified to do so.

TABLE 3.6
Percent of Mineral Element in Typical Mineral Salts

Mineral salt	Percent mineral element	Mineral salt	Percent mineral element
Copper		Zinc	
Cupric carbonate	53.0	Zinc carbonate	52.1
Cupric oxide	80.0	Zinc Sulfate	22.7
Cupric sulfate	25.5	Zinc oxide	80.3
Cobalt		Iron	
Cobaltous carbonate	49.5	Iron oxide	69.9
Cobaltous sulfate	24.8	Ferrous sulfate ($7H_2O$)	20.1
Cobaltous oxide	73.4	Ferrous sulfate (no water)	36.7
Manganese		Ferrous carbonate	41.7
Manganous carbonate	47.8		
Manganous sulfate	32.5	Iodine	
Manganous oxide	77.4	Potassium iodide	76.4
		Calcium iodate	60.0

Table 3.7 gives information on converting parts per million into other terms or vice versa. It can be used as a guide in converting from one term to another.

IX. IODINE

Iodine is needed by the pig for growth and reproduction. A hormone called thyroxine controls the rate of metabolism in the body. Iodine is needed for thyroxine production. If the pig does not have enough iodine in the diet, the thyroid gland in the neck will enlarge in an attempt to make more thyroxine, then goiter develops. Over half of the iodine in the animal's body is in the thyroid gland. Older animals rarely show any symptoms of a lack of iodine. The goiters are usually found in the young at birth; this is a result of a deficiency of iodine in the diet of the mother during the gestation period.

TABLE 3.7
Table of Equivalents

ppm	%	gm/kilo	gm/lb	gm/100 lb	gm/ton	oz/100 lb	oz/ton	lb/100 lb	lb/ton
1.0	0.0001	0.001	0.00045	0.0453	0.907	0.0016	0.032	0.0001	0.002
10,000.0	1.0	10.0	4.53	453.6	9072.0	16.0	320.0	1.0	20.0
1,000.0	0.1	1.0	0.45	45.3	907.0	1.6	32.0	0.1	2.0

Iodine deficiencies occur in certain areas of the world where soil and water are deficient in iodine. In the United States, the main iodine-deficient areas are in the Great Lakes region and westward toward the Pacific Coast. However, it is possible that other borderline iodine-deficient areas may exist in the United States.

A. Effect of Iodine Deficiency in the Pig

A sow which lacks iodine in her diet will give birth to "hairless pigs" (Fig. 3.11). These pigs are bloated and have thick skins and puffy necks. Some of

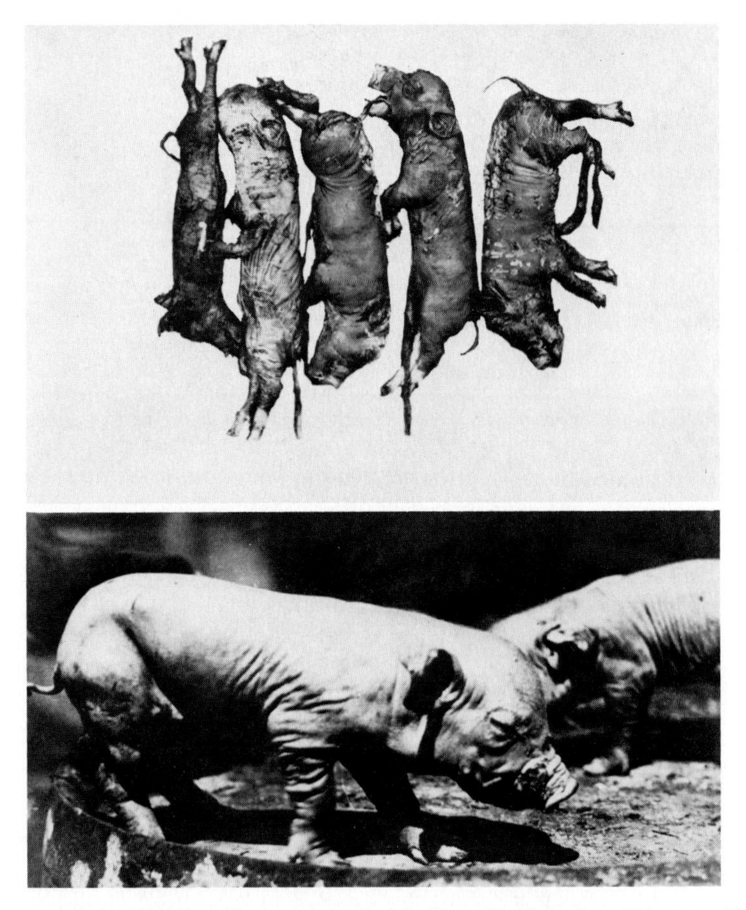

Fig. 3.11. Iodine deficiency. The top photo shows a litter of hairless pigs which were stillborn. The other shows a live hairless pig. (Courtesy of the late J. W. Kalkus, Western Washington Exp. Sta.)

them are born dead; others alive, but weak (17). Thus, an iodine deficiency causes severe losses in pigs.

B. Iodine Requirements of the Pig

The 1973 National Research Council report (9) set the iodine requirement at 0.2 ppm in the diet. A stabilized source of iodine should be used since iodine is easily destroyed in the diet unless it is protected against oxidation. The iodine requirements of the pig can be met by using stabilized, iodized salt containing 0.007% iodine at a level of 0.5% of the diet. In Florida studies, pregnant sows were fed 2500 ppm of iodine as potassium iodide for 30 days before farrowing without harmful effects (18). Studies at North Carolina with the growing-finishing pig showed that the minimum toxic level of iodine (from calcium iodate) appears to be between 400 to 800 ppm in the diet (19). A level of 400–1600 ppm of iodine depressed liver iron concentration. Growth rate, feed intake, and hemoglobin levels were depressed when pigs received diets containing either 800 or 1600 ppm added iodine. This study showed that the availability of iron or its metabolism was interfered with by excessive iodine. Iron supplementation reversed most of the effect of excess iodine. But, there is a good safety factor between the level of iodine needed (0.2 ppm) and that which causes a harmful effect (above 400 ppm). To be on the safe side, however, one should not use much excess iodine in the diet. It is never a good idea to feed any mineral much in excess of the requirements.

X. IRON AND COPPER

Both iron and copper play an important part in hemoglobin formation. They are thus essential in preventing nutritional anemia (20,21). Copper is not contained in hemoglobin, but a trace of it is necessary before the body can utilize iron for hemoglobin formation (22). Both iron and copper are also constituents of different enzymes. Thus, iron and copper are very important, and affect every organ and tissue of the body.

A. Copper Deficiency in Young Pigs

The Hormel Institute (23) produced a copper deficiency in pigs on a raw whole milk diet. Besides anemia, an unusual leg condition developed. The animals lacked rigidity in the leg joints. The hocks became excessively flexed and this forced the animal to a sitting position. The forelegs showed various types and degrees of crookedness. In the extreme state, the animals lost the use of the forelegs and, although not paralyzed, they remained in a prone position (Fig.

3.12; also see Fig. 3.13). In some cases, the administration of copper completely cured the symptoms.

B. Iron and Copper Deficiency in Suckling Pigs

Milk is very low in iron and copper. Feeding iron and/or copper to pregnant or lactating sows has not been effective in preventing anemia in suckling pigs because these minerals are not secreted in the sow's milk in amounts large enough to prevent anemia.

Pigs that are farrowed indoors, away from pasture or soil, should be watched for nutritional anemia. A pig with anemia lacks a healthy pink color and its blood looks watery. The pigs lose appetite, become weak and inactive, exhibit a rough hair coat and develop a mild diarrhea. The pig's breathing becomes heavy or labored in an effort to get enough oxygen to the various tissues of the body. A little exercise leaves the pig exhausted and trying to catch its breath. This is brought about by the lack of hemoglobin in the blood (the hemoglobin carries oxygen from the lungs to all parts of the body). The pig tries to make up for this lack by breathing harder to make the available hemoglobin carry as much oxygen as possible throughout the body. Figure 3.14 shows a pig with anemia (44).

An anemic pig may never do too well and turn out to be a runt. The most severe cases will die by 3 to 4 weeks of age. Anemic pigs are also more susceptible to stress factors and diseases which cause the main losses due to secondary effects such as stunted growth, scouring, and pneumonia. In advanced anemia, one may note a "thumping" condition which is due to pneumonia or

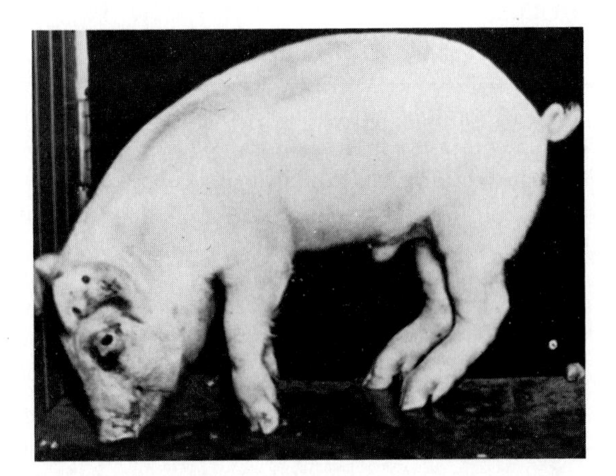

Fig. 3.12. Copper deficiency. Note the drawing under of the rear legs and crookedness of the forelegs, the swelling in the region of the hocks, a turning of the rear legs and the extreme weakness of the carpal joints in the foreleg. (Courtesy of L. E. Carpenter, H. S. Teague and Hormel Institute.)

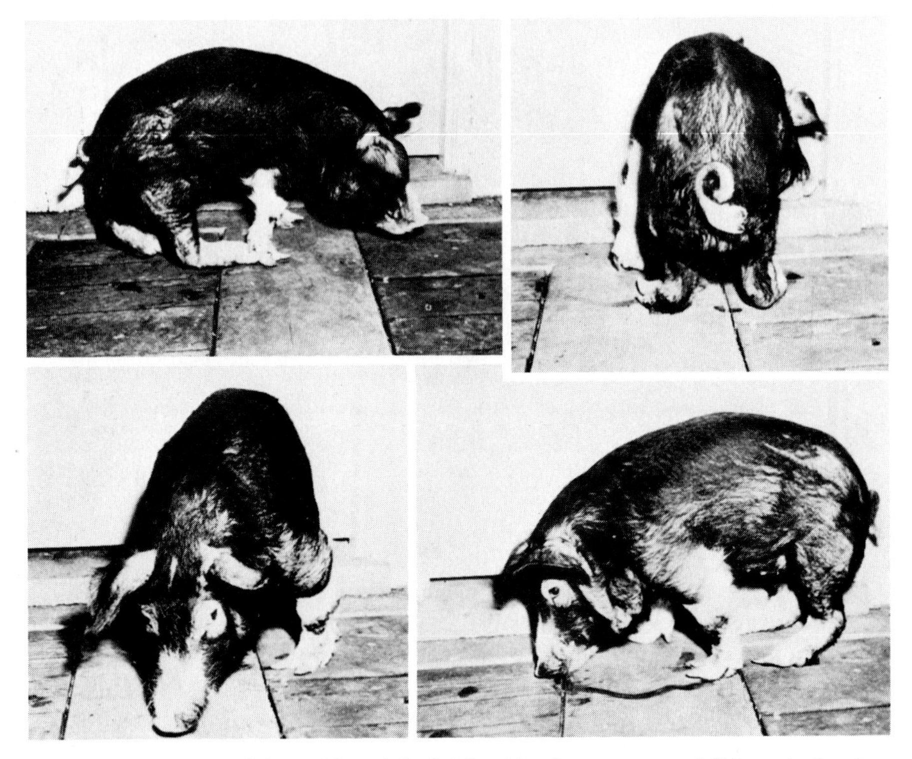

Fig. 3.13. Copper deficiency. Note skeletal deformities due to a copper deficiency in the pig. Both hind and forelegs have been affected severely. [Courtesy of G. E. Cartwright and M. M. Wintrobe, University of Utah. With permission of *Blood* **7,** 1058 (1952).]

fluid in the lungs. Many swine producers do not realize that anemia can be the cause of this problem. Thus, an iron deficiency should always be avoided if at all possible. It is surprising how much anemia still occurs in the United States in spite of many ways to provide iron for the pig.

C. Preventing Anemia in Suckling Pigs

Anemia can be prevented by swabbing the sow's udder with copperas (ferrous sulfate) or another soluble iron salt until the pigs are 4–6 weeks old (24). To make copperas solution, dissolve 1 lb of ferrous sulfate in a quart of hot water. Put this solution on the udder with a clean paint brush, or by a swab made by tying a piece of cloth to a stick. Copperas contains enough copper as an impurity to supply the needs of the pig, but not all personnel working with hogs will swab the sow's udder regularly, and as a result the suckling pigs become iron-

Fig. 3.14. Iron deficiency. Left, anemic pig. Note listlessness and wrinkled skin. These symptoms, along with paleness about eyelids, ears and nose, as well as low hemoglobin, are typical of baby pig anemia. Right, normal pig given iron. (Courtesy of H. D. Wallace, University of Florida.)

deficient. If this were not the case, anemia would have been eliminated many years ago since information on its prevention has been available since 1923.

Placing small amounts of clean soil in a sow's farrowing pen is still followed by some swine producers. Using soil, however, may present a problem in that it may be contaminated with diseases and/or parasites. Moreover, many will not provide the soil frequently enough or will forget about doing it until the pigs develop anemia.

Pigs farrowed on pasture and dirt will get the iron they need there. There are oral tablets and pastes which can be used. A disadvantage with tablets is that some of the pigs will cough them up. One can also supply iron-rich sweetened feeds or pellets which can be fed on the floor of the sow's pen.

The extra daily labor involved in supplying iron at frequent intervals and the undependability of some workers have led to the widespread use and adoption of injectable iron preparations. It is one of the most popular and dependable means of preventing anemia. One injection of 150 mg (or some prefer 200 mg) of iron at 2–3 days of age will be enough to take care of the pig's needs until it is about 3 weeks old. By this time, the pig should be consuming enough feed to get its iron needs from it.

D. Iron and Copper Requirements

The newborn pig will contain about 47 mg of iron in body stores. However, there is considerable variation in iron stores, even among pigs of the same litter. These iron stores will last the pig about a week since it is estimated the pig will need 7 mg of iron absorbed daily for normal growth. In some cases, however, the iron stores are dangerously low by 3 or 4 days of age. Sow's milk is very low in iron and contains only 1 mg of iron/liter (1.057 quart). This is equivalent to about 1 ppm of iron in milk, whereas the pig's requirement for iron is 80 ppm in the total diet consumed. Therefore, unless the pig is given extra iron, it soon becomes anemic. Baby pigs depend quite heavily on milk for their food supply, especially during the first 2–3 weeks of life. The baby pig may nearly double its weight by one week of age and redouble it again in another week or 10 days. So, the pig rapidly outgrows its iron supply. The faster a pig grows, the quicker its iron reserve is used up. So, iron supplementation is very important for the baby pig. The copper needs of the pig are 6 ppm in the total diet. As one adds a source of iron to the baby pig, there is usually enough copper as a contaminant in the iron to provide the copper needed for hemoglobin formation.

E. Hemoglobin in Blood

The hemoglobin level in the blood is an indicator of the iron status of the pig. The 1973 National Research Council report on the nutrient requirements of the pig (9) gives the information tabulated below on hemoglobin level and anemia.

Hemoglobin (gm/100 ml of blood)	Comments
10 or above	Normal level; adequate iron and optimum performance
9	Minimum level required for average performance and the dividing line between normality and anemia
8	Borderline anemia; iron treatment needed
7 or below	Anemic condition that has been shown to retard growth
6 or below	Severe anemia accompanied by marked reduction of performance
4 or below	Severe anemia that can be expected to result in increased mortality rate

F. Effects of Excess Iron

Studies at Purdue (25) and Japan (26) have shown that 5,000 ppm of iron in the diet resulted in toxic effects. This level, however, is over 62 times the recommended level of 80 ppm of iron in the diet. So, there is a big safety factor in the use of iron.

Excess iron interferes with phosphorus utilization by combining with it and making both unavailable to the animal. The sources of phosphorus used will influence the results obtained since they vary in their ability to precipitate iron as an iron-phosphorus complex which is not absorbed from the digestive tract (25). Purdue studies showed that 5000 ppm of iron in the diet decreased rate of gain, the level of phosphorus in the blood, and the ash content of the bone (25). The pigs developed phosphorus-deficiency symptoms. Purdue also showed that excess iron was more toxic with a level of 0.3% phosphorus in the diet when compared to 0.6 or 1.2% phosphorus levels. So, higher levels of phosphorus will help counteract the effect of excess iron in the diet. Therefore, the ratio of iron to phosphorus in the diet is important.

G. Iron Protects against Gossypol Toxicity

The use of some cottonseed meal sources as the only protein supplement in swine diets is limited because of the possibility of gossypol toxicity. However, low gossypol cottonseed meal combined in a 50:50 ratio with soybean meal gives excellent results in swine feeding. University of North Carolina studies have shown that 400 ppm of iron will counteract the toxic effects of 400 ppm of gossypol in the diet (27). This 1:1 ratio of iron to gossypol is needed. Less than one part of iron/one part of gossypol was not as effective. The level of 400 ppm of gossypol in the diet will kill pigs after 37–57 days. However, the addition of 400 ppm of iron prevented the deaths or any harmful effects from gossypol toxicity. Thus, using 400 ppm of iron in the diet is an effective means of preventing any harmful effects from cottonseed meal which would add a total of 400 ppm of gossypol to the diet. It has been postulated that gossypol reacts with iron in the liver, and the iron-gossypol complex is then excreted via the bile (28).

H. Iron and Copper Availability

Copper sulfate, copper carbonate and copper oxide appear to be about equally effective for the pig (9).

Iron sulfate and ferric ammonium citrate are the most effecitve in preventing anemia. Iron carbonate is less effective than iron sulfate (9). Iron oxide has a low solubility and is virtually ineffective for the animal. At best, it is only 4–5% available (29)

I. High Levels of Copper Feeding

Some farmers are using copper at levels of 125–250 ppm in the diet as an antimicrobial compound. No toxic effects have been reported when levels of 125–150 ppm of copper are used in the diet. But, occasional toxic effects have been reported when 250 ppm of copper is used. These harmful effects are probably due to a lack of zinc and iron. The addition of 130 ppm of zinc and 150 ppm of iron has prevented harmful effects from using 250 ppm of copper in the diet. With the baby pig, however, a higher level of 270 ppm of iron is needed to counteract 250 ppm of copper in the diet. In U.S. trials, in about 75% of the cases when an antibiotic is combined with copper an additive effect on growth and feed efficiency is obtained. Therefore, while copper may be beneficial in certain situations, it does not eliminate the use of effective antibiotics. Both play a role in swine feeding.

XI. COBALT

Cobalt is a constituent of vitamin B_{12}. It has been shown that the cobalt fed to sheep is used by the microorganisms in the rumen to make vitamin B_{12}. The need for cobalt by the pig has not been definitely established. Several workers (30–33) have reported that the addition of cobalt to practical swine diets increased the rate of growth and efficiency of feed utilization. North Dakota workers (34) obtained a good growth response from pigs fed added cobalt carbonate at a level of 885 mg of cobalt/lb of diet. However, it is doubtful if cobalt benefits growth if the diet contains adequate amounts of vitamin B_{12} (9,35).

The intestinal microorganisms of the pig are capable of producing some vitamin B_{12} in the lower part of the intestinal tract when cobalt is present. How much of it would be absorbed is not known. But it would be available in the feces, and it is known that pigs will consume some feces. Thus, if vitamin B_{12} is limiting or low, the addition of cobalt may be beneficial. No requirement for cobalt for the pig has been determined, but levels of about 1 ppm in the diet are often added to swine feeds. This level should be adequate for those wishing to use cobalt.

Recent studies at North Carolina (61) show that excess cobalt at levels of 400–600 ppm in the diet causes toxic effects in the pig. The symptoms of cobalt toxicity produced were anorexia, growth depression, stiff-leggedness, incoordination and extreme muscular tremors (Fig. 3.15). The pigs were easily exhausted by minimal exercise. They had a "hump" in their back that was also observed in their dressed carcass. A microcytic, hypochromic anemia developed when a high level (600 ppm) of cobalt was fed. The addition of 200 or 400 ppm of iron to the diet did not have an effect on the toxicity caused by the feeding of cobalt. When iron, manganese and zinc were added in combination, the toxicity produced by the feeding of the 400 ppm level of cobalt was completely alleviated, and was partially overcome when 600 ppm of cobalt was added. The addition of 0.5 or 1.0% methionine to a diet containing 600 ppm of cobalt completely alleviated the toxicity.

A cobalt × iron interrelationship at the absorptive level was demonstrated by the observed difference in serum levels and tissue retention patterns of cobalt and

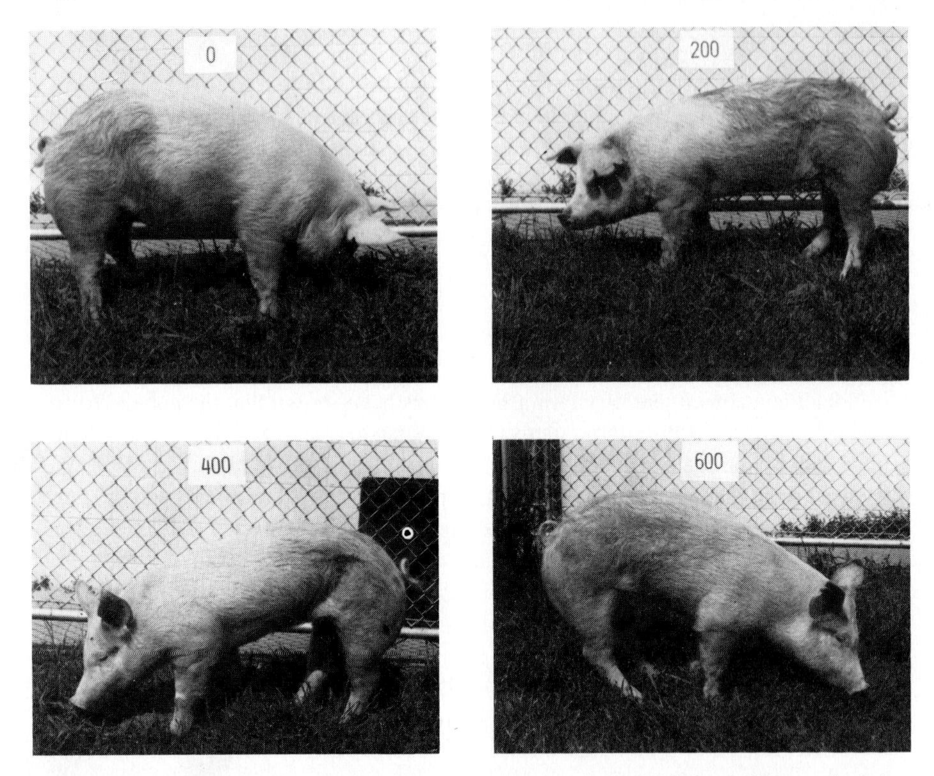

Fig. 3.15. Cobalt toxicity in the pig. Note the smaller size of pigs fed 400 and 600 ppm of cobalt in the diet. Other effects of the toxicity were anorexia, muscular tremors, incoordination, stiff-leggedness, hump back (spinal curvature also seen in carcasses) and growth depression. (Courtesy of A. J. Clawson and D. W. Huck, North Carolina State University.)

iron when various increments of cobalt and iron were fed to the pig. It appears that cobalt and iron may share a common absorptive pathway and are mutually antagonistic.

The addition of methionine to the diet completely alleviated the toxic effects of cobalt, and serum iron levels were restored to near normal, but serum cobalt levels remained elevated. Thus, it appears that methionine may not interfere with cobalt absorption but relieves the antagonistic effect of cobalt on iron absorption.

XII. MANGANESE

A deficiency of manganese causes decreased growth, feed efficiency, and reproduction. It also affects fat deposition and the sense of balance in newborn pigs (Fig. 3.16).

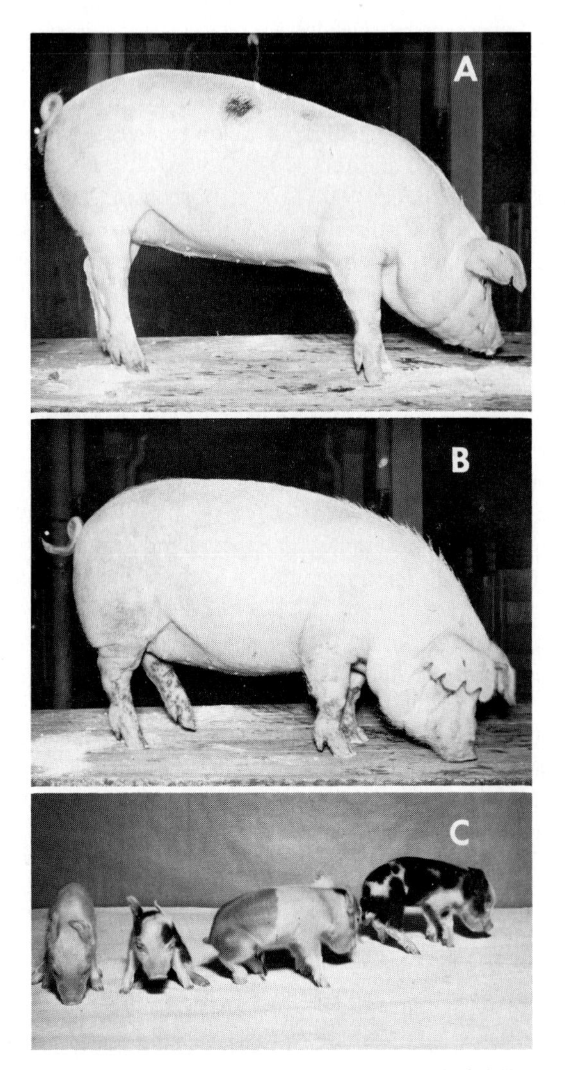

Fig. 3.16. Manganese deficiency. Top photo shows 132 day-old gilt fed 40 ppm manganese since she weighed 8.5 lb. Middle photo shows a littermate gilt started at same weight and 132 days old fed 0.5 ppm of manganese. Note increased fat deposition due to low manganese diet. Lower photo shows a litter from sow fed 0.5 ppm of manganese. The pigs showed weakness and poor sense of balance at birth. (Courtesy of W. M. Beeson, Purdue University.)

Pigs definitely need manganese in the diet, but conflict exists in the published data as to how much. Experimental work on the manganese requirements of the pig has varied a great deal. Some of this is due to differences in the amount of manganese stored in body tissues. It also appears that the higher the ash content of the diet, the more manganese is needed (36,37). This accounts for some of the variability in manganese requirements determined by different scientists. Early work in Arkansas showed that pigs thrived well on a diet containing less than 0.5 ppm of manganese (38,39), and reproduction was satisfactory when the manganese was increased to 6 ppm (39). Work in Wisconsin, however, showed that growth and efficiency of feed utilization were improved by increasing the level of manganese in the diet from 12 to 40 ppm (40). Increasing the level of manganese above 40 ppm did not improve performance. No significant benefits in reproduction and lactation were obtained by adding manganese to a diet containing 12 ppm. However, sows fed 40 ppm of manganese showed a trend toward slightly better performance (40).

The National Research Council Committee on Swine Nutrient Requirements (9) recommends 20 ppm of manganese in the diet as adequate for normal growth and reproduction.

Excess manganese is harmful as was shown by the Wisconsin workers (40). A level of 500 ppm of manganese retarded appetite and growth, especially during the latter part of the trial. The pigs showed stiffness of limbs and stilted gait toward the end of the experiment.

North Carolina studies showed that excess manganese in the diet increased iron needs by decreasing the hemoglobin level in the pig (41). Levels of 50–125 ppm of manganese in the diet had no effect on hemoglobin level in the blood. A supplement of 400 ppm of iron in the diet overcame the depressing effect of 2000 ppm of manganese in the diet on the hemoglobin formation of baby pigs.

Georgia studies showed that adding either 1000 or 2000 ppm of manganese to the diet decreased rate of gain and feed efficiency (42).

Michigan studies have shown that manganese sulfate, manganese carbonate, and manganese oxide are equally well utilized by the growing pig (43).

XIII. ZINC

Zinc is widely distributed throughout the body and plays an essential role in body processes. Alabama workers showed in 1955 that "swine dermatitis" or "parakeratosis" could be cured or prevented by zinc (45). Since then, a great deal of research has been conducted on this problem. But this condition is still found throughout the United States. Most of it occurs because of excess calcium in certain diets which increases the need for zinc. The symptoms obtained are reduced growth rate and appetite, skin lesions which look like mange, a dark watery exudate, diarrhea, vomiting, and in severe cases death. With borderline deficiencies, appetite and growth decrease in some animals, and they may show a fading or bleaching of the hair coat. The thymus gland atrophies. A decrease in litter size occurs with the sow.

A pig with parakeratosis responds very quickly and dramatically to zinc (Figure 3.17). Appetite increases immediately and an improvement in weight gain and skin condition is obvious in a week. Pigs soon recover from the skin lesions and other symptoms and may be completely recovered in one month.

A. Factors Affecting Zinc Needs

The need for zinc is increased by excess calcium in the diet (46,47). The level of calcium which causes parakeratosis, however, will vary considerably. Sometimes a high calcium level diet which supposedly should cause parakeratosis does not do so, and sometimes parakeratosis occurs with a low level of calcium. The zinc in soybean meal, cottonseed meal, sesame meal and other plant protein supplements is less available to the pig (9). This is because these supplements are high in phytic acid, which combines with zinc to form zinc phytate, which is insoluble in the intestinal tract and thus cannot be absorbed by the pig. Therefore, the zinc in plant protein concentrates is less available than that in animal protein supplements such as meat meal and fish meal which contain no phytic acid.

A high level of copper in the diet tends to deplete the stores of zinc in the liver. Some cases of toxicity in pigs fed 250 ppm of copper in the diet are prevented by adding 130 ppm of zinc to the diet. Pigs which are deficient in iron are more susceptible to zinc toxicity. A diet fed in a dry form is more apt to produce parakeratosis than if fed as a wet mash. Autoclaving of the feed also seems to increase the availability of the zinc in it. The level of phosphorus may also affect zinc needs. The requirement for zinc by the male pig appears to be higher than the female. Thus, there are many factors which influence zinc requirements and hence there is a variable need for it by the pig.

B. Zinc Requirement

The requirement for zinc by the pig is 50 ppm in the diet (9). If the calcium level in the diet is excessive, the addition of 50–100 ppm of zinc to the diet will not always completely prevent the growth depression and poor feed conversion associated with parakeratosis, although it will prevent the typical skin lesions (45–47). Therefore, under some conditions, a level of 100–150 ppm of zinc is needed. But in most cases, a level of 100 ppm of zinc should be adequate. This level should provide enough zinc to (1) protect against a modest excess of cal-

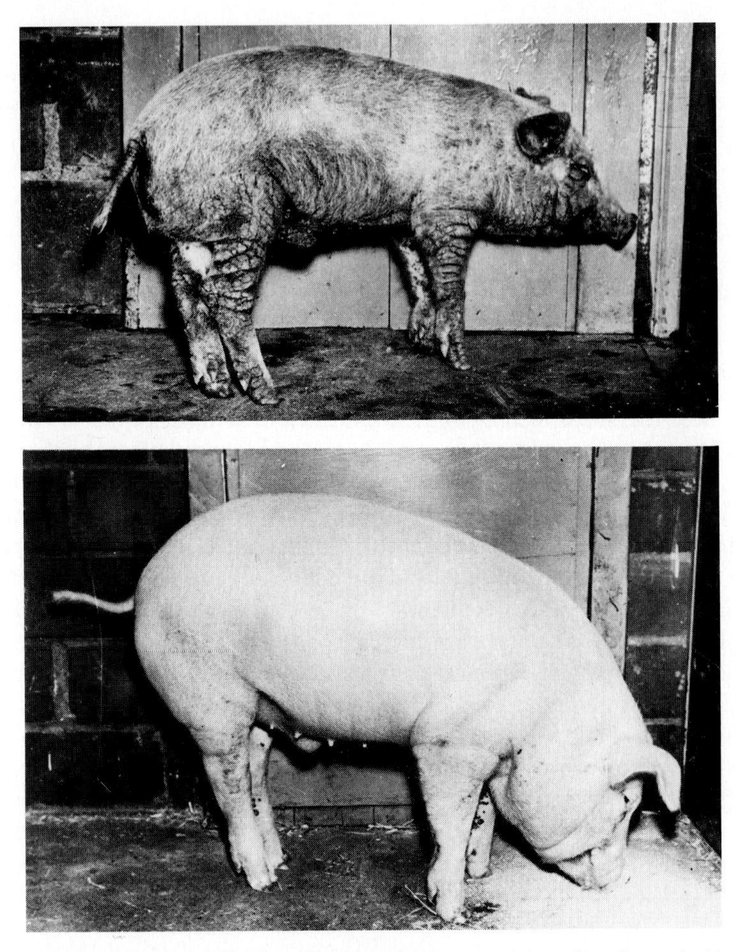

Fig. 3.17. Zinc deficiency. Top photo shows a pig with a typical case of parakeratosis. Note same pig in lower photo 41 days later after zinc was added to the diet. Improvement was noted within a week after zinc was added. (Courtesy of L. E. Hanson, University of Minnesota.)

cium; (2) protect against possible toxicity in the presence of 250 ppm of copper, if used in the diet; and (3) protect against a modest amount of phytic acid (which ties up zinc) in the diet.

C. Excess Zinc

There is no danger in using 150 ppm of zinc since 1000 ppm have been fed without any harmful effects. A level of 2000 ppm of zinc is necessary to produce toxicity (48). Thus it is recommended that a level of 100–150 ppm of zinc be used. This higher level is low cost insurance against the many factors which can increase zinc needs.

The zinc compounds commonly used are zinc oxide, zinc carbonate, and zinc sulfate. They appear to be equally available for the pig.

XIV. MAGNESIUM

Magnesium is closely associated with calcium and phosphorus. It is an essential constituent of bones and teeth and is needed in many body functions. The symptoms of a magnesium deficiency in order of appearance are hyperirritability, muscular twitching, reluctance to stand, weak pasterns, loss of equilibrium, tetany, and finally death. There is a high mortality rate in baby pigs fed low magnesium diets. The magnesium-deficient pigs also exhibit the "stepping syndrome" (Figure 3.18) which causes it to keep stepping or lifting its hind leg almost continuously while standing up (49).

To date, it is thought that magnesium supplementation of swine diets is not needed. But some feed industry nutritionists feel the addition of magnesium may help in preventing hyperirritability and tail biting in pigs under confined conditions. This needs to be verified in more research studies to determine if true. Purdue (50) and a number of other universities have recently studied the addition of magnesium to swine diets, but none have reported any beneficial effects.

Fig. 3.18. Magnesium deficiency. Five-week-old pig showing stepping syndrome; the pig keeps stepping almost continuously while standing. Weakness of pasterns is apparent. (Courtesy of E. R. Miller and D. E. Ullrey, Michigan State University.)

A. Magnesium Requirement

The 1973 National Research Council report (9) states that the exact magnesium requirement of the pig is not known. It suggests, however, that swine diets contain a minimum of 400 ppm of magnesium since this level is adequate for the 3-week-old pig.

B. Excess Magnesium

Formerly it was thought that excess magnesium would cause a large loss of calcium. Recently, however, it has been found that unless the excess of magnesium is very large, no harmful effects occur, provided the diet contains an ample supply of calcium and phosphorus. Thus, a small excess of magnesium will not disturb calcium retention, though it may tend to increase the requirements for calcium and phosphorus by the pig. Proof of this is the fact that dolomitic limestone is a satisfactory source of calcium for bone formation despite its magnesium content, and it is used widely in swine diets (5).

XV. POTASSIUM

Potassium plays some very important roles in the body; thus it is necessary that diets be adequate in this mineral. However, the usual feeds used in swine diets appear to contain adequate amounts of potassium to meet the needs of the pig. A deficiency of potassium has not been observed in pigs under practical farm conditions.

Work at Wisconsin (51) showed that pigs should receive a diet containing 100–120 mg potassium/kilogram body weight per day. This amounts to 0.23–0.28% of the diet. The average daily intake of potassium in their experiments amounted to 5.15 gm/100 lb of liveweight. This figure is about double that obtained in 1942 by California workers (52); their figure was 2.36 gm. As far as can be ascertained, the pigs in the California experiment gained at one-half the rate of the pigs in the Wisconsin trial. It is possible that the decrease in rate of gain caused a lower requirement for potassium in the pigs in the California study. An Illinois study showed the optimum total level of potassium in the diet to be approximately 0.26% (60).

There is still some question concerning the old theory that an excess of potassium causes the animal to excrete more sodium, and this in turn increases the need for more salt. Present opinion indicates that a high potassium intake may result in increased sodium excretion by the body.

No information is available on potassium requirements for reproducing and lactating sows. But since feeds contain more than is needed for growth, it is likely they also contain enough for reproduction and lactation.

XVI. SULFUR

Sulfur is very important since it plays an essential role in many body functions. It occurs in the body mostly in the form of the sulfur-containing amino acids cystine and methionine. It is also present in insulin and in the vitamins thiamin and biotin. Sulfur is present in the body almost entirely in such organic forms. The pig has little ability to use inorganic forms of sulfur (53).

XVII. MOLYBDENUM

Excess molybdenum in the soil occurs in certain areas of the United States and the world. Excess molybdenum in the forage affects ruminants. The use of copper sulfate counteracts most, if not all, the harmful effects of the excess molybdenum in the forage for cattle. However, this condition is more complex than a simple copper-molybdenum relationship. Studies at the Florida Station (54) have shown that the pig is not affected by excess molybdenum at the levels usually found in feeds. No toxicity occurred from feeding 1000 ppm of molybdenum in the diet to pigs for 3 months. The molybdenum fed to the pigs is excreted rapidly since only 0.2% of a dose given by mouth remained in the blood after 30 hours. It was also found that little or no molybdenum is transferred from the sow to the developing fetuses (55). Evidently, the placenta serves as a barrier, preventing molybdenum transfer from the sow to the developing young. As far as is known, excess molybdenum is not a problem in practical swine diets.

XVIII. SELENIUM

Selenium is an essential mineral element. Although its function is closely related to that of vitamin E, the exact biological mechanisms underlying the interrelationship between vitamin E and selenium are unknown.

Pigs deficient in selenium show the following symptoms: sudden death, liver necrosis, edema, waxy degeneration of muscle tissue, yellowish-brown discoloration of body fat, slow growth and reduced feed intake (56). A deficiency of selenium also decreases reproduction. It causes stillborn and small pigs (59). Some reports indicate that selenium or vitamin E supplementation will not increase growth rate but will prevent death and clinical signs of a deficiency. Figure 3.19 shows a vitamin E–selenium deficiency as produced by A. L. Trapp and others at Michigan State University. Most of the symptoms for the pig have also been associated with vitamin E deficiency. The exact roles of selenium and vitamin E are not yet entirely understood. Most scientists refer to a selenium and/or vitamin E deficiency since it is not clear which one is involved or whether

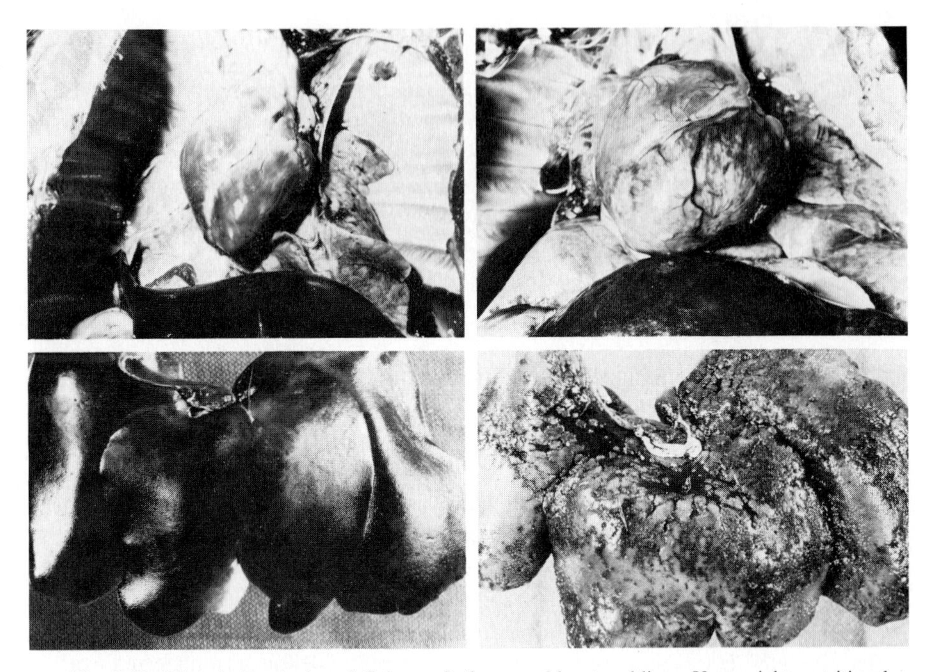

Fig. 3.19. Vitamin E–selenium deficiency. Left, normal heart and liver. Upper right, nutritional muscular dystrophy of myocardium. Lower right, dietary hepatic necrosis and fibrosis. (Courtesy of A. L. Trapp, Michigan State University.)

both of them are. The level of vitamin E in the diet will influence the need for selenium and vice versa. Although vitamin E and selenium influence the need for each other, they both have a nutritional role in the body. Both also have an antioxidant effect. The level of protein and the sulfur-containing amino acids (cystine and methionine) will also influence the need for vitamin E and/or selenium. It is thought this may be due to the antioxidant effect of cystine and methionine. The kind of diet used, its level of biologically active form of vitamin E, and the availability of its selenium content will also have an effect on the level of vitamin E and selenium used. The selenium level in some feeds may be low in availability. When selenium is added to the diet, it is recommended that both vitamin E and selenium be used since neither one can replace the other entirely.

The U.S. Food and Drug Administration (FDA) approved the addition of selenium to swine feeds in February, 1974. Selenium-deficient areas in the United States have been shown to occur in 40 states (56). Since feeds produced in one state may be shipped and fed in others, one needs to consider the possibility of selenium deficiencies occurring throughout the United States. Analyses of corn in 10 midwestern states showed that there are selenium-deficient areas in 7 of the states.

A. Selenium Requirement

The 1973 report of the National Research Council recommends a level of 0.1 ppm of selenium in the total diet as the requirement for the pig (9). This is the level which FDA allows to be added to the diet of the pig. FDA approved the use of either sodium selenite or sodium selenate as the source of selenium. Both will do a good job in animal feeding.

B. Excess Selenium

Selenium in excess is toxic, so one needs to be careful of its use. But selenium must be added at 50–80 times the required amount to be toxic (9). So, there is a big safety factor between the level needed and that which may cause harmful effects. Selenium is no more toxic than many other trace minerals which have been used successfully in swine diets for many years.

With the young growing pig, an excessive level of 5–8 ppm of selenium in the diet causes emaciation, loss of hair, cirrhosis, atrophy of the liver and hooves separate from coronary bands (9). Figure 3.20 shows symptoms obtained from excess selenium in the diet by South Dakota scientists R. C. Wahlstrom, A. L. Moxon, and others. With sows, a level of 10 ppm of selenium causes reduced conception rate and pigs that are small, weak or dead at birth (58).

South Dakota scientists found that 0.02% arsanilic acid and 0.005% 3-nitro-4-hydroxyphenylarsonic acid in the diet counteracted the effects of selenium toxicity (up to 10 ppm of selenite selenium in the diet) (57). However, care must be used with arsenic compounds, since they are also toxic if used in excess.

XIX. FLUORINE

Excess fluorine fed to an animal causes pitting, erosion and wearing down of the teeth, softening and overgrowth of the bones, loss of appetite, poor gains, harmful changes in the kidneys, and finally, death. In severe cases, it causes the teeth to wear down, and the animals then have difficulty chewing feed. Sometimes the pulp cavities of the teeth become exposed. This causes animals to be reluctant to drink cool water because it pains them to do so.

It takes time for an animal to show the symptoms of excess fluorine in the diet. If fluorine is consumed only for a short time, it accumulates in the bones and teeth without any harmful effects. This protects the pig for awhile. But, once the bones become saturated with fluorine, that which is consumed starts producing its poisonous effect in the body. So one may feed high levels of fluorine for a long time without realizing the animal is accumulating it; but, eventually the animal will be affected severely. For example, in one Wisconsin experiment, it

Fig. 3.20. Selenium toxicity. Top pig, note emaciation and hooves separate from coronary bands. Note loss of hair in bottom photo. (Courtesy of R. C. Wahlstrom and A. L. Moxon and South Dakota Agr. Exp. Sta.)

took 3 years before a borderline intake of fluorine affected reproduction in dairy cows.

The majority of feed grade phosphates which originate from rock phosphate deposits have fluoride levels ranging from 2 to 5% and average about 3.5%. In the United States, a phosphate which is to be classified as defluorinated phosphate must contain no more than one part of fluorine to 100 parts of phosphorus.

The 1974 National Research Council report on "Effects of Fluorides in Animals" recommended the levels of fluorine shown in the following tabulation as being safe for the pig (8).

Safe fluorine level in total diet	Fluorine from sodium fluoride or other soluble fluorides		Fluorine from defluorinated natural rock phosphates	
	Sows	Finishing pigs	Sows	Finishing pigs
As ppm in diet	100–150	150	150–225	225
As percent in diet	0.010–0.015	0.015	0.015–0.0225	0.0225

The figures in the tabulation should be used as a guide in determining the level of fluorine which should not be exceeded in the total diet fed to swine.

XX. OTHER MINERALS

There are other mineral elements which are found in animal tissues. They may have some metabolic effect but, to date, no requirement has been demonstrated for swine or has a need for supplementing swine diets with them been shown. These mineral elements include molybdenum, fluorine, bromine, aluminum, nickel, chromium, arsenic, boron, cadmium, tin, vanadium, silicon, cesium, lead, lithium, strontium, barium, radium, and germanium. As soils decline in fertility and as the productivity level of swine increases, there may be a need for adding some of these minerals to swine diets in the future.

REFERENCES

1. Aubel, C. E., and J. S. Hughes, *Am. Soc. Anim. Prod., Rec. Proc. Annu. Meet.* p. 334 (1937).
2. Aubel, C. E., J. S. Hughes, and H. F. Lienhardt, *Kans., Agric. Exp. Stn., Tech. Bull.* **41** (1936).
3. Bohstedt, G., *Ohio, Agric. Exp. Stn., Bull.* **395** (1926).
4. Bohstedt, G., *Am. Soc. Anim. Prod. Rec. Proc. Annu. Meet.* **32,** 14 (1939).
5. Bohstedt, G., *Wis., Agric. Exp. Stn., Circ.* **297** (1940).
6. Combs, G. E., Jr., G. C. Ashton, J. Kastelic, V. C. Speer, M. Emmerson, and D. V. Catron, *J. Anim. Sci.* **14,** 1198 (1955).
7. Loeffel, W. J., R. R. Thalman, F. C. Olson, and F. A. Olson, *Nebr., Agric. Exp. Stn., Res. Bull.* **58** (1931).
8. Shupe, J. L., C. B. Ammerman, H. T. Peeler, L. Singer, and J. W. Suttie, *N.A.S.—N.R.C.* publication (1974).
9. Cunha, T. J., J. P. Bowland, J. H. Conrad, V. W. Hays, R. J. Meade, and H. S. Teague, *N.A.S.—N.R.C.* publication (1973).
10. Pond, W. G., and J. H. Maner, "Swine Production in Temperate and Tropical Environments." Freeman, San Francisco, California, 1974.
11. Anonymous, *Purdue, Agric. Exp. Stn., Mimeo* No. 18 (1945).
12. Vestal, C. M., *Purdue Univ., Agric. Exp. Stn., Circ.* **18** (1945); **20** (1946); **23** (1947); and **28** (1947).

13. Combs, G. E., Florida, Agricultural Experiment Station, Gainesville, personal communication, 1975.
14. Hagsten, I., and T. W. Perry, *J. Anim. Sci.* **39,**182 (1974).
15. Bohstedt, G., and R. H. Grummer, *J. Anim. Sci.* **13,**933 (1954).
16. Sutton, A. L., and V. B. Mayrose, *Purdue, Agric. Exp. Stn., Mimeo* pp. 35–36; *Purdue Swine Day Rep.,* p. 26 (1974).
17. Kalkus, J. W., *Wash., Agric. Exp. Stn., Bull.* **156** (1920).
18. Arrington, L. R., R. N. Taylor, Jr., C. B. Ammerman, and R. L. Shirley, *J. Nutr.* **87,**394 (1965).
19. Newton, G. L., and A. J. Clawson, *J. Anim. Sci.* **39,**879 (1974).
20. Braude, R., *J. Agric. Sci.* **35,**163 (1945).
21. Cartwright, G. E., and M. M. Wintrobe, *J. Biol. Chem.* **176,** 571 (1948).
22. Wintrobe, M. M., G. E. Cartwright, and C. J. Gubler, *J. Nutr.* **50,** 395 (1953).
23. Teague, H. S., and L. E. Carpenter, *J. Nutr.* **43,**389 (1951).
24. Hamilton, T. S., G. E. Hunt, and W. E. Carroll, *J. Agric. Res.* **47,**543 (1933).
25. O'Donovan, P. B., R. A. Pickett, M. P. Plumlee, and W. M. Beeson, *J. Anim. Sci.* **22,** 1075 (1963).
26. Furugouri, K., *J. Anim. Sci.* **34,**573 (1972).
27. Clawson, A. J., and F. H. Smith, *J. Nutr.* **89,**307 (1966).
28. Buitrago, J. A., A. J. Clawson, and F. H. Smith, *J. Anim. Sci.* **31,**554 (1970).
29. Ammerman, C. B., and S. M. Miller, *J. Anim. Sci.* **35,**681 (1972).
30. Noland, P. R., J. P. Willman, and F. B. Morrison, *Cornell, Agric. Exp. Stn., Farm Res. Reprint* **188** (1951).
31. Speer, V. C., D. V. Catron, P. G. Homeyer, and C. C. Culbertson, *J. Anim. Sci.* **11,**112 (1952).
32. Willman, J. P., and P. R. Noland, *Cornell, Agric. Exp. Stn., Farm Res. Reprint* **151** (1949).
33. Klosterman, E. W., W. E. Dinusson, E. L. Lasley, and M. L. Buchanan, *Science* **112,**168 (1950).
34. Dinusson, W. E., E. W. Klosterman, E. L. Wasley, and M. L. Buchanan, *J. Anim. Sci.* **12,**623 (1953).
35. Kline, E. A., J. Kastelić, G. C. Ashton, P. G. Homeyer, L. Quinn, and D. V. Catron, *J. Nutr.* **53,**543 (1954).
36. Johnson, S. R., *J. Anim. Sci.* **3,**136 (1944).
37. Keith, T. B., R. C. Miller, W. T. S. Thorp, and M. A. McCarty, *J. Anim. Sci.* **1,**120 (1942).
38. Johnston, S. R. *Am. Soc. Anim. Prod., Rec. Proc. Annu. Meet.* **33,** 34 (1940).
39. Johnson, S. R., *J. Anim. Sci.* **2,**14 (1943).
40. Grummer, R. H., O. G. Bentley, P. H. Phillips, and G. Bohstedt, *J. Anim. Sci.* **9,**170 (1950).
41. Matrone, G., R. H. Hartman, and A. J. Clawson, *J. Nutr.* **67,**309 (1959).
42. Hale, O. M., R. S. Lowery, and W. C. McCormick, *J. Anim. Sci.* **32,**12 (1971).
43. Kayongo-Male, H., *Mich., Agric. Exp. Stn., A.H.S.W.* **7325,** 107–110 (1970).
44. Wallace, H. D., Florida, Agricultural Experiment Station, Gainesville, unpublished data (1956).
45. Tucker, H. F., and W. D. Salmon, *Proc. Soc. Exp. Biol. Med.* **88,**613 (1955).
46. Lewis, P. K., Jr., W. G. Hoekstra, R. H. Grummer, and P. H. Phillips, *J. Anim. Sci.* **14,**1214 (1955).
47. Luecke, R. W., J. A. Hoefer, and F. Thorp, Jr., *J. Anim. Sci.* **14,**1215 (1955).
48. Brink, M. F., D. E. Becker, S. W. Terrill, and H. H. Jensen, *J. Anim. Sci.* **18,**836 (1959).
49. Miller, E. R., D. E. Ullrey, C. L. Zutaut, B. V. Baltzer, D. A. Schmidt, J. A. Hoefer, and R. W. Luecke, *J. Nutr.* **85,**13 (1965).
50. Krider, J. L., J. L. Albright, M. P. Plumlee, J. H. Conrad, L. Underwood, W. I. Arnold, and R. G. Jones, *J. Anim. Sci.* **35,**1092 (1972).

51. Meyer, J. H., R. H. Grummer, P. H. Phillips, and G. Bohstedt, *J. Anim. Sci.* **9,**300 (1950).
52. Hughes, E. H., and N. R. Ittner, *J. Agric. Res.* **64,**189 (1942).
53. Pfirter, H. P., J. Landis, and A. Schurch, *J. Anim. Sci.* **27,**1158 (1968).
54. Kulwich, R., S. L. Hansard, C. L. Comar, and G. K. Davis, *Proc. Soc. Exp. Biol. Med.* **84,**487 (1953).
55. Shirley, R. L., M. A. Jeter, J. P. Feaster, J. T. McCall, J. C. Outler, and G. K. Davis, *J. Nutr.* **54,**59 (1954).
56. Oldfield, J. E., W. H. Allaway, H. H. Draper, D. V. Frost, L. S. Jensen, M. L. Scott, and P. L. Wright, *N.A.S.—N.R.C.*, publication (1971).
57. Wahlstrom, R. C., L. D. Kamstra, and O. E. Olson, *J. Anim. Sci.* **14,**105 (1955).
58. Wahlstrom, R. C., and O. E. Olson, *J. Anim. Sci.* **18,**141 (1959).
59. Wastell, M. E., R. C. Ewan, E. J. Becknell, and V. C. Speer, *J. Anim. Sci.* **29,**149 (1969).
60. Jensen, A. H., S. W. Terrill, and D. E. Becker, *J. Anim. Sci.* **20,**464 (1961).
61. Huck, D. W., and A. J. Clawson, *J. Anim. Sci.* **43,**253 (1976).

4

Vitamin Requirements of the Pig

I. INTRODUCTION

Most swine producers have heard the word vitamin at some time or another. To many of them, it is some mysterious substance which they think is beyond their control. Research at various experiment stations, however, has shown that a farmer can, by following sound feeding practices, prevent vitamin deficiencies which can cost him hard-earned dollars.

Most of the work on vitamins has been conducted since 1911, when the term vitamine was coined by Casimir Funk (1). At that time, Funk was working at the Lister Institute in London. Later the letter e was dropped, and the present term "vitamin" adopted.

Vitamins are organic compounds which perform many essential functions in the body and are needed only in very small amounts. They are not related to each other as are proteins, carbohydrates, and fats. All are different in structure and also perform different functions. Thus, it is very important that all vitamins be supplied in adequate amounts in swine diets.

Pigs synthesize some of the vitamins in large enough quantities to supply their daily needs. Most vitamins, however, must be supplied in the diet, since the pig does not synthesize them or does not synthesize them in large enough amounts to supply its needs. Below is a list of the known vitamins. It must be stated, though, that there are still unidentified factors and a vitamin could be one of them.

II. LIST OF VITAMINS

It is well recognized that vitamins are as important as protein, minerals, and other nutrients in the diet. We know that unless a feed contains the proper amount and balance of the various necessary vitamins, it is nutritionally inadequate. Naturally, livestock men, research workers, the feed manufacturer, and many others are vitally concerned with the role vitamins play in making diets more

Water-soluble vitamins	B-Complex vitamins (con't)
Vitamin C	Biotin
B-Complex vitamins	Folacin
Thiamin	*p*-Aminobenzoic acid
Riboflavin	B_{12}
Niacin	Fat-soluble vitamins
Pantothenic acid	Vitamin A
B_6	Vitamin D
Choline	Vitamin E
myo-Inositol	Vitamin K

nutritionally adequate and thus more efficient and economical to the livestock producer.

III. WHAT IS KNOWN AND NOT KNOWN

In spite of the tremendous amount of information available on vitamins, present knowledge of the vitamin requirements of the pig for growth, reproduction, and lactation is still inadequate. Although there is considerable information available on the vitamin requirements of the pig during growth, the vitamin requirements of the sow during gestation and lactation are still not well known.

The information available on the vitamin needs of the pig during growth has been obtained mostly with pigs which have suckled the sow for at least 2–4 weeks. The determination of the vitamin needs of the pig immediately after birth, before the pig has nursed, is a challenging area for research. Using such pigs, investigators can determine vitamin and other nutrient needs before the pig has obtained storage of many vital factors from the mother's milk. Studies such as these, which also take into account the nutrition of the mother sow during gestation, and even during her growing and developing period, will show the function that vitamins and other nutrients fulfill in preventing small pig losses that occur, to a large extent, during the first few days after birth. In other words, too much research work has been done on a "piecemeal" basis. Long-term studies are needed which consider the whole life cycle of the pig and which delve into the interrelationships and effects of the various stages on the whole of swine nutrition.

Some vitamins were studied years ago when some of the vitamins now needed were not included in the diet. It is possible that many vitamin deficiency syndromes which have been determined were complicated by a lack of some of the vitamins now being added to swine diets. The role of many of the vitamins in diets used today should be restudied with pigs.

IV. BORDERLINE DEFICIENCY MAY EXIST

A borderline deficiency of any vitamin may exist without the pig showing any of the known symptoms for each individual vitamin deficiency. In that case, poor gains and expensive gains will be made by the pig. In other words, the fact that the pig does not show any deficiency symptoms is no excuse for feeding a poor diet. Borderline deficiencies in swine diets are difficult to detect and so they cost the swine producer considerable money.

Work at Washington State (2) showed that pigs fed a purified diet lacking in all the B-complex vitamins did not show any specific vitamin deficiency symptoms. The pigs failed to grow and looked like runts. When all the vitamins were lacking, the nutritional level of the pig was so low that it failed to grow and thus failed to develop individual vitamin deficiency symptoms. So a deficiency of a group of vitamins may exist on the farm without the pig exhibiting deficiency symptoms such as have been described by various experiment stations for single vitamin deficiencies. This is especially so if the nutritional level is so low that growth is poor.

V. SINGLE VITAMIN DEFICIENCIES RARELY FOUND

Under farm conditions, one will not usually find a single vitamin deficiency. In almost every case, a multiple vitamin deficiency will exist. In other words, the deficiency symptoms may be a combination of symptoms described for the various single vitamins or they may be due to something entirely different. Conditions such as unthriftiness, reduced appetite, and poor growth are common to malnutrition in general. More studies are needed to determine the symptoms and performance obtained with pigs which have multiple deficiencies of the type which may be encountered on the farm under varying conditions of feeding and management.

VI. NATURAL VERSUS PURIFIED DIETS

Some data have been obtained with the pig showing that responses to vitamins may differ depending on whether the vitamins are being added to a purified or to a natural diet. Michigan workers (3–5) showed that the requirements of the pig for niacin, riboflavin, and pantothenic acid were considerably higher on a natural diet than the requirements established earlier from experiments using purified diets. They used a corn, oats, soybean oil meal, meat scraps, alfalfa meal, and mineral diet for young pigs. Their diet supposedly contained enough of the three B vitamins for normal growth when based on the recommendations of the National Research Council (NRC) (6). However, their results showed the diet to be deficient in niacin, pantothenic acid, and riboflavin. This shows that the results obtained with purified diets must also be verified with natural diets.

VII. VITAMIN NEEDS BECOMING MORE CRITICAL

Vitamin needs have become more critical in recent years as the trend has increased toward confinement feeding of pigs. Pasture is no longer depended upon as a source of vitamins for balancing diets for the growing-finishing pig. Moreover, pigs are growing faster with less feed and with increased carcass leanness and quality. These and other factors are making it more important to

Fig. 4.1. Note effect of B vitamin supplementation on deficient pigs obtained from farms in Michigan. Top group of pigs were about 80 days old and averaged 20 lb in weight. Note these same pigs after 35 days of vitamin supplementation. (Courtesy of R. W. Luecke, Michigan State University.)

evaluate vitamin needs to make sure they are adequate in swine diets (Fig. 4.1).

Thirteen factors are listed below which, because of trends in recent years, have had an effect on the increased need for vitamins in swine diets.

1. Selection for meatier and faster growing pigs.
2. Increased use of crossbreeding which can increase pork production per sow about 40%.
3. Genetic differences in animals which can alter vitamin needs.
4. Trend toward complete confinement and slotted floors which lessens opportunity for coprophagy (eating of feces). Feces contain many vitamins.
5. The depletion of certain nutrients in soils which affects nutrient needs.
6. Newer methods of handling and processing feeds and their effect on nutrient level and availability.
7. Various nutrient interrelationships which can affect vitamin needs.
8. Changing environmental conditions in swine units.
9. Trend to earlier weaning which increases need for vitamins in these diets.
10. Increased stress and subclincial disease level conditions because of closer contact between animals in confinement.
11. Presence of antimetabolites in feeds which can increase certain vitamin needs.
12. Molds in feeds which can increase the need for vitamins.
13. Less use of pasture, alfalfa meal, fish meal, animal by-products, distillers' solubles and other feeds which are good sources of vitamins.

These thirteen factors, and others, have resulted in an increased need for adding vitamins to swine diets. Almost all swine diets in the United States are now being fortified with vitamins A, D, B_{12}, riboflavin, niacin, pantothenic acid and choline. An increasing number of feed manufacturers are adding vitamins E and K, and many are adding biotin and B_6 to diets. Diets are being fortified with these vitamins even though not all experiments indicate a need for everyone of them. Most feed manufacturers use them as a precaution to take care of stress factors, subclinical disease level and other conditions on the average farm which may increase vitamin needs.

It must also be emphasized that many farmers buy commercial protein supplements (fortified with vitamins and minerals) to mix with their grain. Unfortunately too many farmers try to cut corners and use less than the recommended protein supplement level and thus a lack of vitamins and other nutrients occur. Hence the reason why many feed manufacturers add levels of vitamins which are higher than NRC recommendations (7). Moreover, they want to provide a safety factor against vitamin losses which can occur in the feed during storage and which may be affected by temperature, rancidity, moisture level and other conditions.

VIII. SUPPLEMENTING DIETS WITH VITAMINS

Many studies have shown that supplementing commonly used swine diets with vitamins will increase rate of gain and decrease feed required per pound of gain. Therefore, it is pretty well established that vitamin supplementation of swine diets is necessary (Fig. 4.2).

IX. UNIDENTIFIED FACTORS

High quality alfalfa meal and pasture, animal protein concentrates, liver, soil, dried distillers' solubles, fish solubles, grass juice concentrate, dried whey, and other feeds have been shown to contain a factor or factors useful either for the growing pig or for the sow during gestation and lactation. Work at Washington State (8) and Florida (9) showed that soil supplies a factor or factors for the young growing pig. Some very interesting and worthwhile studies lie ahead in determining the factors in these feeds, their relationships, and their value in supplementing swine diets.

There is some evidence that both organic and inorganic materials contribute beneficial factors to the growing pig and to the sow during gestation and lactation (7). To what extent one factor or several factors are involved is not clear.

Fig. 4.2. A lack of B vitamins causes small, weak pigs at birth. (Courtesy of T. J. Cunha and University of Florida.)

Furthermore, correction of an imbalance of nutrients could be involved (7). It is suggested that an open mind be kept on the possibility of unidentified factors until more exact information is obtained.

X. PASTURE WILL DECREASE VITAMIN NEEDS

The use of short, lush, green, leafy pastures will minimize vitamin deficiencies in swine. Unfortunately, what many people consider a pasture is, in many instances, no more than an exercise yard for pigs. Pasture conditions must be considered by the feed manufacturer in compounding diets. In areas where the pasture season is of short duration, green, leafy alfalfa meal is an excellent pasture substitute. Work at Utah State (10) showed that alfalfa in the field supplies some factor (or factors) destroyed in the process of drying the alfalfa to hay. It will be interesting to know what this factor (or factors) turns out to be.

XI. THIAMIN

A. Names Used Previously

Vitamin B_1, antiberiberi vitamin, thiamine antineuritic vitamin, oryzamin, torulin, polyneuramin, and aneurin are names previously applied to this vitamin. It is now called thiamin. Its use in phrases such as "thiamin activity" and "thiamin deficiency" is acceptable.

B. Not Apt to Be Deficient in Swine Diets

Thiamin is widely distributed in feeds. For this reason a deficiency is not likely to occur in swine diets. Some good sources of thiamin are brewers' yeast, cereal grains, wheat bran, cull peas, rice bran, and plant protein concentrates. Pork itself is one of the richest sources of thiamin. See Table 11.5 for the thiamin content of feeds.

C. Effects of Deficiency

(a) During growth, a lack of thiamin in weanling pigs causes diarrhea, vomiting, lack of appetite (anorexia), interruption of growth, slight staggering, enlarged and flabby hearts (see Fig. 4.3), a slight reduction in rectal temperature, heart beat, and respiration during the final stages of the deficiency, and finally death (11–15). Pigs can store thiamin and can use it for a long period of time.

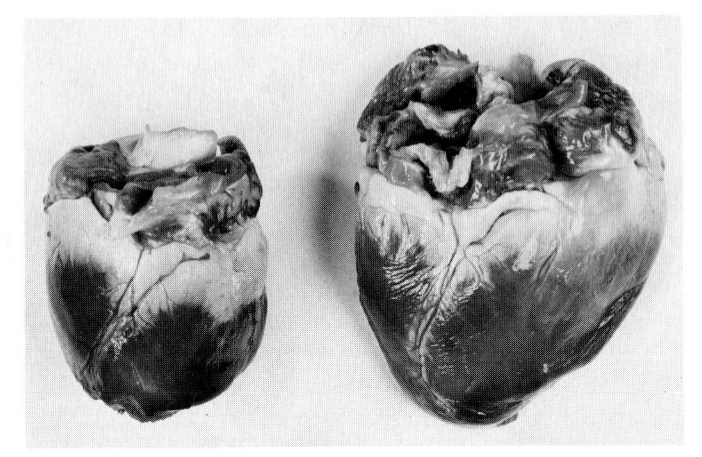

Fig. 4.3. Enlarged heart on right due to a thiamin deficiency. Heart on left is from pig fed same diet plus thiamin. (Courtesy of T. J. Cunha and Washington State University.)

Proof of this is that in Washington State experiments (12) it took at least 56 days for weanling pigs to lose their appetite after being fed a thiamin-deficient diet.

(b) During reproduction and lactation, gilts fed a thiamin-deficient diet at the Washington Station (16) had the following symptoms: (1) some loss of appetite, but not consistent; (2) parturition 9 and 11 days prematurely in two gilts; (3) high birth mortality in litters; (4) weak leg condition in pigs at birth; and (5) unthrifty pigs and subnormal weaning weights. The more depleted the gilts became in thiamin, the more severe the deficiency symptoms. From the forty-second to the fifty-fourth day after birth, three pigs in each litter were given 9 mg thiamin every second day. No apparent benefit on growth or external appearance was obtained from these injections. Tissue analyses showed that gilts on the thiamin-deficient diets were very low in stored thiamin. Low thiamin storage also occurred in their young pigs.

D. Requirements

The NRC (7) recommends 0.5–0.7 mg of thiamin per pound of total diet for swine during growth, reproduction, and lactation (see Tables 2.1 and 2.2). This is slightly higher than the levels worked out by investigators and allows a small margin of safety (13, 17–19). Increasing the level of fat in the diet lowers the thiamin needs of the pig, as is the case with some other animal species. This means that fats have a "thiamin-sparing action" and that the fat content of the diet will affect requirements for this vitamin (17).

E. General Information

A positive relationship exists between thiamin intake and the deposition of this vitamin in the tissues of swine. This makes it possible to increase the amount of thiamin in pork by using feeds high in this vitamin (12,20–22). Usually loin and ham have the highest thiamin content, followed by other cuts in the following order: shoulder, heart, liver, and kidney. A program of enriching pork by feeding diets high in this vitamin would increase the supply of thiamin in the American diet. This is important, because the average American diet is low in thiamin.

A California study has shown that the thiamin required in the diet for the pig is increased at higher temperatures (23).

XII. RIBOFLAVIN

A. Names Used Previously

Vitamin B_2, vitamin G, lactoflavin, ovoflavin, riboflavine and uroflavin are previously used names for riboflavin. Its use in phrases such as "riboflavin activity" and "riboflavin deficiency" is acceptable. See Table 11.5 for the riboflavin content of feeds.

B. May Be Deficient in Swine Diets

Many swine diets are borderline in supplying this vitamin, and many are deficient in riboflavin (Fig. 4.4). Serious consideration should be given to adding a source of riboflavin to practical swine diets to make sure this vitamin is adequately supplied. Good natural sources of riboflavin are lush, green pasture, high quality leafy, green alfalfa meal, yeast, milk, milk by-products, dried distillers' solubles, dried brewers' yeast, plant protein concentrates, and high quality meat scraps and fish meal. It must be emphasized that grains are poor sources of riboflavin. The trend toward confinement feeding and the use of less pasture and/or alfalfa in swine diets has increased the need for supplementation with riboflavin. Both were excellent sources of riboflavin.

C. Effects of Deficiency

(a) During growth, a lack of riboflavin results in alopecia (loss of hair), anorexia (loss of appetite), poor growth, rough hair coat, dermatitis, scours, ulcerative colitis, inflammation of anal mucosa, vomiting, light sensitivity, eye lens opacities, unsteady gait, and many abnormal internal complications (24–27).

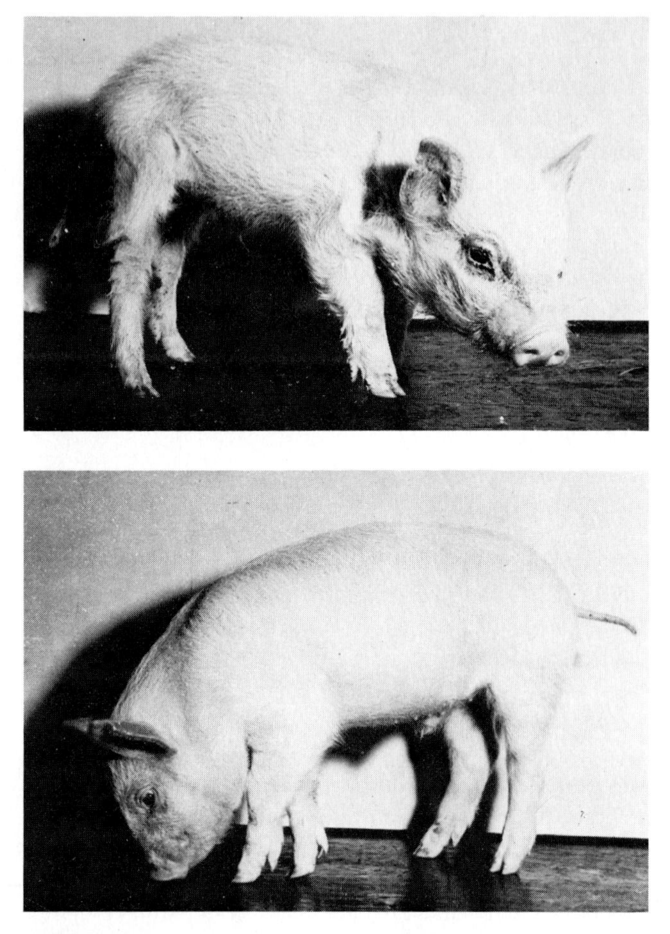

Fig. 4.4. Riboflavin deficiency. The pig on top received no riboflavin in the diet, while the animal below received 3 mg riboflavin/2.2 lb of solids. (Courtesy of R. W. Luecke, Michigan State University.)

(b) During reproduction and lactation, gilts fed a riboflavin-deficient diet exhibited the following symptoms: (1) erratic or, at times, complete loss of appetite; (2) poor gains; (3) parturition 4–16 days prematurely; (4) one case of death of fetus in advanced stage with resorption in evidence; (5) all pigs either were dead at birth or died within 48 hours thereafter; (6) enlarged front legs in some pigs, due to gelatinous edema in the connective tissue and generalized edema in many others; and (7) two hairless litters. The longer the period of riboflavin depletion, the more severe the deficiency symptoms became.

Riboflavin-deficient gilts had low riboflavin storage and gave birth to pigs very low in riboflavin. The above information was obtained at the Washington station (16) (Fig. 4.5), but USDA studies (28) also showed poor conception and reproduction with riboflavin-deficient diets.

D. Requirements

The NRC (7) recommends 1–1.4 mg riboflavin/pound of feed during growth and 1.6–1.8 mg/pound of feed for reproduction and lactation (see Tables 2.1 and 2.2). A USDA study (28) shows that a level of 1.25 mg riboflavin/pound of feed is the practical minimum level for breeding gilts and sows. They also found that a level of 1.65 mg riboflavin/pound of feed gave a slight improvement over the 1.25 mg level, but the results were not conclusive. Thus, the riboflavin need of the sow may be above the 1.25 mg level. An Illinois study (29) would also confirm this, since it was found that the gestation-lactation performance of sows was significantly improved by feeding diets containing 2.3 mg riboflavin/pound as compared to 1.2 mg. These figures are a little higher than the 1.6—1.8 level recommended by the NRC (7).

The NRC recommendations on riboflavin levels for weanling pigs seems to be in agreement with those of most investigators (7). An Illinois (30) study showed that 1.4 mg riboflavin/pound of diet was the minimum for weanling pigs fed in dry lot. A later study by the USDA (31) showed that 0.83 mg riboflavin/pound of diet is adequate for growing swine. Another Illinois study (32) was in better agreement with the USDA results. They found that at 42°F the riboflavin requirement of the pig is 1.04 mg/pound of feed and at 85°F the requirement is lowered to 0.54 mg/pound of feed. This showed that the riboflavin requirement of the pig is higher at a lower temperature. Work at the Illinois Station (33) showed that the riboflavin requirement of the growing pig is between 0.4 and 0.65 mg/pound of the diet when the mean environmental temperature is 53°F. They also found that, under the conditions of their experiment, B_{12} and aureomycin did not significantly influence the riboflavin requirements of the pig.

Studies at Michigan (26) indicate that the baby pig, shortly after birth, requires approximately 1.36 riboflavin/pound of diet. This is a higher riboflavin requirement than that of the weanling 30–40 lb pig. A recent study at Iowa State (34) showed that the riboflavin requirement of the baby pig is between 1.4 to 1.8 mg/pound of diet. They also found no marked differences in the feed efficiency or gains of pigs housed at temperatures ranging from 61 to 90°F. At environmental temperatures below 52°F, the feed required per unit of gain increased, and rate of gain decreased with the decreasing temperature. Though the pigs were tolerant of a rather wide range of temperatures, they suggested an optimum temperature of near 68°F.

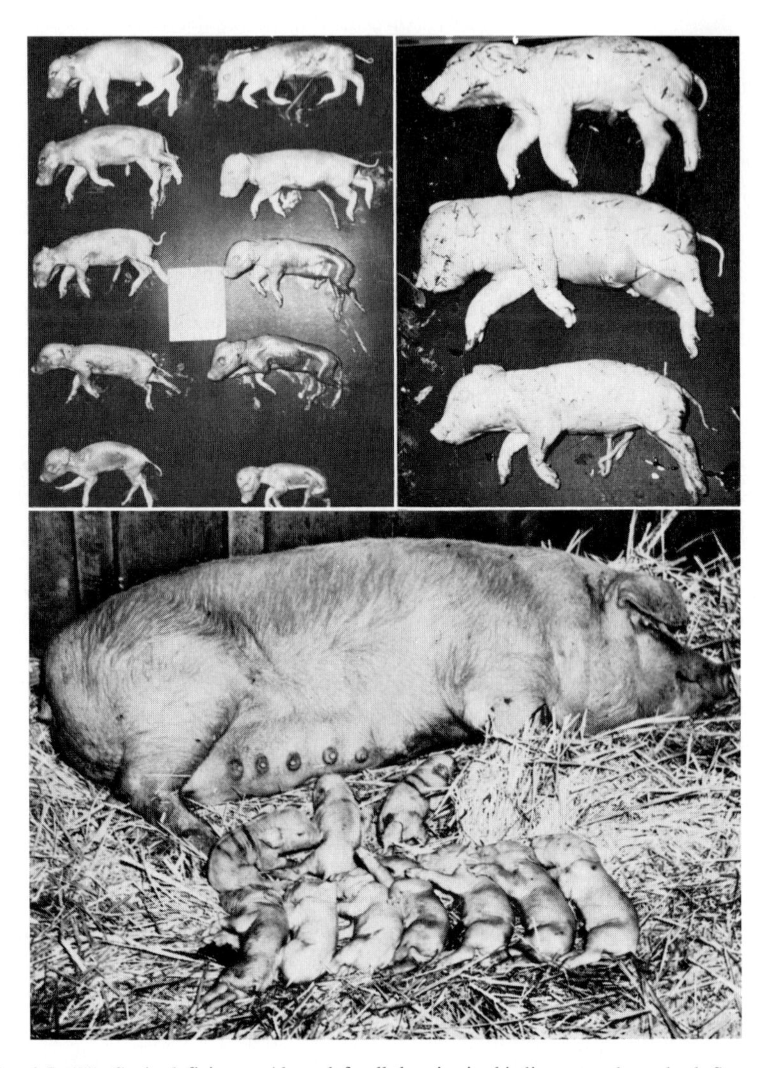

Fig. 4.5. Riboflavin deficiency. Above left: all the pigs in this litter were born dead. Some were in the process of resorption. A few had edema and enlargement of front legs as a result of gelatinous edema. Above right: pigs from a litter in which gelatinous edema was more pronounced. Below: Seven of the ten pigs farrowed were born dead and the other three were dead within 48 hours. The sow received a riboflavin-deficient diet for a shorter period than the sows farrowing the other two litters. (Courtesy of T. J. Cunha and Washington State University.)

This discussion indicates that more information is still needed on the riboflavin requirement of the pig at various stages in its life cycle and under varying conditions. Evidently, low temperatures affect riboflavin requirements. It has been shown that increased fat and lowered protein in the diet increase the riboflavin requirements of the rat. Possibly other factors may influence riboflavin requirements. More studies are needed with the pig to determine the exact requirements and why so much variation has occurred in the riboflavin requirements reported to date.

E. General Information

Riboflavin is a component of enzyme systems essential to normal metabolic processes. Pork liver is higher in riboflavin than the other pork cuts and is followed, in order, by kidney, heart, ham, shoulder, and loin. Pork muscle contains much less riboflavin than thiamin. The pork heart, liver, and kidney, however, contain more riboflavin than the muscle tissue and more riboflavin than thiamin. The dietary level of riboflavin has much less effect on the tissue content of this vitamin than is true with thiamin. This means that one can enrich pork to a much higher level of thiamin than that of riboflavin by feeding diets higher in these two vitamins (11,12).

XIII. NIACIN

A. Names Used Previously

Pellagra preventive factor, PP factor, pellagramine, vitamin PP, and niamid are names previously used for niacin. Phrases such as "niacin activity" and "niacin deficiency" are acceptable usage.

B. May Be Deficient in Swine Diets

A deficiency of niacin is likely to occur when corn is a large part of the diet. Corn is a poor source of niacin. Barley and wheat have about twice as much niacin as corn. Many diets are borderline in niacin content. Good sources of niacin are brewers' yeast, wheat bran, rice bran, rice polish, peanut meal, and dried distillers' solubles. See Table 11.5 for the niacin content of feeds.

C. Effects of Deficiency

(a) During growth a deficiency of niacin in the weanling pig causes poor appetite, slow growth (Fig. 4.6), occasional vomiting, an exfoliative type of

Fig. 4.6. Niacin deficiency. Note retarded growth of pig on right that received the same diet as pig on left except for niacin. (Courtesy of M. M. Wintrobe, University of Utah and *Journal of Nutrition*.)

dermatitis, loss of hair, rough haircoat, normocytic anemia, diarrhea, and a high incidence of necrotic lesions in the colon and cecum (3,4,35–39). Some workers have noted paresis, spasticity of muscles of the rear limbs, and ulceration of the lips and tongue, but in some instances a deficiency of several nutrients may have existed (7). Wide variation has been observed in the severity of symptoms in pigs with similar breeding and environmental backgrounds. Occasionally animals appear to thrive with no niacin, and other animals appear to vary in their requirement (38).

(b) During reproduction and lactation, it was not possible to produce niacin deficiency with sows fed a purified diet when either 18 or 26.1% casein was included in the diet in Washington State experiments (40). Evidently, the diet contained enough tryptophan to supply niacin needs, or the sows were not fed the niacin-deficient diet long enough to develop a deficiency.

D. Requirements

The NRC (7) recommends a level of 4.5–10 mg niacin/pound of feed for growing pigs and 8–10 mg niacin/pound of feed for sows during reproduction and lactation (see Tables 2.1 and 2.2). This recommendation is in good agreement with niacin levels found by various investigators.

More work is needed on determining niacin needs of swine, since the requirement for this vitamin is influenced by the protein and tryptophan content of the diet. For example, Utah workers (39) were not able to produce niacin deficiency in pigs on a diet containing 26% casein. But when the casein content was lowered to 10%, niacin deficiency symptoms were produced. This is because of an interrelationship between niacin and tryptophan (3,4).

Pigs can use the tryptophan to make niacin. But, niacin cannot be reconverted back to tryptophan by the pig. Thus, if tryptophan is high enough in the diet (which occurs with a high-casein diet), it is difficult to produce a niacin deficiency. This means that the protein and tryptophan level in the diet is very important in determining niacin requirements. It also means that if a diet contains enough niacin, tryptophan is not used up in the process of niacin synthesis. In practice, therefore, one should make sure swine diets are adequate in niacin since it is very low in price, whereas tryptophan is very expensive. There is considerable variation in niacin requirements depending on the diet used, the level of protein and tryptophan, and the weight of the pig, since it has been shown that the niacin requirement of the pig decreases as the pig grows older and larger (41).

E. General Information

Niacin, or a derivative of niacin, is required by all living cells, and it is an essential component of important enzyme systems involved in glycolysis and tissue respiration. Of great importance is the fact that the niacin in cereal grains (corn, wheat, milo and others) and their by-product feeds is in a bound form which is essentially unavailable to the pig (44,45). In mixing diets therefore, the niacin values for corn, and other cereal grains and their by-product feeds should be disregarded. It is safe to assume that these feeds provide no niacin.

A study at Illinois showed that 0.01% of either DL- or L-tryptophan in the diet had the conversion equivalent to at least 6 mg niacin/pound of diet (42). Illinois workers (43) showed that the tryptophan in corn gluten feed is largely unavailable to the pig. Pelleting of the corn gluten feed did increase the tryptophan availability, however. So both niacin and tryptophan need to be considered when corn gluten feed is used.

XIV. PANTOTHENIC ACID

A. Names Used Previously

The names antidermatitis factor, liver filtrate factor, and yeast filtrate factor have been used for pantothenic acid. Its use in phrases such as "pantothenic acid activity" and "pantothenic acid deficiency" is acceptable.

B. May Be Deficient in Swine Diets

Many swine diets are borderline in supplying pantothenic acid and many are deficient in this vitamin. Corn and soybean oil-meal diets are apt to be deficient in pantothenic acid. Good sources of pantothenic acid are alfalfa meal, fish solubles, liver meal, cane molasses, peanut meal, dried whey, dried brewers' yeast, wheat bran, and distillers' solubles. See Table 11.5 for the pantothenic acid content of feeds.

C. Effects of Deficiency

(a) During growth, a deficiency of pantothenic acid causes poor growth (Fig. 4.7), excess lacrimation, coughing, decrease in appetite, dermatitis, incoordinated movements of the hind legs, a spastic gait (goose stepping) (Fig. 4.8), rough hair coat and skin, dark brown exudate around the eyes, excessive nasal secretion, diarrhea, loss of suckling reflexes and control of the tongue, loss of hair (alopecia), rectal hemorrhages, ulcerative colitis, low urinary excretion of

Fig. 4.7. Pantothenic acid deficiency. The smaller pig received the same diet as the larger one except for pantothenic acid. [Courtesy of M. M. Wintrobe, University of Utah and *Bull. Johns Hopkins Hosp*. Pantothenic acid deficiency in swine. **73,** 313 (1943). © The Johns Hopkins University Press.]

Fig. 4.8. Note high stepping due to pantothenic acid deficiency in the pig. (Courtesy of R. W. Luecke, Michigan State University.)

pantothenic acid, and moderate normocytic anemia (46–51,62). It has been shown that the inclusion of biotin in the diet of a pantothenic acid-deficient pig was effective in prolonging the life of the pig, but caused the pantothenic acid deficiency symptoms to appear in half the time. This may be due to some interrelationship of biotin and pantothenic acid (46).

(b) During reproduction and lactation, a study at the Washington Station (40) showed that lack of pantothenic acid resulted in loss of appetite, reduced water intake, "goose-stepping" with hind legs, diarrhea, and rectal hemorrhages. Although the gilts became pregnant, they did not farrow or show any signs of pregnancy. An autopsy of the gilts revealed macerating feti in the uterine horns in all cases. Hemorrhagiconecrotic cecocolitis, gastroenteritis, and catarrhal of the stomach and small intestine were also observed. Thus a lack of pantothenic acid resulted in complete reproduction failure. A study at Illinois (52) showed that a dietary intake of 0.7 or 2.7 mg pantothenic acid/pound of diet was not sufficient to prevent development of deficiency symptoms. Gilts fed the 2.7 mg level conceived, the gestation was supported to term. Abnormal pigs, however, were farrowed. No pigs were born to gilts receiving the 0.7 mg level of pantothenic acid, although all but one individual exhibited estrus and were bred. Pantothenic acid-deficient sows exhibited soft feces, diarrhea, bleeding from the anus, and locomotor incoordination which gradually progressed to "goose-stepping" and to eventual inability to rise. The pigs born to the sows fed 2.7 mg pantothenic acid/pound of diet showed a reduced desire to nurse. Severe muscular weakness, incoordination, and diarrhea increased until most of the pigs died.

An Ohio study (59,60) reported an abnormal locomotion in suckling pigs from sows which had undergone a long period of inadequate pantothenic acid intake (Fig. 4.9).

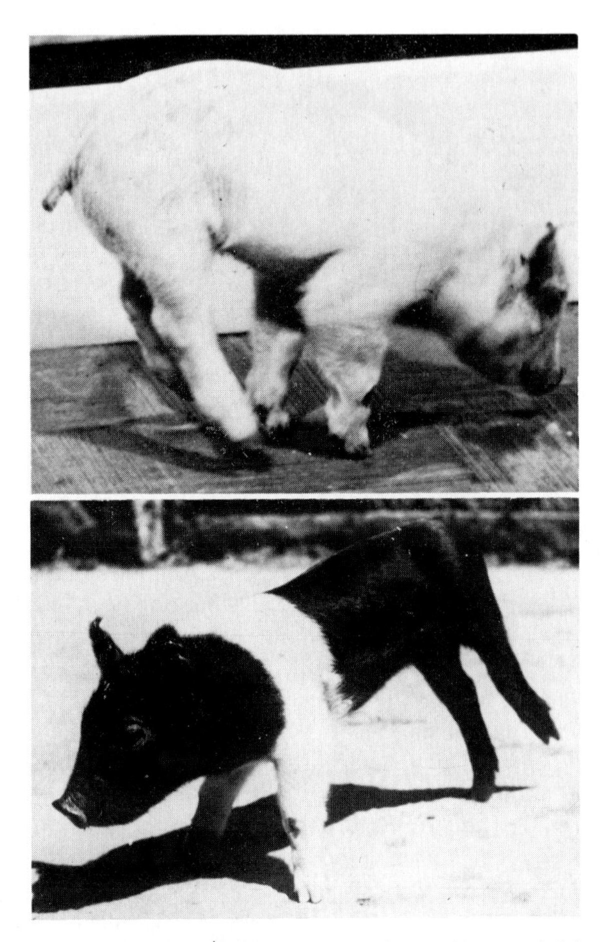

Fig. 4.9. Pantothenic acid deficiency. These pigs were farrowed by sows deficient in pantothenic acid. Abnormal locomotion with little or no flexion of leg joints was the primary symptom observed. (Courtesy of H. S. Teague and Ohio State University.)

D. Requirements

The NRC (7) recommends a level of 5.0–6.0 mg pantothenic acid/pound of total diet for the growing pig and 6.0–7.5 mg/pound of feed for the sow during reproduction and lactation (see Tables 2.1 and 2.2). The Michigan Station found that for one-half of the pigs a level of 4.15 mg pantothenic acid/pound of feed was sufficient for normal growth, whereas for the remaining half of the pigs it was inadequate (53,54). Under their conditions, the requirement of the pig for pantothenic acid was greater than 4.15 mg but less than 6.15 mg/pound of feed. Evidently there is a wide variation in the requirements for pantothenic acid among the animals. This should be taken into consideration in determining the vitamin needs of the pig. Vitamins supplied must be adequate for all pigs and not for just some of them. A Michigan study (55) showed that the pantothenic acid requirement of the baby pig for optimum growth and feed efficiency approximates 5.7 mg calcium pantothenate/pound of solids. A study at Illinois (52) showed that this same level of pantothenic acid, 5.7 mg/pound of diet, appeared to be adequate to support normal reproduction of sows. A USDA study (63) showed that a minimum of 5.4 mg pantothenic acid/pound of diet is required for reproduction in swine.

Another important aspect to consider in determining pantothenic acid needs is the finding by Michigan workers (53) that adding a mixture of synthetic B vitamins to a corn-soybean meal diet resulted in very severe symptoms of pantothenic acid deficiency in weanling pigs. However, if the same corn-soybean meal diet was fed alone, without any vitamins added, no symptoms of incoordination appeared. The reason for this may be that the pigs without any B vitamins grew so poorly that the level of pantothenic acid in the diet was sufficient to prevent symptoms of incoordination. By adding the B vitamins to the diet, the researchers increased the growth rate. This, in turn, increased the dietary requirements for pantothenic acid to the point where deficiency symptoms appeared.

This indicates that B vitamin balance is very important from a practical standpoint. If one is going to add B vitamins to a diet, then all the vitamins which are apt to be lacking should be added. If one, or more than one vitamin, is omitted, a deficiency may be produced which otherwise would not occur unless the diet was partially fortified with other vitamins. Work at the Iowa Station (56) has shown that pantothenic acid and vitamin B_{12} exert a "sparing" action on one another in the absence of aureomycin in the diet. They also found that aureomycin appears to "spare" both pantothenic acid and vitamin B_{12}. They also state that with healthy, undepleted pigs weighing 35–45 lb a 14% protein corn-soybean meal diet balanced in other respects and containing adequate amounts of vitamin B_{12} and aureomycin need not be supplemented with pantothenic acid for optimum growth. A Florida trial (61) also showed that aureomycin at a level of 10 mg/pound of diet had a sparing effect on the pantothenic acid requirement of the weanling pig.

On the other hand, Michigan workers (53,57,58) showed that pantothenic acid supplementation benefited pigs fed corn-soybean oil meal diets. More work is needed on this problem and on determining the requirements for pantothenic acid of the pig under varying conditions and with different kinds of diets.

E. General Information

Pantothenic acid is a component of coenzyme A, an important enzyme in intermediary metabolism. Pantothenic acid is available as the calcium salt (calcium pantothenate) and, because of its stability and crystalline nature, the salt is commonly used by the feed industry to add to swine diets. Calcium pantothenate is frequently marketed as the racemic mixture (both D and L forms) of DL-calcium pantothenate. Only the D isomer of pantothenic acid has vitamin activity for the pig. Therefore the following amounts should be used: (a) 1000 mg D-calcium pantothenate = 920 mg pantothenic acid; (b) 1000 mg DL-calcium pantothenate =460 mg pantothenic acid.

XV. VITAMIN B_6

A. Names Used Previously

Vitamin H, factor Y, yeast eluate factor, adermin, antiacrodynia rat factor, and antidermatitis rat factor all have been used as names for vitamin B_6. Vitamin B_6 includes three compounds which have B_6 activity: pyridoxine, pyridoxal and pyridoxamine. See Table 11.5 for the pyridoxine content of feeds.

B. May Be Deficient in Swine Diets in a Few Situations

A deficiency of vitamin B_6 is not thought to occur in well-balanced diets. But it could be deficient in a few situations. Good sources of B_6 are dried brewers' yeast, wheat bran, beef liver, cereal grains, fish, and meat.

C. Effects of Deficiency

(a) During growth a deficiency of B_6 with the young pig causes poor appetite, microcytic hypochromic anemia, incoordination of the muscles, spastic gait, poor growth, fatty infiltration of the liver, epileptiform fits, coma, rough hair coats, a brown exudate around the eyes, low urinary excretion of pyridoxine, internal abnormalities, and impairment of vision (49,64–68).

(b) During reproduction and lactation, a Colorado study (69) showed that sows fed a corn-milo-soybean meal diet responded to vitamin B_6 supplementation at a level of 2.0 mg/pound of feed. The B_6 also increased rate of gain and feed efficiency when added to the same diet for growing-finishing pigs (69). German workers (70) reported that suboptimal vitamin B_6 levels may occur under practical feeding conditions and affect protein synthesis in the pig. They also indicated that vitamin B_6 requirements may be increased under field conditions.

In Australia, Dr. Dudley Smith, a Ph.D. in swine nutrition and former Head of swine reasearch at the Ruakura Station in New Zealand, stated that adding 0.5 mg vitamin B_6/pound of feed to the sows' diets prevented their offspring from having epileptic-like fits at 3 weeks of age (Fig. 4.10) (71).

D. Requirements

The NRC (7) recommends a level of 0.7 mg pyridoxine/pound of total feed for 11- to 22 lb pigs (see Tables 2.1 and 2.2). It recommends 0.5 mg/pound of feed for 22–44 lb pigs. These are the only recommendations given. A Michigan study (72) showed that 0.23 mg vitamin B_6/pound of feed was nearly adequate for growth and feed consumption. However, data on blood hemoglobin, red blood cell and lymphocyte counts indicate that the minimum requirement is not less than 0.34 mg/pound of feed. With respect to urinary xanthurenic acid, however, the level of B_6 needed was greater than 0.34 mg but probably less than 0.45 mg/pound of feed. If histopathology data had been obtained, the requirement might still be higher. This study shows that nutrient requirements may be higher if criteria in addition to growth and feed efficiency are used in experimental studies. This fact may account for certain nutrient deficiencies showing up under

Fig. 4.10. Vitamin B_6 deficiency. This pig is having an epileptic-like fit. (Courtesy of the late E. H. Hughes, University of California.)

farm conditions where stress and other factors increase nutrient needs. A Michigan study (73) showed that 0.45 mg vitamin B_6/pound of diet met the swine gestation requirement for normal reproduction. A Georgia study (74) showed a vitamin B_6 requirement of 0.53 mg/pound of feed with 2% fat in the diet, and a 0.31 mg/pound of feed requirement with 12% fat in the diet. Increasing the fat content (corn oil) in the diet decreased the B_6 requirement of 3-week-old pigs.

E. General Information

Vitamin B_6 includes three compounds which have B_6 activity. Also, there may be other forms of pyridoxine. Pyridoxine, pyridoxal, and pyridoxamine are equal in activity for animals under many conditions. Under others, however, pyridoxal and pyridoxamine may show slightly less activity than pyridoxine. The three forms, however, show very different activities for many microorganisms. In plant tissues, all three forms of the vitamin occur in similar amounts. In yeast, glandular organs, and meats, however, most of the B_6 is present as pyridoxal and pyriodxamine, with only traces of pyridoxine. Thus, vitamin B_6 studies must consider the form of the vitamin in the feed as well as the effectiveness of each in animal or microorganism response.

XVI. CHOLINE

A. Names Used Previously

Bilineurine was the name previously used for choline. Its use in phrases such as "choline activity" and "choline deficiency" is acceptable.

B. May Be Deficient in Swine Diets

A choline deficiency may occur in practical, well-balanced swine diets. Good sources of choline are liver, wheat germ, cottonseed meal, soybean meal, peanut meal, tankage, and fish meal. Corn is low in choline; wheat, barley, and oats contain two to three times as much choline as corn. See Table 11.5 for the choline content of feeds.

C. Effects of Deficiency

(a) During growth, a lack of choline in the diet of the young pig results in unthriftiness, poor conformation (short-legged and pot-bellied), lack of coordination in movements, a characteristic lack of proper rigidity in the joints (particularly the shoulders), fatty infiltration of the liver, characteristic renal glomerular occlusion, and some tubular epithelial necrosis (75–79). These symptoms have been obtained with baby pigs on a diet containing 0.8% methionine (78). Doubling the methionine content was effective in preventing these symptoms with the baby pig (77). This means that the methionine level in the diet is important when studying a choline deficiency.

(b) During reproduction and lactation, a lack of choline in the diet resulted in poor reproduction, lactation, and survival of young. Subnormal weaning weights and fatty livers were obtained with the young. Some of the pigs had spraddled hind legs (Fig. 4.11). No benefit, as far as appetite or increased gains, occurred from injecting half of the pigs of choline-deficient sows from the forty second to fifty sixth days of lactation. Pigs from choline-deficient sows were rough in appearance and became increasingly so with age (16).

D. Requirements

The NRC (7) recommends a level of 500 mg choline/pound of feed for pigs 11–22 lb and 410 mg/pound of feed for pigs 22–44 lb in weight. Recommendations for pigs of other weights or for sows are not given. Illinois workers (78) showed that the baby pig (2 days old when started on trial) requires 0.1%

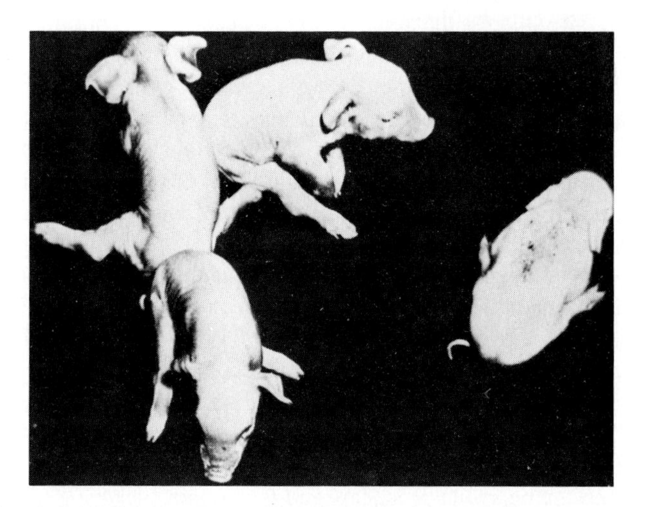

Fig. 4.11. Choline deficiency. The condition shown here—spraddled hind legs—is produced with a purified ration and is prevented by choline supplementation. Other factors may be involved in this condition. (Courtesy of T. J. Cunha and Washington State University.)

choline (454 mg choline/pound of feed) when fed a synthetic milk containing approximately 0.8% methionine (30% casein).

The level of methionine in the diet is important in determining choline needs. Methionine can furnish methyl groups for choline synthesis. Choline, however, is effective only in sparing methionine which otherwise would be used to make up for a choline shortage. Methionine is not used up for choline synthesis if there is an adequate level of choline in the diet. Thus, the establishment of a choline requirement is complicated by the methionine level in the diet.

Proof of this is the finding that when 1.6% methionine was included in the diet it was not possible to produce a choline deficiency in baby pigs (77). The data with baby pigs on a synthetic diet, however, may not apply entirely to older pigs fed natural feeds. Thus, other studies are needed to determine choline needs for growing pigs as well as for sows during reproduction and lactation. A level of 20 mg choline/kg body weight daily was used in Washington experiments with sows and prevented choline deficiency symptoms (16). This level, however, has not been determined as the requirement for the sow.

A report from the NCR-42 Committee on Swine Nutrition (80) showed that sows fed diets supplemented with choline farrowed more pigs per litter (10.54 versus 9.89), more live pigs per litter (9.33 versus 8.64) and weaned more pigs per litter (7.72 versus 7.29) than the sows fed the control diets. These results were obtained from 551 sows in 22 trials in 9 university studies. Choline was added at a level of 350 mg/pound of diet. It is interesting to note that in four of the nine university studies, the number of pigs weaned with choline supplementation did not increase, but in the other five studies the benefits derived in some trials were much higher than the averages given above. This would seem to indicate that under many conditions the need for choline supplementation is considerable. Why choline was not beneficial in certain trials must be determined to better understand when and why it is needed. In this extensive study, choline supplementation did not aid the spraddled hind leg condition. The cause of this condition is a difficult one to determine.

In the past 8–10 years, I have observed a spraddled hind leg condition in the United States, Canada, Europe, Australia, New Zealand, Japan, the Philippines and Latin America. I first produced this condition in a Washington State University study (16) with a purified diet deficient in choline. However, it now seems it may involve more than choline when practical or farm diets are used.

In 1967, I suggested in a *Hog Farm Management* article that anyone encountering the spraddled hind leg condition try choline to determine if it would help. Since that time, many swine producers and feed companies have reported to me that choline supplementation of sow diets either eliminated or greatly alleviated the problem of spraddled hind legs.

After one large producer encountered the problem with 80 sows (all sows had almost all pigs with spraddled legs), choline in the diet was increased from 2200 mg to 3000 mg daily per sow. During the next farrowing of these same 80 sows, only an occasional sow delivered a pig with spraddled legs. I saw these sows at both farrowings, so was actually able to observe them.

The spraddled leg condition started to appear as swine producers began to decrease the amount of feed given sows during gestation from 6–7 lb daily to 3–4.5 lb per day. But in doing so, they did not increase the level of choline or methionine in the diet. As a result, the diet may not supply enough choline or methionine daily to meet the needs of the sow at the lower feed intake level.

The spraddled hind leg condition may be more complex than a choline deficiency. Opinions and reports indicate that thiamin, vitamin E, selenium, methionine, a virus and even heredity factors might be involved. I have seen the spraddled legs occur most frequently with crossbred sows. This may involve heredity factors or possibly may be that a crossbred sow produces at a higher level and thus requires more choline and other nutrients. My observations and those of others in practice indicate that sows need 3000–4500 mg of choline daily to prevent the spraddled hind legs. However, Dr. H. W. Newland at the Ohio Station (81) and Dr. K. J. Dobson of the Department of Agriculture (82) in Adelaide, South Australia found that adding 4500 mg choline per sow per day did not help the spraddled hind leg condition.

Even more difficult to explain is the fact that many swine producers and university reseachers feed diets with less than 2200 mg choline daily per sow and have no problem with spraddled hind legs. Moreover, there is the NRC-42 report (80) in which choline supplementation was of benefit in litter size but did not aid the spraddled hind legs (even though the incidence in the unsupplemented diets was low). Part of the answer might be B_{12} and folic acid. In poultry, according to Dr. M. L. Scott of Cornell, the requirement for choline depends to a large extent on folic acid and vitamin B_{12}. He states that marked increases in choline requirement are noted with folic acid and/or vitamin B_{12} deficiency. Both of these vitamins should be studied in the pig to determine if they could be involved with choline and the spraddled leg condition. Ordinarily, folic acid supplementation is not needed in swine diets but it may be now, in certain situations, since the use of pasture and alfalfa meal in swine diets has decreased. Higher levels of B_{12} may also be needed in certain situations.

To further complicate the choline problem, Dr. Larry Erickson of Schreiber Mills wrote to me that he has not encountered spraddled hind legs with corn diets but has had many cases with barley diets and some with milo diets (83). This is the opposite of what would be expected on the basis of choline levels, since barley has 467 mg choline, milo has 310 mg and corn the least with 200 mg/lb.

Dr. Erickson also stated that their sow supplement added 2000 mg of choline to the sows' daily diet. Their sow supplement consisted of soybean meal, meat meal, tankage, dehydrated alfalfa meal, vitamins and minerals. By adding an additional 2000 mg of choline daily to the sow, the spraddled leg condition was overcome. He also found that the higher level of choline the last 30 days prior to farrowing also prevented spraddled hind legs. A number of other people have also told me this.

In spite of the conflicting results, it would be worthwhile for anyone experiencing the spraddled hind leg condition in pigs to consider choline supplementation since it has been beneficial to many farmers and feed industry nutritionists.

The Nebraska Station reported that the addition of 37 mg choline/pound of an 11% protein corn-soybean meal diet (to which methionine, lysine and tryptophan had been added to equal that in a 14% protein corn-soybean meal diet) increased rate of gain 8.8% and increased feed efficiency 11% (84). The pigs weighed 81–100 lb at the initiation of the trial. The choline supplementation caused the pigs on the 11% protein diet to gain as well as those fed the 14% protein diet. In addition, carcass quality was the same.

This work indicated that it might be possible to use a lower protein level if an adequate level of choline and amino acids is present. The interesting aspect of this study is that chemical analyses of the feeds indicated that the choline level was adequate. This raises the possibility of an effect of processing or storage or other factors on the choline availability of feeds. It could also affect the methionine availability of feeds which in turn could affect choline needs.

Studies at Cali, Columbia in South America (85) showed that some death losses occur due to a spraddled leg condition. But, some of the pigs recuperated after the first 7–10 days after birth. This might indicate that in borderline deficiencies the condition can be corrected through the sow's milk. A Minnesota study (86) showed that sows fed a diet without supplemental choline had a significantly lower conception rate expressed as a percentage of total sows (57 versus 73%), had a significantly lower farrowing rate expressed as a percentage of bred sows (62 versus 78%), farrowed significantly fewer total pigs per litter (9.3 versus 10.1), farrowed fewer live pigs per litter (8.0 versus 9.1) and weaned significantly fewer pigs (6.6 versus 7.3) compared with sows given supplemental choline. The incidence of spraddled legs was lower (2.1 versus 5.7%) with the sows supplemented with choline. A supplemental level of 187 mg choline/pound of diet was adequate in this study. A Virginia study (87) showed that choline had little effect through the first four lactations with sows but the response occurred during the fifth and sixth lactations. The results indicated that choline helped keep older sows in the producing herd longer.

Research studies to date indicate that choline supplementation of swine diets may benefit growth, litter size at weaning, and may keep older sows in the producing herd longer. More studies are needed, however, on the value of choline supplementation of swine diets to more adequately understand the role of this vitamin. Undoubtedly, this will not occur until the various interrelationships of choline to other nutrients are better understood.

XVII. BIOTIN

A. Names Used Previously

The names vitamin H, coenzyme R, factor W, bios II, bios IIB, factor X, and anti-egg white injury factor have been used for biotin. Its use in phrases such as ''biotin activity'' and ''biotin deficiency'' is acceptable.

B. May Be Deficient in Certain Situations

The 2-day-old suckling pig evidently develops a biotin deficiency when fed a diet lacking in this vitamin. Evidently, the very young suckling pig does not synthesize enough biotin for its needs (88). Good sources of biotin are liver, yeast, milk, and fish. Corn and wheat products are poor sources of biotin. See Table 11.5 for biotin levels in feeds.

C. Effects of Deficiency (See Figures 4.12–4.16)

(a) During growth, a deficiency of biotin in the diet results in alopecia, spasticity of the hind legs, transverse cracking of the soles and tops of the hooves, dermatosis of the skin characterized by dryness, roughness, a brownish exudate, and ulceration of the skin, and inflammation of the mouth mucosa (88–90). In the 8-week-old pig, the addition of 30% desiccated egg white or sulfathalidine to a purified diet resulted in a biotin deficiency (89,90). The avidin in egg white ties up biotin in the intestinal tract, making it unavailable to the pig. Sulfathaladine decreases the intestinal synthesis of biotin. Sulfaguanidine feeding did not produce a biotin deficiency in the 8-week-old pig fed a purified diet (90).

(b) The addition of biotin to a purified diet was of no benefit to sows in a Washington experiment (40). The supplementation of biotin to a practical diet in a Colorado trial (91) was of some benefit in reproduction.

D. Requirements

The requirements of the pig for biotin have not been determined. In 8-week-old pigs, biotin deficiency symptoms were prevented, however, by feeding biotin at a level of 20 μg/100 gm of feed (90) or by injecting 100 μg biotin per

Fig. 4.12. Biotin deficiency. The two pigs in the middle are deficient in biotin. Note loss of hair and dermatosis. The other two are control pigs. (Courtesy of T. J. Cunha and Washington State University.)

pig daily (89). These figures might be used as guides in the use of biotin until definite requirements are determined.

E. General Information

Until recently it was assumed that the pig did not need the addition of biotin to its diet. Biotin synthesis by microorganisms in the intestinal tract plus the biotin in feeds supposedly took care of its needs. The exception would be the 2-day-old pig fed a purified diet devoid of biotin, since it does not synthesize enough biotin to meet its body needs. Therefore, very early weaned pigs would need biotin in the diet until about 3 or 4 weeks of age when biotin synthesis becomes more adequate in the pig.

A trial in Colorado with SPF pigs averaging 15 lb body weight showed that adding 50 μg of D-biotin/lb of corn-milo-soybean meal diet increased rate of gain and feed efficiency 15% during the first 53 days (91). During the entire 122-day trial, it increased rate of gain 7.5% and feed efficiency 5% (see Table 4.1). None of the pigs used in this trial showed biotin deficiency symptoms. However, about 10% of their mothers showed signs of biotin deficiency. Studies at Nebraska (92) and by Dr. G. E. Combs of Florida in 1967 showed that biotin supplementation had little effect at their University Research Farms (see Table 4.1). Of importance in the Nebraska study was that the addition of chlortetracycline plus penicillin, with or without sulfamethazine, did not affect biotin needs. This is important to know since these three compounds are used as Aureo SP-250 in the diets of many pigs during early growth.

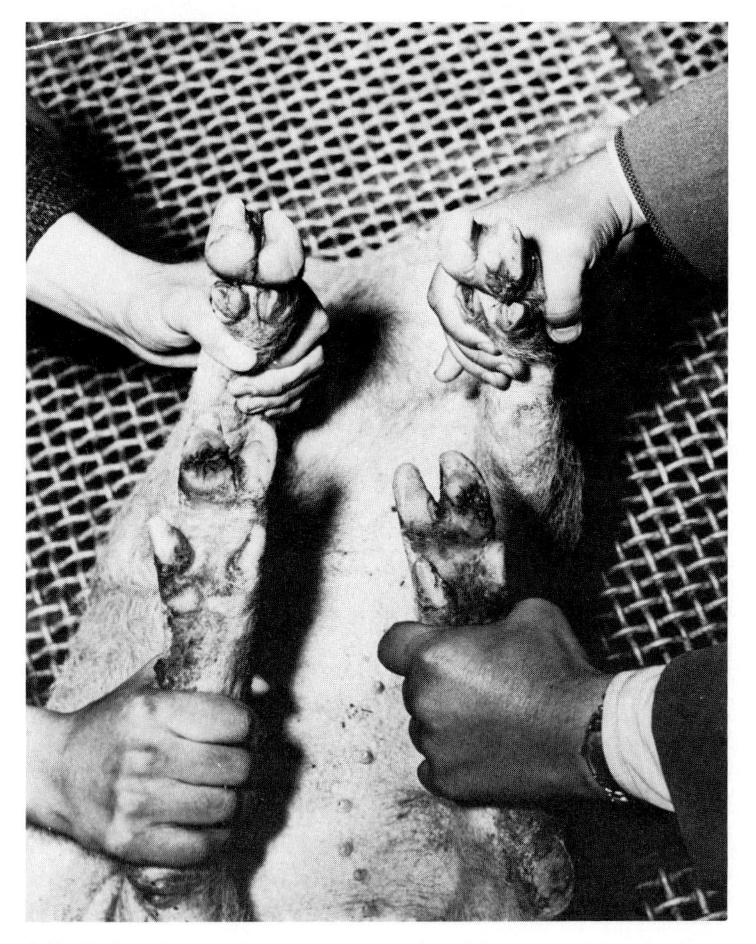

Fig. 4.13. Biotin deficiency. Note transverse cracking of the soles and the tops of the hooves of this biotin-deficient pig. (Courtesy of T. J. Cunha and Washington State University.)

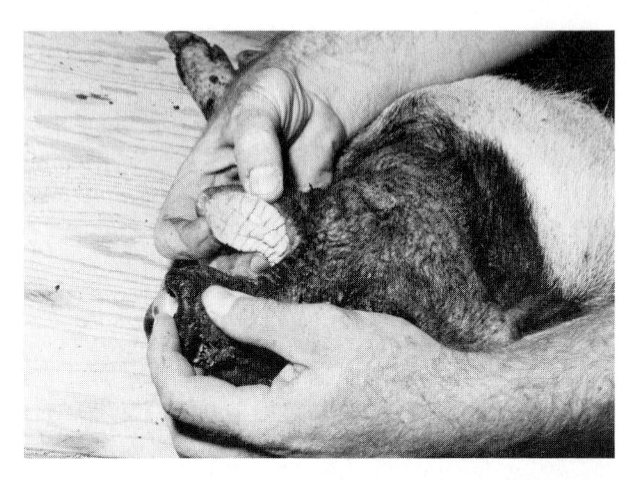

Fig. 4.14. Biotin deficiency. Note sore tongue which bled at times. (Courtesy of G. E. Combs, University of Florida.)

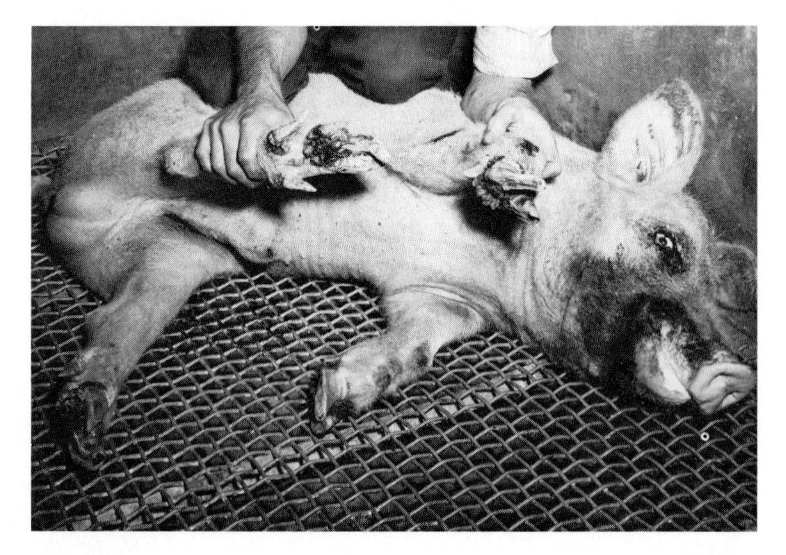

Fig. 4.15. Biotin deficiency. Note cracking and bleeding in feet, dermatosis and hair loss. (Courtesy of M. M. Wintrobe, University of Utah.)

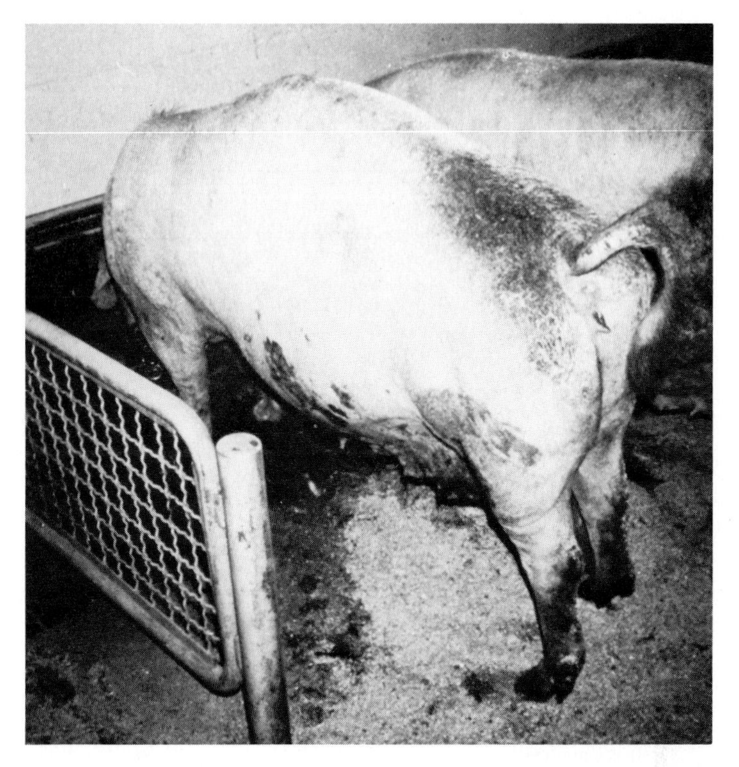

Fig. 4.16. Biotin deficiency. Note arched back which is observed with advanced biotin deficiency. Also note hair loss and dermatosis. (Courtesy of T. J. Cunha, University of Florida.)

The fact that the biotin data (Table 4.1) obtained at the Universities of Florida and Nebraska differ from those obtained at the American Hog Company is not discouraging. In a somewhat comparable situation, Hoffman-La Roche personnel observed a biotin deficiency with turkeys in the field. They then shipped the same turkeys and feed to their research laboratories at Nutley, New Jersey but were unable to produce the biotin deficiency there. This would seem to indicate that something is either increasing the need for biotin or decreasing its synthesis in the intestinal tract under certain conditions which may be difficult to duplicate under laboratory conditions. It also means that more attention must be paid to what is happening on the farm with regard to biotin needs.

Dr. W. M. Beeson of Purdue University in 1970 observed two large commercial swine operations, one in the United States and the other in Spain, where biotin supplementation was beneficial. In the Spain operation, he found that a combination of alfalfa meal plus biotin was needed for best results. Dr. Beeson

TABLE 4.1
Effect of Biotin Supplementation at Florida, Nebraska and American Hog Farms

	Biotin (μg/lb)				
	0	25	50	0	50
University of Florida					
Av. daily gain (lb)	1.36	1.44	1.36	0.66	0.72
Feed/lb gain	2.10	2.03	2.18	1.95	1.86
	Biotin (μg/lb)				
	0	200	400		
University of Nebraska					
Av. daily gain (lb)	0.69	0.75	0.76		
Gain/feed ratio	0.48	0.49	0.49		
	Biotin (μg/lb)				
	First 53 days			During 122 day trial	
	0	50		0	50
American Hog Farms, Colorado					
Av. daily gain (lb)	0.75	0.86		1.16	1.25
Feed/lb gain (lb)	2.38	2.02		3.34	3.17

feels that a biotin deficiency is complicated by other nutrient deficiencies which are apparently taken care of by feeding about 0.5 lb of dehydrated alfalfa pellets daily to the sows. A number of commercial feed companies in the United States and in other countries have also encountered biotin deficiencies under field conditions. Usually only 10–20% of the sows in the herd exhibit biotin deficiency symptoms, therefore, one will not observe the symptoms in too many sows. Moreover, the little pigs may show no deficiency symptoms and will still respond to biotin.

Just why a biotin deficiency occurs under certain conditions is still not known. The use of confinement and slatted floors may be a factor since this lessens the opportunity for coprophagy (eating of feces). It is possible that the pig meets some of its biotin needs by rooting the manure, bedding and/or soil. Recent studies also indicate that there is considerable difference in the availability of biotin in certain feeds. The biotin in many feeds may be less than 50% available to the animal. Therefore, feeds being processed and fed today may have different biotin levels and availability than previously.

The kind of pig being selected today may also have some effect on biotin needs. Its biotin requirements may be higher. Oxidation rancidity in feeds could also be a factor since biotin is destroyed by rancidity. Biotin antagonists (antibiotin compounds) and biotin interrelationships with other nutrients could also affect biotin needs.

The trend to corn-soybean diets which do not contain dehydrated alfalfa, dried whey and fermentation by-products (all good sources of biotin) could also be contributing to biotin deficiencies. Another cause might be molds in the feed which could affect the availability of the biotin for the pig. Streptavidin, isolated from *Streptomyces,* is a biotin-binding substance. It acts similar to avidin in raw egg white in tying up biotin. *Streptomyces* are molds found in soil, and possibly in litter and moldy feeds. All of these factors must be considered in reevaluating the need for biotin by the pig.

In the past 9 years, I have seen pigs with hair loss, dermatitis and cracked feet, similar to those observed with a biotin deficiency, in Germany, England, Scotland, Spain, Italy, Canada, Honduras, Mexico, New Zealand, Australia, the Philippines, Brazil, Argentina, Chile, Uraguay, Peru, Colombia, Venezuela and the United States. I have also received many reports from persons who have tried biotin and achieved a response with its use. Therefore, it is apparent that biotin deficiencies occur under certain conditions and that biotin supplementation is beneficial under certain situations. But it is still not known under what conditions biotin supplementation is beneficial.

The exact level of biotin needed in supplementing practical diets is not known. The level needed for growth will probably range from 25 to 100 μ/lb of feed. For reproduction, the range will probably vary from 50 to 200 μ/lb of feed. Anyone encountering symptoms of a biotin deficiency and wishing to try it might experiment with different levels of biotin since beneficial results may depend on having the correct level plus other nutrients which may also be needed in a proper balance and level with biotin.

Biotin supplementation may be quite complex since it interrelates with so many other nutrients, such as pantothenic acid, pyridoxine, ascorbic acid, B_{12}, and folacin. Therefore, simply adding one level of biotin may or may not be beneficial, nor may it answer the question as to whether the diet is adequate in biotin.

The information obtained to date indicates one should consider the possibility of the need for biotin under certain conditions. It is hoped this discussion will stimulate nutritionists to (1) look for biotin deficiencies and (2) try biotin if they suspect a deficiency in their area.

XVIII. *myo*-INOSITOL

A. Names Used Previously

The names *i*-inositol, inositol, *meso*-inositol, inosite, nucite, and dambose have been used. Today it is called *myo*-inositol. Its use in phrases such as "*myo*-inositol activity" and "*myo*-inositol deficiency" is acceptable.

B. Not Apt to Be Deficient in Swine Diets

Myo-inositol is not apt to be deficient in swine diets. It is widely distributed in feeds. Good sources of *myo*-inositol are wheat germ, barley, oats, wheat, liver and molasses.

C. Effects of Deficiency

(a) During growth, the addition of *myo*-inositol to a purified diet was of no benefit to 8-week-old pigs. This would indicate either that the pig synthesizes enough *myo*-inositol for its needs or that it does not need the vitamin added to the diet (90,95). In that trial, however, it was shown that if a biotin deficiency was produced by using sulfathalidine in the diet, *myo*-inositol alleviated to a large extent the deficiency symptoms which were prevented entirely by biotin. A possible explanation is that *myo*-inositol acted indirectly by stimulating the intestinal synthesis of biotin. Work with 1- and 4-day-old pigs in an Illinois study (76) showed no evidence for a PABA or *myo*-inositol need when their combined deficiencies were superimposed on a choline deficiency, although their omission from the diet appeared to accentuate the degree of fatty infiltration of the liver.

(b) During reproduction and lactation, the addition of *myo*-inositol to a purified diet for the sow was of no benefit (40). Wisconsin data, however, showed that *myo*-inositol was beneficial in lactation when added to a corn-soybean diet for brood sows and rats (93,94). These studies need more confirmation, however. Other types of diets should be studied as well.

D. Requirements

The requirements for *myo*-inositol by the pig have not been determined. Levels of 0.1–0.3% *myo*-inositol have been used in the diet (90,93,94). None of these levels, however, are regarded as definite requirements. Many investigators have been able to obtain good growth with the pig and small laboratory animals without adding *myo*-inositol to purified diets. Other groups of workers, however, under different conditions, have shown that *myo*-inositol benefits the pig and small laboratory animals. Thus, under certain conditions, a need for *myo*-inositol can be shown. The reason for the need under those conditions is not known. Since certain laboratory animals can synthesize *myo*-inositol, it is logical to assume that under some conditions the intestinal synthesis of *myo*-inositol is altered, thus changing the requirements of the animal. Interrelationships of nutrients also occur; thus the diet fed and its content of various nutrients may influence the need for *myo*-inositol. Substances, antimetabolites for example, in natural feeds may have an effect on the *myo*-inositol needs of the animal. There may also be some other explanation for this. Regardless of the reason, considerably more work is needed to clear up the picture on the *myo*-inositol needs and requirements of the pig.

XIX. *p*-AMINOBENZOIC ACID (PABA)

A. Names Used Previously

BX factor, vitamin Bx, chromotrichia factor, anti-gray hair factor, and trichochromogenic factor have been used as names for this vitamin. Its use in phrases such as "*p*-aminobenzoic acid activity" and "*p*-aminobenzoic acid deficiency" is acceptable.

B. Not Apt to Be Deficient in Swine Diets

p-Aminobenzoic acid is not apt to be deficient in swine diets. *p*-Aminobenzoic acid or PABA (as it is often abbreviated) is widely distributed in feeds. Good sources of PABA are dried brewers' yeast, liver, alfalfa meal, wheat germ, and wheat middlings.

C. Effects of Deficiency

(a) During growth, experiments at the Washington Station (96) with 8-week-old pigs fed for 7 weeks with either PABA or folacin added alone or in combination with *myo*-inositol and biotin to a purified diet showed no beneficial effect on growth, efficiency of feed utilization, or external appearance. The addition of PABA or folacin alone to the purified diet did, however, stimulate hemoglobin formation to a small extent. Illinois experiments (76) with 1- to 4-day old pigs showed no definite evidence for a requirement for PABA or *myo*-inositol when their combined deficiencies were superimposed on a choline deficiency, although their omission from the diet appeared to accentuate the degree of fatty infiltration of the liver.

(b) During reproduction and lactation, the addition of PABA to the diet of sows fed a purified diet was of no benefit (40).

D. Requirements

The requirement for PABA by the pig has not been determined. Levels of 10 mg/100 gm of feed (96) and 2.6 mg/1000 gm of milk (76) have been used for growing pigs and 20 mg/100 gm of feed for sows (40). None of these levels, however, are regarded as requirements. More studies are needed to determine the possible need and requirement for PABA by the pig.

XX. FOLACIN

A. Names Used Previously

Folic acid, pteroylglutamic acid, vitamin Bc, vitamin M, L casei factor, norite eluate factor, SLR factor, and factor U have all previously been used to designate this vitamin. The term folacin should be used for folic acid and related compounds exhibiting qualitatively the biological activity of folic acid. Thus, phrases such as "folacin activity" and "folacin deficiency" are preferred usage.

B. Usually Not Deficient in Swine Diets

So far as is known, a deficiency of folacin is not thought to occur in practical swine diets. One study, however, has shown some benefit during early growth from adding folacin to a wheat-barley-tankage diet fed to 8-week-old pigs (96). This study needs to be repeated and other practical diets supplemented before it can definitely be stated that folacin is beneficial when added to practical swine diets. Good sources of folacin are liver, peanuts, and soybeans. The cereal grains are fair sources of folacin. See Table 11.5 for the folacin level in feeds.

C. Effects of Deficiency

(a) During growth, baby pigs receiving colostrum for 24 hours were successfully raised to 8 weeks of age on a synthetic milk diet (97). A clear-cut deficiency of folacin was not consistently produced on this synthetic milk diet even with the addition of a bacteriostatic agent (sulfathaladine). In one of the trials, however, the lower growth rate and lighter hair coats of the pigs indicated the beginning of a folacin deficiency.

Washington workers (98) found that folacin was needed for normal hematopoiesis if pigs (8 weeks old at the start) were fed a purified diet for a long period of time (21 weeks). Normocytic anemia was produced in pigs by adding sulfasuxidine to the diet for 21 weeks. Anemia was prevented by folacin and to a lesser extent by an antipernicious anemia liver extract. A more severe anemia was produced by using a crude folacin antagonist. A combination of folacin and biotin was more effective than folacin in counteracting the effect of the crude folacin antagonist.

Various other workers (99–106) have produced folacin deficiencies in the pig by a folacin antagonist or sulfonamide. All studies showed that folacin is concerned with blood cell formation. A more severe anemia (99) was produced with a 10% casein diet than with a 26% casein diet. This would indicate, as has been also shown with rats, that folacin requirements are decreased at higher protein levels.

Since almost all the swine work on folacin has been conducted with high protein diets, studies are needed with lower levels of protein. One study with a natural diet containing about 17% protein showed that folacin supplementation benefited growth during the first 4 weeks and caused the pigs to be cleaner in appearance and have bigger appetites at the end of the experimental period (96). This study indicates that work is needed with practical diets to determine whether folacin supplementation is needed.

(b) During reproduction and lactation, folacin supplementation was tried with a purified diet containing 26.1% casein in Washington experiments (40). Indications were that folacin appeared to improve reproduction somewhat. Although it also seemed to aid in lactation, the effect was not so apparent as it was in reproduction. More studies are needed to determine the effect of a folacin deficiency in the sow, especially with lower protein diets.

D. Requirements

The requirements for folacin by the pig have not been determined. Levels of 50–100 μg folacin/100 gm of feed (96,98) for 8-week-old pigs, 0.05 mg/liter of synthetic milk for 2-day-old pigs (97), and 200 μg/100 gm of feed for sows (40) have been used successfully. None of these levels, however, are regarded as requirements.

E. General Information

The folacin molecule contains *p*-aminobenzoic acid (PABA). Some think PABA functions as a precursor of folacin; under some conditions this is true, but it is not known if this is always the case. This should be verified since it would clarify the role of folacin and PABA in animal nutrition.

There is some relationship between leucovorin (synthetic citrovorum factor) and folacin. For certain microorganisms and for chicks and turkeys, leucovorin has some folacin activity. There is some evidence that before folacin can carry out some of its functions it must first be converted to leucovorin. It has also been shown that ascorbic acid is needed for this conversion.

In the future, some very interesting information will be obtained in unraveling the interrelationship of these vitamins and what role they play in the nutrition of the animal. This incomplete information is being given since it may help the reader in understanding the complexity of certain interrelationships of these nutrients in the diet. It also helps explain why a certain balance between nutrients in the diet is needed for best results.

XXI. VITAMIN B_{12}

A. Names Used Previously

The names zoopherin, animal protein factor, erythrotin, factor X, and physin have been used for B_{12}. The term vitamin B_{12} should be used for all compounds exhibiting the biological activity of cyanocobalamin, hydroxycobalamin and nitrocobalamin. Thus, phrases such as ''vitamin B_{12} activity'' or ''vitamin B_{12} deficiency'' are preferred usage.

B. May Be Deficient in Swine Diets

Many swine diets are borderline or deficient in supplying this vitamin and can be benefited by B_{12} supplementation. Good sources of vitamin B_{12} are liver, fish meal, fish solubles, peanut meal, meat scraps, and milk. The cereal grains are poor sources of B_{12}. See Table 11.5 for the vitamin B_{12} level in feeds.

C. Effects of Deficiency

(a) During growth, symptoms of a B_{12} deficiency show up in slower growth rate, rough hair coat, dermatitis (Fig. 4.17), a tendency to roll over onto the side or back, hyperirritability, posterior incoordination, voice failure, and pain in the rear quarters. Hematopoiesis is not normal as shown by high total erythroid counts in the bone marrow. Slight normocytic anemia also results (107–112).

(b) During reproduction and lactation, vitamin B_{12} has been shown to increase birth weights and survival of young pigs (113).

Under some conditions the reproductive performance of sows has been improved by the inclusion of higher than recommended levels of vitamin B_{12} in the diet. The response is evidenced by an increase in the number and birth weight of pigs. Response to such elevated levels of B_{12} is not consistent and may relate to variable synthesis by intestinal bacteria or utilization of the vitamin (7,122–124). Ohio Station studies (122) showed that feeding 1000 mg of B_{12}/ton of feed (almost 100 times NRC recommended level of B_{12}) resulted in an increase of 2.3 and 2.0 live pigs farrowed and 2 and 1.1 more pigs weaned in the third and fourth

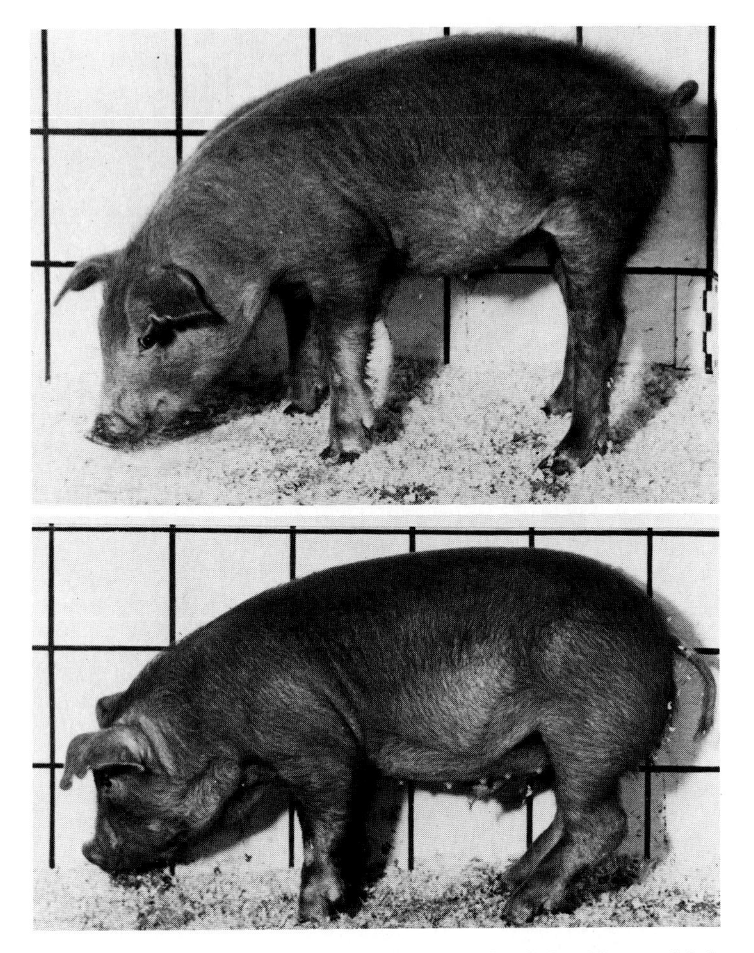

Fig. 4.17. Vitamin B_{12} deficiency. Above, pig deficient in vitamin B_{12}. Note rough hair coat and dermatitis. Below, control pig. (Courtesy of the late D. V. Catron and Iowa State University.)

farrowing, respectively. No apparent increase in reproductive performance occurred in the first and second farrowings from B_{12} supplementation. This high level of B_{12} is not yet recommended but it does indicate that for an extended reproductive period the B_{12} requirement may be higher than presently recommended. Ohio workers (125) showed that 41.6–58% of the radioactive vitamin B_{12} introduced via colon cannulas into the first part of the colon was absorbed by the pig. It is important to know that this high a level of B_{12} is absorbed in this area of the large intestines.

D. Requirements

In a number of studies at the Iowa Station, it was found that the B_{12} requirement of the weanling pig is approximately 4–5 μg B_{12}/lb of total diet when added to a corn-soybean diet containing antibiotics (112,120,121). Studies at Illinois (109) showed that suckling pigs to 6 weeks of age require approximately 9 μg B_{12}/lb of dry matter consumed. When B_{12} was injected instead of being fed orally, the requirement was approximately half this latter amount. The NRC (7) recommends a level of 5–10 μg/lb of feed during growth and 5–6.4 μg/lb of feed during gestation and lactation.

E. General Information

Data obtained with a corn-peanut meal diet at the Florida Station showed that B_{12} alleviated the methionine needs of the pig (114,115). Pseudo-vitamin B_{12} has been shown to be inactive in the baby pig (116). It also appears that B_{12} is not involved in the synthesis of choline from methionine in the pig (117). Wisconsin data (118) showed that pigs receiving a diet with vitamin B_{12} responded with an increased vitamin B_{12} activity of the muscle. The needs for B_{12} are increased as the level of protein is increased in the diet (119). Thus, the level of protein is important when B_{12} requirements are being studied.

Vitamin B_{12} contains the trace element cobalt, and its synthesis by the intestinal flora is dependent on the presence of this mineral in the feed. This may be the major, if not the only, function of cobalt as an essential nutrient (7).

XXII. VITAMIN C

A. Names Used Previously

Vitamin C has been called hexuronic acid, ascorbic acid, cevitamic acid, antiscorbutic vitamin, and scorbutamin. The term vitamin C should be used for all compounds exhibiting qualitatively the biological activity of ascorbic acid. Thus, phrases such as "vitamin C activity" and "vitamin C deficiency" are preferred usage.

B. May Need in Certain Situations

Vitamin C has occasionally been beneficial in a few university studies. It is not known, however, why it is helpful in a few instances and not in others.

C. General Observations

Until recently it was assumed that the pig synthesized all the vitamin C it needed and that vitamin C supplementation of diets was not beneficial. Recent studies, however, raise some doubt about this assumption. These studies indicate that pigs may have borderline deficiencies of vitamin C when a high rate of growth, high stress level or some other conditions (not yet understood) exist on the farm.

It is pretty well established that vitamin C plays some role in bone formation. In poultry, vitamin C supplementation is beneficial on egg shell thickness and strength during hot weather. Dr. A. S. Jones of Scotland found lower levels of vitamin C (0.3 mg% compared to 0.5 mg%) in pigs that were weaned early and were prone to leg weakness. However, he has not been able to relate the leg weakness specifically to a lack of vitamin C (126).

Vitamin C supplementation has also counteracted high environmental temperature (100°F) effects in laying hens. Therefore, high temperatures increase C needs in poultry. This would indicate the need for studying vitamin C needs when the temperature is high either outside or inside swine housing. Purdue reported that vitamin C in the blood plasma of pigs was lowered at a high temperature (127). Dr. J. E. Burnside of Southern Illinois University reported that average daily gain climbed from 1.26 to 1.51 lb when vitamin C was added to the diet of pigs stressed by heat (128). He feels that hogs under stress respond to extra vitamin C in the diet with an increase in rate of gain and feed efficiency.

The effect of stress on vitamin C requirements should be studied since certain strains of pigs are more susceptible to environmental stress than others. For example, in Wisconsin a strain of Poland China pigs with varying degrees of stress-susceptibility was used in adrenal studies since the adrenal gland is affected in stress-susceptible pigs.

Dr. J. E. Burnside, while at Southern Illinois University, did a great deal of work on vitamin C. He and P. R. Moss concluded the following from a study (130) involving seven trials with the pig:

1. Under high level of stress conditions, vitamin C supplementation is beneficial. The higher stress level requires higher levels of vitamin C. Pigs on a low stress level do not respond to vitamin C.

2. Under low stress conditions the blood plasma vitamin C level is low, whereas it is high with a high burden of stress over a long period of time.

3. Average daily gains and blood plasma vitamin C levels are related. This was also confirmed by Purdue studies (127), in which it was shown that plasma vitamin C levels were highest when weight gains were lowest and vitamin C levels were the lowest when weight gains were highest. Both weight gains and vitamin C levels fluctuated in cycles. One explanation offered is that during rapid

gains there is a decrease in blood plasma vitamin C until it reaches a level which no longer supports rapid growth. Then growth rate decreases until blood plasma vitamin C levels build up again.

4. Although some of the pigs developed pneumonia, those getting vitamin C continued to gain well at a rate of 1.35 lb daily as compared to 1.14 lb for those pigs receiving no vitamin C.

5. There may be a relationship between vitamin C, the adrenals and the thyroid.

6. Not all pigs in a pen respond to vitamin C. With some pigs, there is a big response. This has been verified by Texas studies (131) in which it was shown that individual guinea pigs have highly variable needs for vitamin C. A big mistake made by many scientists in evaluating experimental data is not to consider individual pig needs or response. In many trials, only a few pigs develop observable deficiency symptoms or respond to treatment. This means more individual feeding, in addition to group feeding, is needed in evaluating certain treatments.

Studies at the University of Kentucky (132,133) and at Purdue (134) have also shown that vitamin C will occasionally benefit growth and feed efficiency in growing pigs. University of Florida studies showed that 11 lb pigs fed a semi-synthetic diet responded with a 9% increase in daily gain during a 5 week period when 100 mg vitamin C were added per pound of feed. However, no response occurred when the same level of vitamin C was added to a practical corn-soybean meal diet (135).

Cornell workers (136) found that adding 0.5% ascorbic acid to the diet tended to overcome the effects of a high level of copper (250 ppm) in the feed. Ascorbic acid shows fairly good stability in unpelleted feeds, but with pelleting and the stress of heat and moisture, stability is not too good (137). Therefore, an antioxidant must be added to the feed when vitamin C is studied. Moreover, the feed should be mixed every 3 or 4 days and kept in a cool, dry place in order to assure as much vitamin C retention as possible.

The possible need for supplementing swine diets with vitamin C is not established yet. Preliminary information, however, indicates it may be needed under certain situations. Until more information is obtained, a level of 100 mg vitamin C/lb of feed is suggested as one to add.

European workers (189) fed three groups of boars ascorbic acid daily in the morning feed. There was a control group with no vitamin C, one group given 1 gm of ascorbic acid daily and another 2 gm of ascorbic acid per day during the hot summer months when conception rate in sows is lower. Conception rate increased with ascorbic acid feeding. The conception was some 20% higher in the sows served by boars receiving 2 gm ascorbic acid per day as compared with sows served by the untreated boars. This is an interesting report and needs more verification.

Recent Canadian studies suggest that energy restriction inhibits the capacity of the pig to synthesize ascorbic acid (190). In another study, they stated that in swine fed a high level of energy, supplementary vitamin C resulted in increased bone matrix metabolism, probably catabolic in nature (191). In still another study, the Canadian group found that pigs deficient in vitamin E and/or selenium had elevated blood serum vitamin C levels (192). They indicated this could be a reflection of an increased need for vitamin C or an inability to utilize it. They also suggested that swine are probably capable of synthesizing vitamin C in the presence of a vitamin E and/or selenium deficiency. These are interesting results and additional studies are needed to indicate their possible implications in practical swine feeding.

XXIII. VITAMIN A

A. Names Used Previously

Ophthalmin, antiinfective vitamin, biosterol, retinol and fat-soluble A are names that have been applied to this vitamin. Vitamin A should be used for all β-ionone derivatives exhibiting qualitatively the biological activity of retinol. Phrases such as "vitamin A activity" and "vitamin A deficiency" are preferred usage.

B. Often Deficient in Swine Diets

Many swine diets are lacking in vitamin A activity. Thus, in compounding a diet, one must make sure that it contains enough carotene and/or vitamin A to supply the needs of the pig. Good sources of vitamin A are fish liver oils, liver, eggs, and milk. Good sources of carotene are lush, green pasture, silage, and high quality, leafy green alfalfa meal.

C. Effect of Deficiency (Figs. 4.18 and 4.19)

(a) During growth, a lack of vitamin A in the young pig caused a decrease of blood plasma vitamin A levels, a tendency to carry the head tilted to one side (an infection on the inner ear, otitis media, was found in each of these cases), incoordination of movement as exhibited by a swaying gait, paresis of the hind legs, gradual loss of control of the hind legs, weakness of the back, severe tonic spasms which lasted from 2 to 3 minutes, seborrhea characterized by a brown, greasy exudate over the entire body surface, a rise in cerebrospinal fluid pressure, night blindness, constriction and degeneration of the optic nerves, and deaths from pneumonia. An effect of the vitamin A deficiency on appetite or rate

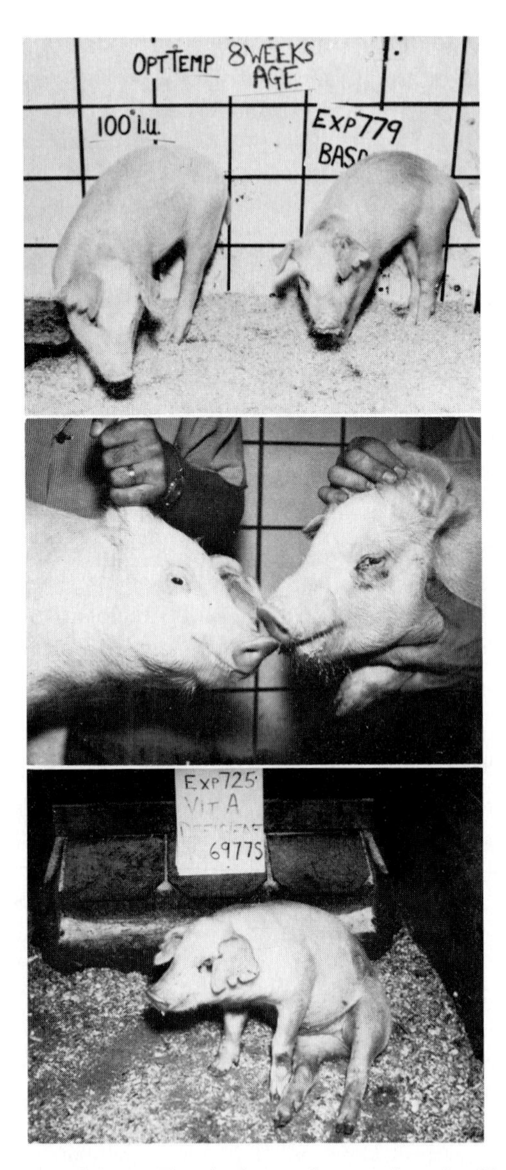

Fig. 4.18. Vitamin A deficiency. The pigs in top photo are 8 weeks old. The one on right is deficient in vitamin A and shows slow rate of growth. The pig on right in the middle photo is deficient in vitamin A and has xerophthalmia. The pig in the bottom photo is paralyzed in the hind limbs. (Courtesy of V. C. Speer, Iowa State University.)

Fig. 4.19. Vitamin A deficiency in growing pig. Top photo shows a pig exhibiting partial paralysis and seborrhea; middle photo shows a pig in initial stage of spasm; bottom photo shows a pig exhibiting lordosis and weakness of hind legs. (Courtesy of J. F. Hentges, R. H. Grummer, and University of Wisconsin.)

of gain was not detected until eventual paralysis and weakness prohibited movement to the feeder (137–142).

(b) During reproduction and lactation, a lack of vitamin A in the sow produced the following symptoms: failure of estrus, resorption of young, wobbly gait, weaving and crossing of the hind legs at the walk, dropping of the ears, curving with head down to one side, spasms, loss of control of hind and fore quarters and thus inability to stand up, and impaired vision. Depending on the degree of severity of the vitamin A deficiency, the fetuses were either resorbed, born dead, or carried to term. Fetuses carried to term showed a variety of defects, including various stages of arrestment of formation of the eyes to a complete lack of eyeballs, harelips, cleft plate, misplaced kidneys, accessory earlike growths, some with one eye and some with one large and one small eye, and bilateral cryptorchidism (141,142,145). Vitamin A is stored for long periods of time, and it takes a considerable length of time to deplete pigs and sows of their reserve (141–145).

D. Requirements

The NRC (7) recommends a level of 590–1000 IU of vitamin A per pound of total diet during growth and 1500–1865 IU of vitamin A per pound of total diet for reproduction and lactation (see Table 2.1 and 2.2). The recommended level for growth is a little higher than the requirements for the growing pig worked out by various investigators (137,138,141,144,148). The level recommended by the National Research Council should supply the needs of the growing pig as well as allow for a certain amount of storage.

E. General Information

It should be stressed that vitamin A is stored for a long period of time and it may take months to produce a deficiency. A Kentucky study (146) showed that sows that have built up stores of vitamin A until they are 8–9 months of age and are then placed on a diet practically devoid of carotene or vitamin A can produce at least two normal litters. Borderline deficiencies may exist without noticeable deficiency symptoms developing. The deficiency symptoms shown for vitamin A indicate that it is very important for growth, reproduction and lactation. It is needed for normal eye function and for the maintenance of the tissues in the respiratory, reproductive, nerve and urinary tract of the pig.

One cannot tell the exact status of vitamin A nutrition or storage in the pig by analyzing the blood for vitamin A. When vitamin A values drop below 10 μg/gm in the liver and below 10 μg/100 ml in the blood, the pig becomes quite deficient in vitamin A. Liver biopsies and determination of liver vitamin A values would give the best indication of the vitamin A storage or status in the pig.

Vitamin A does not occur in plant products; they contain carotene only. The pig can convert the carotene into vitamin A in its intestinal wall and possibly elsewhere in the body. The National Research Council (7) indicates that the pig can convert 1 mg of β-carotene into 500 IU of vitamin A. There is some indication that the conversion rate may be lower. A Michigan study (147) showed that when corn is the only source of carotene, 1 mg carotene had a vitamin A activity of 261 IU. In general, a high intake of nitrates either interferes with the conversion of carotene to vitamin A or in some way increases the need for vitamin A. While the exact status of levels of nitrates and of their effects on the pig are not exactly known, water containing 50–100 ppm of nitrates is potentially unsafe. Water is usually considered harmful when it contains 1000–3000 ppm of nitrates.

A level of 3000 ppm (0.3%) of nitrates on a dry matter basis is usually the safe upper limit in feeds from the standpoint of toxicity. Supplementing with vitamin A is low cost insurance against the extra need caused by nitrates for this vitamin in the feed or water. Since pigs consume about 2.5–3 lb of water per pound of feed, it is a good practice to test their drinking water for nitrates.

Grain in storage loses carotene activity (149). Thus, the use of old corn which has been in storage increases the A needs in the diet.

A report from the Florida Station (150) showed that pigs weighing 8.8 lb at the initiation of the trial consumed an average of 184,000 IU vitamin A daily for 42 days without any harmful effect on daily gain, feed consumption and efficiency of feed utilization. This indicates that there is a big safety factor in the use of vitamin A since the pigs were getting 114,286 IU of vitamin A in each pound of feed.

Vitamin A in excess can be toxic, however. An Iowa State study (151) showed that 100,000 IU of vitamin A per pound of feed was not toxic. However, a level of 200,000, 300,000 and 400,000 IU per pound of feed was harmful and toxic symptoms showed up in 43,32 and 17.5 days, respectively. Others have also shown harmful effects of very high levels of vitamin A (152–154). Thus, one must exercise caution in the levels of vitamin A administered. Most of the harmful effects have been obtained by feeding over 100 times the daily requirements for a period of time. Thus, small excesses of vitamin A for short periods of time should not exert any harmful effects.

Kansas workers (155) and others showed that unit for unit carotene is less effective than vitamin A for growth, gestation, and lactation for swine. This means that the requirements of swine for carotene are higher than those for vitamin A.

International standards for vitamin A are based on the utilization of vitamin A and β-carotene by the rat and are as follows: 1 IU of vitamin A = 1 USP unit = vitamin A activity of 0.300 μg of crystalline vitamin A alcohol corresponding to 0.344 μg of vitamin A acetate or 0.550 μg of vitamin A palmitate.

XXIV. VITAMIN D

A. Names Used Previously

Antirachitic vitamin, rachitasterol, and rachitamin are previously used terms for vitamin D. The term vitamin D should be used for all steroids exhibiting qualitatively the biological activity of cholecalciferol. Phrases such as "vitamin D activity" and "vitamin D deficiency" are preferred usage.

B. May Be Deficient in Swine Diets

Most of the commonly used feeds contain little or no vitamin D. Fortunately, the action of sunlight on the skin of the pig produces vitamin D. The skin contains the provitamin which is converted to vitamin D by the ultraviolet rays of the sun. Summer sunlight is more potent than winter sun. Thus, animals on pasture should not suffer from a vitamin D deficiency.

In the winter, however, hogs are often outside only a part of the day and there are fewer sunny days. So, it is unsafe to rely on winter sun to supply the vitamin D needs of pigs kept inside buildings. This means that the vitamin D must be supplied in the diet. More swine producers are switching to complete confinement, and away from pasture. They must make sure their diets are adequate in vitamin D.

Good sources of vitamin D are sun-cured roughages, fish oils, and irradiated yeast. Certain irradiated sterols are used as commercial sources of vitamin D for animal feeding.

C. Effect of Deficiency

(a) During growth, a deficiency of vitamin D causes poor growth, stiffness and lameness, a general tendency to "go down" or lose the use of the limbs (posterior paralysis), frequent cases of fractures, softness of the bones, bone deformities, enlargement and erosion of joints, and unthriftiness (53,155–160). Fractures of the bones occur frequently and bones become softened and deformed by the weight of the animal and the pull of the body muscles. Enlargement and erosion of joints and beading of the ribs (a growth at the junction of bone and cartilage) are other deficiency symptoms. Vitamin D deficiency symptoms are very similar to those of a calcium or phosphorus deficiency (Fig. 3.6). This is because all three are concerned with proper bone formation.

(b) During reproduction and lactation the sow needs vitamin D.

D. Requirements

The NRC (7) recommends a level of 57–100 IU of vitamin D/lb of total diet for growth and 100–125 IU/lb of feed for reproduction and lactation. Vitamin D is needed for efficient utilization of calcium and phosphorus and, consequently, for normal calcification of growing bone. With an adequate intake of calcium and phosphorus and a proper ratio between them, less vitamin D is needed by the pig (95,155,157,159). No amount of vitamin D, however, will compensate for severe deficiencies of either calcium or phosphorus in the diet.

Minnesota workers (160) showed that white pigs resisted vitamin D deficiency symptoms about twice as long as colored pigs. The white pigs, however, still required vitamin D, even though in lesser amounts than colored pigs. These same workers in a later study (159) showed that an average of 45 minutes daily exposure to January sunshine for 2 weeks was sufficient to cure rickets in Minnesota. They found, however, that feeding pigs indoors and allowing them voluntary access to an outside pen was not a satisfactory way to cure rickets. The pigs did not go outside enough, presumably because of the cold or otherwise inclement weather during December and January. They also found that the need for vitamin D was inversely proportional to the calcium and phosphorus content of the diet.

Vitamin D is stored in the body for a long period of time, but not to the same extent as vitamin A. Thus, the pig resists a deficiency of vitamin D for a considerable time. Minnesota data (159) showed that it can take as long as 4 months for a pig on a vitamin D-deficient diet to develop signs of a deficiency.

E. General Information

There are at least ten sterol derivatives which have vitamin D activity. Vitamin D_2 and D_3, however, are the important ones in swine feeding. Vitamin D_2 is the form found in plant products such as hay and in irradiated yeast. Vitamin D_3 is the animal form found in fish oils, in irradiated milk, and in the body after irradiation. Both D_2 and D_3 have the same value for the pig, but D_3 is much more effective for chicks and turkeys. Vitamin D_2 (or calciferol) is obtained from ergosterol by irradiation. Vitamin D_3 (or activated 7-dehydrocholesterol) is obtained from 7-dehydrocholesterol by irradiation. Irradiated ergosterol (vitamin D_2) is sold for human use under the trade name Viosterol.

In a Florida study pigs were weaned at 3 weeks of age and fed corn-soybean meal diets supplemented with 0, 900, 9,000 or 90,000 IU of vitamin D/lb of feed for 17 weeks. The pigs were housed in the absence of sunlight and fed *ad libitum* in self-feeders. Rate of gain and efficiency of feed utilization did not differ significantly with the different vitamin D levels (162). Wisconsin workers showed no harmful effects from feeding of 100,000 IU of vitamin D/lb of diet to weanling pigs for 21 weeks (163). These two studies, and others, have shown there is a wide range between the recommended levels of vitamin D and a level which is harmful. Nevertheless, since harmful effects can be obtained from excessive D, it is best to avoid using very high levels. For example, in Scotland, pigs fed 500,000 IU or more of vitamin D daily ceased eating or died in less than a month, while pigs fed 250,000 IU daily developed a reduced appetite and growth rate. They also developed hypercalcemia and fully calcified, partially calcified and noncalcified vascular lesions. There was evidence of rapid recovery, when excess vitamin D was eliminated and the blood content declined to normal levels. The pathogenesis of vitamin D toxicity is still not fully understood (164).

Recent Florida data (165) and that of many other investigators have shown that vitamin D is needed for calcium absorption. A Michigan trial (161) showed that the minimal vitamin D_2 requirement of the baby pig on a purified casein protein diet is no greater than 45 IU/lb of diet. Another study (166) showed that the practical minimum vitamin D_2 needed in a purified, isolated soybean protein diet by the baby pig is higher, between 114 to 228 IU/lb of diet. This indicates that vitamin D needs are higher when soybean protein diets are used as compared to casein as a source of protein. Data by Michigan workers (167) indicated no evidence for β-carotene being rachitogenic. Previous studies by others had shown the possibility that β-carotene increased vitamin D needs.

A new development is the finding that 1,25-dihydroxycholecaliciferol (25-OH-CC) may be the active form of vitamin D_3 in the intestines (168). This compound which is produced in the kidney from vitamin D_3, may be 4–13 times as active as D_3 in stimulating calcium transport. It is the most biologically active form of vitamin D found to date and it may possibly be reclassified as a steroid hormone. Developments in this area should be followed closely since it may change vitamin D use and supplementation.

Of interest are a number of recent reports indicating that molds in feeds interfere with vitamin D_3 (169). When corn contains a mold *Fusarium roseum,* a metabolite of this mold keeps the vitamin D_3 in the intestinal tract from becoming miscible and thus not absorbed by the chick. Other molds may also be involved. This results in a pseudo-rickets condition and a large percentage of birds with bone disorders. A number of flocks have been successfully treated by adding water-dispersible forms of vitamin D to the drinking water at 3–5 times normally recommended vitamin D levels. This development should be followed closely since field cases of unexplained bone and leg problems with swine could be due to the same cause.

One international unit of vitamin D is defined as the biological activity of 0.025 μg of crystalline vitamin D_3.

XXV. VITAMIN K

A. Names Used Previously

Coagulation vitamin, antihemorrhagic vitamin, 2-methyl, 1,4-naphthoquinone, prothrombin factor, and phylloquinones are names previously given to vitamin K. The term vitamin K should be used for 2-methyl, 1,4-naphthoquinone and all derivatives exhibiting qualitatively the biological activity of phytylmenaquinone (phylloquinone). Thus, phrases such as "vitamin K activity" and "vitamin K deficiency" are preferred usage.

B. Apt to Be Deficient in Swine Diets

Until a few years ago, vitamin K supplementation of swine diets was not practiced. It was thought that the pig synthesized most, if not all, the vitamin K it needed. Moreover, some of the feeds used supplied additional vitamin K.

Now, vitamin K supplementation of swine diets is widely practiced. Vitamin K is needed for blood clotting. A lack of vitamin K causes hemorrhages in the body because of the failure of blood to clot properly.

After returning from a trip to Australia and New Zealand, I reported in *Feedstuffs* (June 7, 1969) that a number of persons, including veterinarians, had observed cases in these countries of hemorrhaging in the navel with newborn pigs. Dr. Dudley Smith, manager of the Huntley 2200-sow operation in Australia), found that adding vitamin K to the sow's diet prevented the hemorrhagic condition from occurring.

Just why vitamin K deficiencies are now occurring in the United States is not entirely known. The important thing is that they are and there is need to protect against them.

C. Reports of Hemorrhaging

In 1969, Minnesota (170) reported that vitamin K helped control hemorrhaging which occurred with pigs in a growth experiment. In 1970, a Nebraska (171) university study reported on a hemophiliac-like condition, also referred to as the hemorrhagic syndrome, in 26 lb pigs. The condition occurred 9 days after the pigs were fed a standard university formulated 18% protein starter diet. The pigs were then divided into three groups. One group was kept as a control with no treatment, whereas the other two groups were fed either 2.5% dehydrated alfalfa meal or 2 gm menadione sodium bisulfate (vitamin K) per ton of feed. The pigs fed the alfalfa or vitamin K did not show visible symptoms of the disease. Symptoms were observed in 6 days in the control group. Gross visible symptoms were large subcutaneous hemorrhages, blood in the urine, and abnormal breathing. Clotting time was normal in the pigs fed alfalfa or vitamin K (about 4–5 minutes) whereas it was slow in the controls (average of 10.43 minutes). A study in Missouri in 1970 reported on vitamin K supplementation benefiting a hemorrhagic condition severe enough to cause death in pigs (172). Hawaii studies (173) in 1971 showed that vitamin K supplementation was beneficial to pigs fed a high sugar ration. A more complete report (174) on sugar-induced heart lesions and hemorrhagic syndrome in the pig was published in 1973. Two Florida reports also showed vitamin K supplementation of swine diets to be beneficial (175,176). Therefore, these trials, and others, have shown that vitamin K supplementation is needed under certain conditions.

D. Reason for Vitamin K Need

The exact cause for a vitamin K need is not definitely known. Some of the causes which have been postulated are as follows:

1. Hemorrhaging gastric ulcers, which occur frequently, may increase vitamin K needs.
2. A mycotoxin produced by certain molds which may be present in the feed might cause the disease.
3. An antimetabolite (antivitamin K) may be in the feed and thus increase vitamin K needs.
4. As confinement feeding has increased, less pasture and alfalfa, good sources of vitamin K, are used, so diets are now lower in this vitamin.
5. We may be breeding strains of pigs which require more vitamin K. Increased litter size and rate of gain may also be increasing the need for vitamin K.
6. The use of slatted floors lessens the opportunity for coprophagy (eating of feces). Since vitamin K is synthesized in the intestinal tract the feces are a source of this vitamin.

There are undoubtedly other possibilities for vitamin K deficiencies now showing up, but those listed above are probably the main causes.

E. Field Observations

Field reports which have appeared in various magazines and communication with other nutritionists indicate this condition to be quite prevalent. Some of this hemorrhaging occurs at birth in the navel and following castration. Some of it occurs without causing death. Some pigs will develop enlarged blood-filled joints and become lame. Others may have swellings along the body wall which are filled with unclotted blood. Hematomas (or blood swellings) in the ears also occur.

Fortunately, the use of vitamin K has prevented the condition. Dr. Dudley Smith of Australia prevented the hemorrhagic condition at birth in the navel with 0.5 gm vitamin K per ton of sow feed. University reports in the United States indicate that 2 gm vitamin K per ton is adequate. I have heard of instances in which levels of 2 to as high as 16 gm vitamin K per ton of feed were needed because the lower levels were not effective under certain farm conditions. Until more research data are obtained it is difficult to recommend the exact level needed. The level required is probably somewhere between 2 to 16 gm vitamin K per ton of feed.

F. Requirement

The 1973 Swine Nutrient Requirements Publication (7) recommends a level of 2.2 gm menadione (vitamin K) per ton of diet. This is a minimum requirement and it should be increased if field conditions indicate a higher level is needed. Fortunately, vitamin K is inexpensive and so supplementation with it is not costly, especially if lower levels can be used successfully.

Until more is known about vitamin K, the safe course is to supplement with this vitamin if symptoms of a deficiency are occurring on the farm.

G. General Information

When vitamin K is deficient, the coagulation time of the blood is increased and the prothrombin level is decreased. Thus, vitamin K is needed for prothrombin formation, which in turn is needed for blood coagulation. Vitamin K is synthesized by microorganisms in the intestinal tract of the pig.

XXVI. VITAMIN E

A. Names Used Previously

Vitamin E has also been called antisterility vitamin and factor X. The term vitamin E should be used for all tocol and tocotrienol derivatives exhibiting qualitatively the biological activity of α-tocopherol. Thus, phrases such as "vitamin E activity" or "vitamin E deficiency" are preferred usage.

B. Apt to Be Deficient in Swine Diets

In 1965 it was thought that vitamin E was of no value in supplementing swine diets. By 1975, there was no question about the need for vitamin E fortification of swine diets in most situations. Good sources of vitamin E are wheat germ oil, cereal grains (particularly in the germ and the by-products containing the germ), alfalfa hay and meal, green forage, and liver, as well as soybean, peanut, and cottonseed oils.

C. Effect of a Deficiency

A deficiency of vitamin E causes the following: slow or interrupted growth, reduced feed intake, liver necrosis, brownish-yellow discoloration of adipose tissue, waxy degeneration of muscle tissue, edema and sudden death. Feeding sows a diet deficient in vitamin E during gestation and lactation may cause increased embryonic mortality and muscular incoordination in suckling pigs. There is limited evidence that some baby pigs from sows deficient in vitamin E become sensitive to iron administration. Such pigs will soon die if they are given iron orally or by injection (7). Alabama workers (177) were able to produce a fatal liver necrosis in growing pigs by feeding a diet deficient in vitamin E and containing 2% cod liver oil. The authors stated that without cod liver oil in the diet, the liver necrosis would probably not have occurred since the oil is needed as a stress factor to develop the condition.

During reproduction and lactation, a limited study with gilts (178) showed that a diet deficient in vitamin E caused lowered reproductive performance because of fetal death. Pigs from the sows reared on the deficient diet exhibited muscular incoordination caused by the disintegration of the muscle fibers. A study at the Hormel Institute (179) showed that supplementing the diet of the sow with vitamin E during gestation did not affect the size or apparent health of pigs at birth, but did favorably affect their survival rate and the growth of the nursing pigs under environmental conditions in which so-called baby pig disease was enzootic.

Dr. W. M. Beeson and others at Purdue University reported in 1950 that vitamin E supplementation was beneficial to sows during reproduction. The sows were being fed corn, soybean meal, alfalfa meal and mineral diet. The addition of vitamin E increased the birth weight of the pigs, increased the number of strong pigs, reduced baby pig losses for the first 14 days after farrowing and increased weaning weights.

An Iowa study (180) with a vitamin E and/or selenium deficiency showed that the stillborn pigs and those that died shortly after birth were small in size. Another Iowa study (188) showed that the supplementation of a diet with vitamin

E or selenium, or both, did not significantly affect growth rate. It did, however, reduce mortality from 54% down to 7%. A Cornell report (181) showed that pigs born from dams on low vitamin E diets were unable to stand or nurse, appeared disoriented, showed tremors and incoordination and died within minutes or a few days of birth. On necropsy, the dams showed typical vitamin E deficiency symptoms in the heart and liver, in addition to severe edema in many cases. This is similar to information I obtained on a visit to Dr. R. G. Grey of the Dookie Station (Australia) where selenium-deficient sows farrowed prematurely at 106 days of the gestation period. The baby pigs were very small and weak and most of them died shortly after birth.

Dr. Duane E. Ullrey in an article in *Feedstuffs* (November 12, 1973) reported that a Michigan State University study showed that 54 sows supplemented with vitamin E and selenium weaned 1.1 more pigs per litter than 54 sows receiving no E or selenium. Dr. Ullrey also stated that the symptoms of a lack of vitamin E or selenium, or both, show up more with an increase in environmental stress. For example, when pigs are weaned and mixed with pigs of other litters, the fighting which occurs to establish social order (the peck order or who is boss in the pen) precipitates the symptoms or lesions of a vitamin E and/or selenium deficiency. This has been observed experimentally by a number of other scientists. If pigs become excessively chilled or overheated in response to very cold or hot weather, the incidence of symptoms also increases.

Dr. Ullrey feels that in the Michigan State herd the pigs fed practical corn-soybean meal diets unsupplemented with vitamin E or selenium showed a 20% death loss due to a deficiency of these nutrients. Yet in living pigs, clinical signs of a deficiency of vitamin E and/or selenium are difficult to see. This would indicate that many unexplained death losses on the farm may be due to vitamin E and/or selenium deficiencies. Evidently stress factors will result in the death of many pigs which are borderline in a deficiency of vitamin E and/or selenium. All of this discussion, plus other studies not cited here, would indicate that vitamin E and/or selenium are involved in reproduction and are needed for maximum results.

D. Vitamin E and Selenium Interrelationship

Most of the vitamin E deficiency symptoms for the pig have also been associated with selenium deficiency. The exact role of vitamin E and selenium is not yet entirely understood. Most scientists refer to a vitamin E and/or selenium deficiency since it is not clear which one is involved or whether both of them are. The level of vitamin E in the diet will influence the need for selenium and vice versa. Although both vitamin E and selenium influence the need for one another, they both play a nutritional role in the body. Both also have an antioxidant effect.

The level of protein and the sulfur-containing amino acids (cystine and methionine) will also influence the need for vitamin E and/or selenium. It is thought this may be due to the antioxidant effect of cystine and methionine. The kind of diet used, its level of biologically active forms of vitamin E and the availability of its selenium content will also have an effect on the level of vitamin E and selenium to add. The selenium level in some feeds may be low in availability. When selenium is added to the diet, it is recommended that both vitamin E and selenium be used since neither one can replace the other entirely.

E. Vitamin E Activity

The α-tocopherol is the preferred form of vitamin E since it has the greatest nutritional value. For example, if α-tocopherol is given a value of 100, the beta and zeta forms of tocopherols have about one-third this value, whereas the gamma, delta, epsilon and eta-tocopherols have less than 1% the activity of α-tocopherol. Therefore, one can be misled by a total tocopherol analysis of feeds. The important thing to know is the α-tocopherol value of feeds.

F. Why Do Vitamin E Deficiencies Occur?

Some of the reasons for the occurrence of vitamin E deficiencies are listed below.

1. More pigs are being raised in confinement without pasture and alfalfa which are good sources of vitamin E.
2. Heating and pelleting grains lower their vitamin E values. The use of high moisture grains increases the need for vitamin E.
3. Rancidity destroys vitamin E. Thus the use of fats or feeds with unsaturated fatty acids, which are susceptible to rancidity, could be increasing vitamin E needs.
4. More selenium deficiences are occurring and this increases the need for vitamin E supplementation.
5. The vitamin E level in feeds has been overestimated. Only α-tocopherol has maximum vitamin E activity.
6. Selenium in feeds may have only 50% availability for the animal. Thus, selenium analyses of feeds may overestimate their value. If selenium is low it increases need for vitamin E.
7. Animals are growing faster and crossbred sows are producing at a higher rate which would increase vitamin E needs.
8. Certain strains or breeds may require more vitamin E than others.
9. Sows are being fed less total feed daily during gestation. This increases the level of vitamin E needed per pound of feed.

There are other reasons for the increase of vitamin E deficiencies, but the nine listed above are among the more important ones.

G. Requirements

The 1973 Nutrient Requirements Publication (7) suggests that 5 IU of vitamin E be added per pound of diet until more specific research information is obtained on both vitamin E and selenium requirements. The level of selenium, protein, cystine, methionine and other factors will influence the exact level of vitamin E needed. Until all these interrelationships are better understood, it is difficult to recommend the level of vitamin E required by the pig during the various stages of its life cycle and under varying conditions.

H. White Pig Disease

In a visit to Australia, I visited R. G. Grey of Dookie Agriculture College near Shepparton, Australia who has obtained experimental information on "white pig disease" (182). This condition occurs with the fastest growing pigs when they usually weigh about 90–100 lb (may be 60–90 lb in other areas). The pigs develop hemorrhages in the cecum and colon. These hemorrhages cause bleeding which results in anemia and then causes the pig to become white or pale in appearance and hence the name "white pig disease."

If the disease is noticed early, one dose of 5–10 mg selenium and 300–600 mg vitamin E will save most of the pigs affected. If not, they will usually die in 24–48 hours after developing a very pale skin condition. Only one to four pigs in each litter are usually affected and only about 20% of the sows, at the most, in the herd will have pigs showing this condition. These are probably the most deficient pigs in vitamin E. Of interest is that these pigs, on necropsy, do not show the liver necrosis or mulberry heart condition which occurs in New Zealand (which is mainly a selenium deficiency and which also responds to selenium and vitamin E). This shows the complexity of the vitamin E–selenium picture and that it is still not well understood.

I. Vitamin E Deficiency and Iron Sensitivity

Studies at the Royal Veterinary College in Stockholm, Sweden showed that vitamin E-deficient sows produced baby pigs that were sensitive to iron administration (183,184). This caused some of these pigs to die shortly after iron was given them orally or by injection. A summary of these studies indicates the following:

1. Only an occasional litter develops iron-sensitive pigs. This may be those that are the most deficient in vitamin E. Only some of the pigs in the litter die. Very seldom will all of the pigs die. The pigs will start dying in 4–6 hours after iron administration. Peak mortality occurs in 10–12 hours. My observations in other countries, however, indicate that deaths may occur shortly after iron administration. In the surviving pigs, the symptoms of poisoning begin to recede about 12 hours after the iron treatment, though a few deaths occur during the next 2 or 3 days.

2. Gilts which farrow for the first time have a higher incidence of pigs sensitive to iron. This is probably because the gilt is still growing (plus carrying a litter) and thus needs more vitamin E.

3. Vitamin E or selenium did not protect against baby pig deaths when given at the same time as the iron. The vitamin E had to be given 1 day before the iron, and selenium 2 or 3 days before the iron. However, ethoxyquin (an antioxidant) was effective in preventing death losses when given with the iron. This raises the possibility of the simultaneous use of iron and ethoxyquin preparations to prevent death losses of baby pigs from E deficient sows. The best recommendation, however, is to make sure the sow diet is adequate in vitamin E since a deficiency of this vitamin can have other harmful effects on the pig. Deaths after larger oral doses of iron as ferrous sulfate (180 mg and 360 mg/lb of body weight) could not be prevented with vitamin E (184).

This work indicates that farmers noticing baby pigs dying shortly after iron administration should take another look at the adequacy of their sow diets in vitamin E and/or selenium.

A Michigan study (185) with pigs weaned at 5 weeks of age showed no evidence of iron toxicity. It may be that 5-week-old pigs from vitamin E-deficient sows are no longer susceptible to iron toxicity.

J. General Information

There is no evidence that vitamin E in excess causes any harmful effects. This does not mean, however, that excessive levels should be used. Vitamin E is a strong antioxidant. Thus, it protects vitamin A, essential fatty acids, and other nutrients from destruction. It has been definitely shown that the vitamin A requirement of animals is partially dependent on the adequacy of the vitamin E content of the diet. Vitamin E is especially helpful when a borderline level of the vitamin A is fed.

There is some information to indicate that vitamin E is concerned with the formation of blood in the newborn pig (186). Pigs with anemia are more susceptible to pneumonia and other diseases. This is another reason to make sure sow diets are adequate in vitamin E.

A study in Norway (187) showed that adding vitamin E to swine diets prior to slaughter reduced the susceptibility of the fat to oxidation and improved the quality of the taste in pork.

REFERENCES

1. Funk, C., *J. Physiol. (London)* **43,** 395 (1911).
2. Cunha, T. J., Washington, Agricultural Experiment Station, Pullman, unpublished data (1948).
3. Luecke, R. W., W. N. McMillen, F. Thorpe, Jr., and C. Tull, *J. Nutr.* **36,** 417 (1948).
4. Luecke, R. W., W. N. McMillen, F. Thorpe, Jr., and C. Tull, *J. Nutr.* **33,** 251 (1947).
5. McMillen, W. N., R. W. Luecke, and F. Thorpe, Jr., *J. Anim. Sci.* **8,** 518 (1949).
6. Beeson, W. M., E. W. Crampton, T. J. Cunha, N. R. Ellis, and R. W. Luecke, *N.A.S.—N.R.C., Publ.* **295** (1953).
7. Cunha, T. J., J. P. Bowland, J. H. Conrad, V. W. Hays, R. J. Meade, and H. S. Teague, *N.A.S.—N.R.C., Publ.* (1973).
8. Cunha, T. J., R. W. Colby, H. W. Hodgskiss, T. C. Huang, and M. E. Ensminger, *J. Anim. Sci.* **7,** 523 (1948).
9. Cunha, T. J., J. E. Burnside, H. H. Hopper, A. M. Pearson, and R. S. Glasscock, *J. Anim. Sci.* **8,** 616 (1949).
10. Rasmussen, R. A., H. H. Smith, R. W. Phillips, and T. J. Cunha, *Utah Agric. Exp. Stn., Tech. Bull.* **302** (1942).
11. Ensminger, M. E., W. W. Heinemann, T. J. Cunha, and E. C. McCulloch, *Wash., Agric. Exp. Stn., Bull.* **468** (1945).
12. Heinemann, W. W., M. E. Ensminger, T. J. Cunha, and E. C. McCulloch, *J. Nutr.* **31,** 107 (1946).
13. Hughes, E. H., *J. Nutr.* **20,** 239 (1940).
14. Wintrobe, M. M., R. Alcayaga, S. Humphreys, and R. H. Follis, Jr., *Bull. Johns Hopkins Hosp.* **73,** 169 (1943).
15. Wintrobe, M. M., R. H. Follis, Jr., S. Humphreys, H. Stein, and M. Lauritsen, *J. Nutr.* **28,** 283 (1944).
16. Ensminger, M. E., J. P. Bowland, and T. J. Cunha, *J. Anim. Sci.* **6,** 409 (1947).
17. Ellis, N. R., and L. L. Madsen, *J. Nutr.* **27,** 253 (1944).
18. Miller, E. R., D. A. Schmidt, J. A. Hoefer, and R. W. Luecke, *J. Nutr.* **65,** 423 (1955).
19. Van Etten, C. H., N. R. Ellis, and L. L. Madsen, *J. Nutr.* **20,** 607 (1940).
20. Miller, C. D., and S. H. Work, *J. Anim. Sci.* **5,** 350 (1946).
21. Miller, R. C., J. W. Pence, R. A. Dutcher, P. T. Ziegler, and M. A. McCarty, *J. Nutr.* **26,** 261 (1943).
22. Pence, J. W., R. C. Miller, R. A. Dutcher, and P. T. Ziegler, *J. Anim. Sci.* **4,** 141 (1945).
23. Peng, C. L., and H. H. Heitman, Jr., *Br. J. Nutr.* **32,**1 (1974).
24. Hughes, E. H., *J. Nutr.* **20,** 233 (1940).
25. Lehrer, W. P., Jr., and A. C. Wiese, *J. Anim. Sci.* **11,** 244 (1952).
26. Miller, E. R., R. L. Johnston, J. A. Hoefer, and R. W. Luecke, *J. Nutr.* **52,** 405 (1954).
27. Wintrobe, M. M., W. Buschke, R. H. Follis, Jr., and S. Humphreys, *Bull. John Hopkins Hosp.* **75,** 102 (1944).
28. Miller, C. O., N. R. Ellis, J. W. Stevenson, and R. Davey, *J. Nutr.* **51,** 163 (1953).
29. Krider, J. L., D. E. Becker, W. E. Carroll, and B. W. Fairbanks, *J. Anim. Sci.* **7,** 332 (1948).
30. Krider, J. L., S. W. Terrill, and R. F. Van Poucke, *J. Anim. Sci.* **8,** 121 (1949).

31. Miller, C. O., and N. R. Ellis, *J. Anim. Sci.* **10,** 807 (1951).
32. Mitchell, H. H., B. C. Johnson, T. S. Hamilton, and W. T. Haines, *J. Nutr.* **41,** 317 (1950).
33. Terrill, S. W., C. B. Ammerman, D. E. Walker, R. M. Edwards, H. W. Norton, and D. E. Becker, *J. Anim. Sci.* **14,** 593 (1955).
34. Seymour, E. W., V. C. Speer, and V. W. Hays, *J. Anim. Sci.* **27,** 389 (1968).
35. Burroughs, W., B. H. Edgington, W. L. Robison, and R. M. Bethke, *J. Nutr.* **41,** 51 (1950).
36. Cartwright, G. E., B. Tanning, and M. M. Wintrobe, *Arch. Biochem.* **19,** 109 (1948).
37. Hughes, E. H., *J. Anim. Sci.* **2,**23 (1943).
38. Powick, W. C., N. R. Ellis, L. L. Madsen, and C. N. Dale, *J. Anim. Sci.* **6,** 310 (1947).
39. Wintrobe, M. M., H. J. Stein, R. H. Follis, Jr., and S. Humphreys, *J. Nutr.* **30,** 395 (1945).
40. Ensminger, M. E., R. W. Colby, and T. J. Cunha, *Wash., Agric. Exp. Stn., Circ.* **134** (1951).
41. Braude, R., S. K. Kon, and E. G. White, *Biochem. J.* **40,** 843 (1946).
42. Harmon, B. G., D. E. Becker, A. H. Jensen, and D. H. Baker, *J. Anim. Sci.* **31,**339 (1970).
43. Jensen, A. H., J. T. Yen, D. H. Baker, and B. G. Harmon, *J. Anim. Sci.* **31,**1023 (1970).
44. Luce, W. G., E. R. Peo, Jr., and D. B. Hudman, *J. Nutr.* **88,**39 (1966).
45. Luce, W. G., E. R. Peo, Jr., and D. B. Hudman, *J. Anim. Sci.* **26,**76 (1967).
46. Colby, R. W., T. J. Cunha, C. E. Lindley, D. R. Cordy, and M. E. Enminger, *J. Am. Vet. Med. Assoc.* **113,** 589 (1948).
47. Hughes, E. H., *J. Agric. Res.* **64,** 185 (1942).
48. Hughes, E. H., and N. R. Ittner, *J. Anim. Sci.* **1,** 116 (1942).
49. Lehrer, W. P., Jr., and A. C. Wiese, *Idaho, Agric. Exp. Stn., Res. Bull.* **21** (1952).
50. Wiese, A. C., W. P. Lehrer, Jr., P. R. Moore, O. F. Pahnish, and W. V. Hartwell, *J. Anim. Sci.* **10,** 80 (1951).
51. Wintrobe, M. M., R. H. Follis, Jr., R. Alcayaga, M. Paulson, and S. Humphreys, *Bull. Johns Hopkins Hosp.* **73,** 313 (1943).
52. Ullrey, D. E., D. E. Becker, S. W. Terrill, and R. A. Notzold, *J. Nutr.* **57,** 401 (1955).
53. Luecke, R. W., W. N. McMillen, and F. Thorpe, Jr., *J. Anim. Sci.* **9,** 78 (1950).
54. Luecke, R. W., J. A. Hoefer, and F. Thorpe, Jr., *J. Anim. Sci.* **12,** 605 (1953).
55. Stothers, S. C., D. A. Schmidt, R. L. Johnston, J. A. Hoefer, and R. W. Luecke, *J. Nutr.* **57,** 47 (1955).
56. Catron, D. V., R. W. Bennison, H. M. Maddock, G. C. Ashton, and P. G. Homeyer, *J. Anim. Sci.* **12,** 51 (1953).
57. Luecke, R. W., R. Thorpe, Jr., W. N. McMillen, and H. W. Dunne, *J. Anim. Sci.* **8,** 464 (1949).
58. McMillen, W. N., R. W. Luecke, and F. Thorpe, Jr., *J. Anim. Sci.* **7,** 529 (1948).
59. Teague, H. S., W. M. Palmer, and A. P. Grifo, Jr., *Ohio, Agric. Exp. Stn., Anim. Sci. Mimeo* **200** (1970).
60. Teague, H. S., A. P. Grifo, Jr., and W. M. Palmer, *J. Anim. Sci.* **33,**239 (1971).
61. McKigney, J. I., H. D. Wallace, and T. J. Cunha, *J. Anim. Sci.* **16,**35 (1957).
62. Godwin, R. F. W., *J. Comp. Pathol. Ther.* **72,**214 (1962).
63. Davey, R. J., and J. W. Stevenson, *J. Anim. Sci.* **22,**9 (1963).
64. Cartwright, G. E., and M. M. Wintrobe, *J. Biol. Chem.* **172,** 557 (1948).
65. Follis, R. H., Jr., and M. M. Wintrobe, *J. Exp. Med.* **81,** 539 (1945).
66. Hughes, E. H., and R. L. Squibb, *J. Anim. Sci.* **1,** 320 (1942).
67. Lehrer, W. P., Jr., A. C. Wiese, P. R. Moore, and M. E. Ensminger, *J. Anim. Sci.* **10,** 65 (1951).
68. Wintrobe, M. M., R. H. Follis, Jr., M. H. Miller, H. J. Stein, R. Alcayaga, S. Humphreys, A. Suksta, and G. E. Cartwright, *Bull. Johns Hopkins Hosp.* **72,** 1 (1943).
69. Adams, C. R., C. E. Richardson, and T. J. Cunha, *J. Anim. Sci.* **26,**903 (1967).

70. Kirchgessner, V. M., and H. Friesecke, *Z. Tierphysiol., Tierernahr. Futtermittekd.* **17**,235,238 (1962).
71. Smith, D., personal visit and communication in Australia (1968).
72. Miller, E. R., D. A. Schmidt, J. A. Hoefer, and R. W. Luecke, *J. Nutr.* **62**,407 (1957).
73. Ritchie, H. D., E. R. Miller, D. E. Ullrey, J. A. Hoefer, and R. W. Luecke, *J. Nutr.* **70**,491 (1960).
74. Sewell, R. F., and M. M. Nugara, *J. Anim. Sci.* **20**,951 (1961).
75. Ellis, N. R., *Nutr. Abstr. Rev.* **16**, 1 (1945–1946).
76. Johnson, B. C., and M. F. James, *J. Nutr.* **36**, 339 (1948).
77. Nesheim, R. O., and B. C. Johnson, *J. Nutr.* **41**, 149 (1950).
78. Neumann, A. L., J. L. Krider, M. F. James, and B. C. Johnson, *J. Nutr.* **38**, 195 (1949).
79. Wintrobe, M. M., M. H. Miller, R. H. Follis, Jr., H. J. Stein, C. Mushatt, and S. Humphreys, *J. Nutr.* **24**, 345 (1942).
80. NRC-42 Committee on Swine Nutrition, *J. Anim. Sci.* **37**,281 (1973).
81. Newland, H. W., Ohio, Agricultural Experiment Station, Columbus, personal communication (1971).
82. Dobson, K. J., Department of Agriculture, Adelaide, Australia, personal communication (1970).
83. Erickson, L., Schreiber Mills Inc., St. Joseph, Missouri, personal communication (1971).
84. Vipperman, P. E., Jr., E. R. Peo, Jr., and P. J. Cunningham, *J. Anim. Sci.* **31**,213 (1970).
85. Maner, J. H., CIAT Cali, Columbia, S. A., personal communication (1970).
86. Stockland, W. L., and L. G. Blaylock, *J. Anim. Sci.* **39**,1113 (1974).
87. Kornegay, E. T., *J. Anim. Sci.* **33**,232 (1971).
88. Lehrer, W. P., Jr., A. C. Wiese, and P. R. Moore, *J. Nutr.* **47**, 203 (1952).
89. Cunha, T. J., D. C. Lindley, and M. E. Ensminger, *J. Anim. Sci.* **5**, 219 (1946).
90. Lindley, D. C., and T. J. Cunha, *J. Nutr.* **32**, 47 (1946).
91. Adams, C. R., C. E. Richardson, and T. J. Cunha, *J. Anim. Sci.* **26**,903 (1967).
92. Peo, E. R., Jr., G. F. Wehrbein, B. Moser, P. J. Cunningham, and P. E. Vipperman, *J. Anim. Sci.* **31**,209 (1970).
93. Cunha, T. J., O. B. Ross, P. H. Phillips, and G. Bohstedt, *J. Anim. Sci.* **3**, 415 (1944).
94. Ross, O. B., P. H. Phillips, and G. Bohstedt, *J. Anim. Sci.* **1**, 353 (1942).
95. Parrish, D. B., C. E. Aubel, J. S. Hughes, and J. D. Wheat, *J. Anim. Sci.* **10**, 551 (1951).
96. Cunha, T. J., L. K. Bustad, W. E. Ham, D. R. Cordy, E. C. McCulloch, I. F. Woods, G. H. Conner, and M. A. McGregor, *J. Nutr.* **34**, 173 (1947).
97. Johnson, B. C., M. F. James and J. L. Krider, *J. Anim. Sci.* **7**, 486 (1948).
98. Cunha, T. J., R. W. Colby, L. K. Bustad, and J. F. Bone, *J. Nutr.* **36**, 215 (1948).
99. Cartwright, G. E., and M. M. Wintrobe, *Proc. Soc. Exp. Biol. Med.* **71**, 54 (1949).
100. Cartwright, G. E., J. Fay, B. Tanning, and M. M. Wintrobe, *J. Lab. Clin. Med.* **33**, 397 (1948).
101. Cartwright, G. E., J. G. Palmer, B. Tanning, H. Ashenbrucker, and M. M. Wintrobe, *J. Lab. Clin. Med.* **36**, 675 (1950).
102. Cartwright, G. E., M. M. Wintrobe, and S. Humphreys, *J. Lab. Clin. Med.* **31**, 423 (1946).
103. Franklin, A. L., E. L. R. Stokstad, M. Belt, and T. H. Jukes, *J. Biol. Chem.* **169**, 427 (1947).
104. Heinle, R. W., A. D. Welch, and H. L. Shorr, *J. Lab. Clin. Med.* **34**, 1763 (1949).
105. Johnson, B. C., A. L. Neumann, R. O. Nesheim, M. F. James, J. L. Krider, A. S. Dana, and J. B. Thiersch, *J. Lab. Clin. Med.* **36**, 537 (1950).
106. Welch, A. D., R. W. Heinle, G. Sharpe, W. L. George, and M. Epstein, *Proc. Soc. Exp. Biol. Med.* **65**, 364 (1947).
107. Anderson, G. C., and A. G. Hogan, *J. Nutr.* **40**, 243 (1950).

108. Johnson, B. C., and A. L. Neumann, *J. Biol. Chem.* **178,** 1001 (1949).
109. Nesheim, R. O., J. L. Krider, and B. C. Johnson, *Arch. Biochem.* **27,** 240 (1950).
110. Neumann, A. L., J. L. Krider, and B. C. Johnson, *Proc. Soc. Exp. Biol. Med.* **69,** 513 (1948).
111. Neumann, A. L., J. B. Thiersch, J. L. Krider, M. F. James, and B. C. Johnson, *J. Anim. Sci.* **9,** 83 (1950).
112. Richardson, D., D. V. Catron, L. A. Underkofler, H. M. Maddock, and W. C. Friedland, *J. Nutr.* **44,** 371 (1951).
113. Vestal, C. M., W. M. Beeson, F. N. Andrews, L. M. Hutchings, and L. P. Doyle, *Purdue, Agric. Exp. Stn., Mimeo* No. 50 (1950).
114. Sewell, R. F., T. J. Cunha, C. B. Shawver, W. A. Ney, and H. D. Wallace, *Am. J. Vet. Res.* **13,** 186 (1952).
115. Cunha, T. J., H. H. Hopper, J. E. Burnside, A. M. Pearson, R. S. Glasscock, and A. L. Shealy, *Arch. Biochem.* **23,** 510 (1949).
116. Firth, J., and B. C. Johnson, *Science* **120,** 352 (1954).
117. Firth, J., S. P. Mistry, M. F. James, and B. C. Johnson, *Proc. Soc. Exp. Biol. Med.* **85,** 307 (1954).
118. Kelly, R. F., R. W. Bray, and P. H. Phillips, *J. Anim. Sci.* **13,** 332 (1954).
119. Cary, C. A., A. M. Hartman, L. P. Dryden, and G. D. Likely, *Fed. Proc., Fed. Am. Soc. Exp. Biol.* **5,** 128 (1946).
120. Catron, D. V., D. Richardson, L. A. Underkofler, H. M. Maddock, and W. C. Friedland, *J. Nutr.* **47,** 461 (1952).
121. Vohs, R. L., H. M. Maddock, D. V. Catron, and C. C. Culbertson, *J. Anim. Sci.* **10,** 42 (1951).
122. Teague, II. S., and A. P. Grifo, Jr., *J. Anim. Sci.* **23,**894 (1964).
123. Teague, H. S., and A. P. Grifo, Jr., *J. Anim. Sci.* **25,**895 (1966).
124. Teague, H. S., and A. P. Grifo, Jr., *Ohio, Agric. Exp. Stn., Res. Summ.* **13** (1966).
125. Henderickx, H. K., H. S. Teague, D. R. Redman, and A. P. Grifo, Jr., *J. Anim. Sci.* **23,**1036 (1964).
126. Jones, A. S., The Rowett Research Institute, Aberdeen, Scotland, personal communication (1968).
127. Riker, J. T., III, T. W. Perry, R. A. Pickett, and C. J. Heidenreich, *J. Nutr.* **92,**99 (1967).
128. Burnside, J. E., Southern Illinois University, Carbondale, personal communication (1969).
129. Judge, M. D., E. J. Breskey, R. G. Cassens, J. C. Forrest, and R. K. Meyer, *Am. J. Physiol.* **214,**146 (1968).
130. Moss, P. R., M. S. Thesis, Southern Illinois University, (1969).
131. Williams, R. J., and G. Deacon, *Science* **156,**543 (1967).
132. Cromwell, G. L., and V. W. Hays, *J. Anim. Sci.* **31,**198 (1970).
133. Cromwell, G. L., V. W. Hays, and J. R. Overfield, *J. Anim. Sci.* **29,**131 (1969).
134. Pickett, R. A., and B. D. Virgin, *Purdue Swine Day Rep.* p. 1 (1965).
135. Combs, G. E., and H. D. Wallace, *Fla., Agric. Exp. Stn., Mimeo Ser.* **AN 71-6** (1971).
136. Gipps, W. F., W. G. Pond, D. R. Van Campen, F. A. Kallfelz, and J. Tasker, *J. Anim. Sci.* **33,**229 (1971).
137. Adams, C. R., Hoffman La Roche, personal communication (1969).
138. Dunlop, G., *J. Agric. Sci.* **24,** 435 (1935).
139. Elder, C., *J. Am. Vet. Med. Assoc.* **87,** 22 (1935).
140. Foot, A. S., K. M. Henry, S. K. Kon, and J. Mackintosh, *J. Agric. Sci.* **29,** 142 (1939).
141. Guilbert, H. R., R. F. Miller, and E. H. Hughes, *J. Nutr.* **13,** 543 (1937).
142. Hentges, J. F., Jr., R. H. Grummer, P. H. Phillips, G. Bohstedt, and D. K. Sorensen, *J. Am. Vet. Med. Assoc.* **120,** 213 (1952).
143. Hale, F., *Tex. State J. Med.* **33,** 228 (1937).

144. Hentges, J. F., Jr., R. H. Grummer, P. H. Phillips, and G. Bohstedt, *J. Anim. Sci.* **11,** 266 (1952).
145. Hughes, J. S., C. E. Aubel, and H. F. Lienhardt, *Kans., Agric. Exp. Stn., Tech. Bull.* **23** (1928).
146. Selke, M. R., C. E. Barnhart, and C. H. Chaney, *J. Anim. Sci.* **26,**759 (1967).
147. Wellenheiter, R. H., D. E. Ullrey, E. R. Miller, and W. T. Magee, *J. Nutr.* **99,**129 (1969).
148. Frape, D. L., V. C. Speer, V. W. Hays, and D. V. Catron, *J. Nutr.* **68,** 173 (1959).
149. McDonald, T. A., W. H. Smith, R. A. Pickett, and W. M. Beeson, *J. Anim. Sci.* **25,**1024 (1966).
150. Combs, G. E., and H. D. Wallace, *Fla., Agric. Exp. Stn., Mimeo Ser.* **AN 70-6** (1970).
151. Anderson, M. D., V. C. Speer, J. T. McCall, and V. W. Hays, *J. Anim. Sci.* **25,**1123 (1966).
152. Hurt, H. D., R. C. Hall, Jr., M. C. Calhoun, J. E. Rousseau, Jr., H. D. Eaton, R. E. Wolke, and J. J. Lucas, *J. Anim. Sci.* **25,**1966.
153. Pryor, W. J., A. A. Seawright, and P. J. McCosker, *Aust. Vet. J.* **45,**563 (1969).
154. Dobson, K. J., *Aust. Vet. J.* **45,**570 (1969).
155. Aubel, C. E., and J. S. Hughes, *Am. Soc. Anim. Rec. Proc. Annu. Meet.* **30,** 334 (1937).
156. Aubel, C. E., J. S. Hughes, and W. J. Peterson, *J. Agric. Res.* **62,** 531 (1941).
157. Bethke, R. M., B. H. Edgington, and C. H. Kick, *J. Agric. Res.* **47,** 331 (1933).
158. Bohstedt, G., *Ohio, Agric. Exp. Stn., Bull.* **395** (1926).
159. Johnson, D. W., and L. S. Palmer, *J. Agric. Res.* **63,** 639 (1941).
160. Johnson, D. W., and L. S. Palmer, *J. Agric. Res.* **58,** 929 (1939).
161. Miller, E. R., D. E. Ullrey, C. L. Zutant, B. V. Baltzer, D. A. Schmidt, B. H. Vincent, J. A. Hoefer, and R. W. Luecke, *J. Nutr.* **83,**140 (1964).
162. Combs, G. E., and H. D. Wallace, *Fla., Agric. Exp. Stn., Mimeo Ser.* **AN 70-5** (1970).
163. Roberts, H. F., W. G. Hoekstra, and R. H. Grummer, *J. Anim. Sci.* **20,**950 (1961).
164. Quarterman, J., A. C. Dalgarno, A. Adams, B. F. Fell, and R. Boyne, *Br. J. Nutr.* **18,**65 (1964).
165. Combs, G. E., T. H. Berry, H. D. Wallace, and R. C. Crum, Jr., *J. Anim. Sci.* **25,**827 (1966).
166. Beeson, W. M., Purdue, Agricultural Experiment Station, LaFayette, Indiana, personal communication (1969); Hoefer, J. A., and R. W. Luecke, *J. Anim. Sci.* **23,**884 (1964).
167. Hendricks, D. G., E. R. Miller, D. E. Ullrey, R. O. Struthers, B. V. Baltzer, J. A. Hoefer, and R. W. Luecke, *J. Nutr.* **93,**37 (1967).
168. Norman, A. W., J. F. Myrtle, R. J. Midgett, H. G. Nowicki, V. Williams, and G. Popjak, *Science* **173,**51 (1971).
169. Naber, E. C., *Feed Manage.* 39 (1975).
170. Meade, R. J., Minnesota, Agricultural Experiment Station, Minneapolis, personal communication (1969).
171. Fritschen, R. O., E. R. Peo, Jr., L. E. Lucas, and O. D. Grace, *J. Anim. Sci.* **31,**199 (1970).
172. Muhrer, M. E., R. G. Cooper, C. N. Cornell, and R. D. Thomas, *J. Anim. Sci.* **31,**1025 (1970).
173. Nakamura, R. M., and C. C. Brooks, *J. Anim. Sci.* **33,**237 (1971).
174. Brooks, C. C., R. M. Nakamura, and A. Y. Miyahara, *J. Anim. Sci.* **37,**1344 (1973).
175. Houser, R. H., E. C. Harland, and H. L. Rubin, *Fla., Agric. Exp. Stn., Mimeo Ser.* **ARCLO71-3** (1971).
176. Neufville, M. H., H. D. Wallace, and G. E. Combs, *J. Anim. Sci.* **37,**288 (1973).
177. Hove, E. L., and H. R. Seibold, *J. Nutr.* **56,** 173 (1955).
178. Adamstone, F. B., J. L. Krider, and M. F. James, *Ann. N. Y. Acad. Sci.* **52,** 63 (1949).
179. Carpenter, L. E., and W. O. Lundberg, *Ann. N. Y. Acad. Sci.* **52,** 269 (1949).
180. Wastell, M. E., R. C. Ewan, E. J. Bicknell, and V. C. Speer, *J. Anim. Sci.* **29,**149 (1969).
181. Aydin, A., W. G. Pond, and D. Kirtland, *J. Anim. Sci.* **37,**274 (1973).

182. Grey, R. G., Dookie Agricultural College, Sheparton, Australia, personal communication (1968).
183. Tollerz, G., and N. Lannek, *Nature* **201,**846 (1964).
184. Tollerz, G., Royal Veterinary College, Department of Medicine, Stockholm, Sweden. Studies on the tolerance to iron in piglets and mice. Bulletin (1965).
185. E. R. Miller, J. P. Hitchcock, K. K. Kuan, P. K. Ku, D. E. Ullrey, and K. K. Keahey, *J. Anim. Sci.* **37,**287 (1973).
186. Baustad, B., and I. Nafstad, *Br. J. Nutr.* **28,**183 (1972).
187. Astrup, H. N., *Acta Agric. Scand.* **19,**152 (1973).
188. Ewan, R. C., M. E. Wastell, E. J. Becknell, and V. C. Speer, *J. Anim. Sci.* **29,**912 (1969).
189. Dolipher, I. J., and C. Muhaxhire, *Vet. Arh.* **41,** 202–216 (1970).
190. Brown, R. G., V. D. Sharma, and L. G. Young, *Can. J. Anim. Sci.* **50,**605 (1970).
191. Brown, R. G., V. D. Sharma, L. G. Young, and J. G. Buchanan-Smith, *Can. J. Anim. Sci.* **51,**439 (1971).
192. Brown, R. G., B. A. Sharp, L. G. Young, and A. Van Dreumel, *Can. J. Anim. Sci.* **51,**537 (1971).

5

Protein Requirements of the Pig

I. INTRODUCTION

A lack of protein is frequently a limiting factor in the diet. This is because farm grains and their by-products are deficient in both quantity and quality of protein for swine. And since protein supplements are expensive feeds, some farmers still tend to feed too little protein.

Animals continually use protein either to build new tissues, as in growth and reproduction, or to repair worn-out tissues. Thus, swine require a regular intake of protein. If adequate protein is lacking in a diet, the pigs suffer a reduction in growth or loss of weight. Ultimately, protein will be withdrawn from certain tissues to maintain the functions of the more vital tissues of the body as long as possible. Protein is needed to form milk, meat, hide, hoof, hair, hormones, enzymes, blood cells and other constituents in the body. Thus, protein affects almost every body function. It has been shown also that animals are more resistant to infections if they are fed an adequate protein diet. The compounds in the bloodstream which help resist disease are proteins. So, adequate protein in the diet is one way of helping to keep animals resistant to diseases.

Proteins are made up of many amino acids combined with one another. Protein in the diet is not required as such. It is needed as a source of essential amino acids and nitrogen for nonessential amino acid synthesis. The amino acids are put together in various combinations to form body proteins, and are many times referred to as the building blocks of proteins. Every protein has a definite amino acid composition and no two are alike.

Amino acids contain nitrogen combined with carbon, hydrogen, oxygen, and sometimes sulfur, phosphorus and iron. The nitrogen is in the form of an amino group (NH_2); it is from this that the name of the amino acid is derived. Amino acids have been commercially synthesized, and a few are available in large quantities. The amino acids occurring in nature are in the L-form. However, both the D- and L-forms are synthesized in the laboratory. Animals can change the D-form to the L-form with some amino acids but not with others.

An excellent review (1) on this subject with the chick indicates the following information. The chick utilizes certain D-amino acids more efficiently under one set of conditions and less under others. The amount of other D-amino acids in the diet influences this. Vitamin supplements or carbohydrate sources will also influence utilization. Adding an excess of an amino acid such as phenylalanine, when D-tyrosine is being evaluated, will influence the apparent degree of utilization. A summary of the activity of the D-forms of the essential amino acids for the chick is tabulated below.

Well utilized	Poorly utilized
Cystine	Arginine
Glutamic acid	Isoleucine
Leucine	Histidine
Methionine	Lysine
Phenylalanine	Threonine
Proline	Tryptophan
Tyrosine	Valine

The nonessential amino acids, aspartic acid and serine, are fairly well utilized in the D-form. Other compounds, mainly keto acids and other analogues, have amino acid activity. The main one is the hydroxy analogue of methionine (MHA), which some scientists think is at least 80% as effective as methionine when supplemented on a methionine equivalent basis. Glyoxylic acid, creatine, citrulline and 3-indolpyruvic acid have activity in the chick.

Information derived from the chick is presented since much more is known about this subject with the chick than with the pig. This information can be used as a guide for the pig, but there is no assurance that all will apply. For the pig, the National Research Council (NRC) (3) states the following: "When amino acid supplements are provided, DL-methionine can replace the L-form in meeting the need for methionine. D-Tryptophan has a biological activity of about 60% of L-tryptophan for the growing pig (23,55). Thus, 0.15% DL-tryptophan is equivalent to 0.12% L-tryptophan in meeting the needs of the growing pig for this amino acid. It is assumed that the pig can utilize D-phenylalanine to some extent in meeting the total need for phenylalanine + tyrosine, but the efficiency of inversion is not known. Similarly, D-histidine may be used to a small extent. Isoleucine, leucine, lysine, threonine and valine are poorly invertible, if at all, and the D-isomer may make no contribution to fulfilling the need for these specific amino acids. Also, the effect of large quantities of the D-isomers of the poorly invertible amino acids on the efficiency of inversion of D-isomers of the more readily invertible amino acids in swine is not established".

II. ESSENTIAL AMINO ACIDS

Animals can make certain amino acids from other amino acids or other nutrients in the diet. These are called nonessential amino acids. Other amino acids, however, cannot be made in the body from other substances or cannot be made fast enough by the animal's body to supply its needs. These are, therefore, known as essential amino acids. It has been shown that the pig requires ten essential amino acids for maximum growth (3,4). (See Table 5.1.)

Swine may obtain the nonessential amino acids in the diet or by synthesis in the body from the essential amino acids. An adequate amount of nonessential amino acids in the diet lessens the need for certain essential amino acids by conserving them. For example, cystine is synthesized from methionine. Thus, if the cystine content in the diet is low, there must be enough methionine in the diet to supply the body needs for methionine as well as for the synthesis of cystine. It is more efficient to supply the nonessential amino acids as such rather than through their synthesis from essential amino acids.

TABLE 5.1
Amino Acid Classification for the Pig

Essential amino acids[a]	Nonessential amino acids[a]
Lysine	Glycine
Tryptophan	Serine
Methionine	Alanine
Valine	Norleucine
Histidine	Aspartic acid
Phenylalanine	Glutamic acid
Leucine	Hydroxyglutamic acid
Isoleucine	Cystine[c]
Threonine	Citrulline
Arginine[b]	Proline
	Hydroxyproline
	Tryrosine[d]

[a]Indispensible and dispensible could replace the terms essential and nonessential respectively. This new terminology has been proposed by Dr. A. E. Harper (2) and others as more accurate and meaningful. Eventually, it may be used instead.

[b]Partially synthesized.

[c]Can replace 50–70% of the methionine requirement.

[d]Can replace 30% of the phenylalanine requirement.

III. QUALITY OF PROTEIN

Feeds which supply the proper proportion and amount of the various essential amino acids supply so-called good quality protein. Those feeds which furnish an inadequate amount of any of the essential amino acids have poor quality protein.

If any one amino acid is lacking in proper amount, it will limit the utilization of the other amino acids in the diet (22). This means that one serious amino acid deficiency will cause the entire diet to be inadequate. For this reason, it is important that feeds low in one or more of the essential amino acids not be fed alone; otherwise, swine will make poor use of the protein supplied by that feed in performing the body functions which require protein.

Protein supplements are usually slightly low in one or more of the essential amino acids. Consequently, combinations of protein supplements with other feeds should be fed to make up this deficiency. In most cases, proper selection of feeds accomplishes this. There may be occasions, however, where supplementation of certain diets with synthetic amino acids may be desirable. Recent findings indicate that lower protein diets than used in the past can successfully be fed to pigs. This will be putting more stress on the quality of protein or the essential amino acid content of the diet.

Table 5.2 shows the order of limitation of amino acids in a few of the more commonly used swine feeds. For example, if lysine and methionine are added to the diet, there is the possibility that deficiencies of tryptophan, threonine, isoleucine and valine will occur. Tryptophan and threonine are often more limiting than methionine in some swine diets. Thus, proper supplementation of diets with amino acids requires expertise and knowledge of the subject.

TABLE 5.2
Limiting Amino Acids for Pigs[a]

	Order of limitation				
Feedstuffs	First	Second	Third	Fourth	Fifth
Soybean meal	Methionine	Threonine	Valine	Lysine	Isoleucine
Meat and bone	Tryptophan	Methionine	Isoleucine	Threonine	Histidine
Corn	Lysine	Tryptophan	Isoleucine	Threonine	Valine
Milo	Lysine	Threonine	Methionine	Isoleucine	Tryptophan
Wheat	Lysine	Threonine	Methionine	Valine	Isoleucine
Barley	Lysine	Threonine	Methionine	Isoleucine	Valine
Corn-soybean meal	Lysine	Methionine	Threonine	Isoleucine	Tryptophan

[a]This information is that of Dr. V. W. Hays, University of Kentucky.

IV. TIME FACTOR IN PROTEIN FEEDING

For efficient use, essential amino acids must be fed at the proper level, at the proper time, and with the proper level of all other essential amino acids. A pig cannot consume an excess of amino acids today to take care of tomorrow's needs—amino acids are not stored or carried over. Actually, animals do better if they are fed all the required essential amino acids at the same time. The lack of any one essential amino acid will decrease the use of all the others. Proteins in the feed are not utilized in the body as such. First they are broken down into amino acids, which are then recombined to form the animal's own body protein. For protein synthesis, therefore, it is necessary that all the required amino acids be present simultaneously. Tests show (5) that pigs could be fed some corn and protein supplement in the morning and only corn in the evening with good results (this was an interval of 24 hours between protein supplement feeding). This finding was verified by later research abroad (22). But when the interval of supplemental protein feeding was increased to 36 hours (two feedings of corn only, then one feeding of corn and protein supplement) or 48 hours (three feedings of corn, then one feeding of corn and protein supplement), a decrease in rate of gain, efficiency of feed utilization, and nitrogen retention resulted (see Table 5.3).

This information indicates that the interval between grain and protein supplement feeding must not be too great. Many swine producers are not aware of this fact; they let their pigs suffer from a protein deficiency because they do not fill their protein feeders regularly or do not feed protein supplement to their pigs every day. This also explains why pigs self-fed grain and supplement free-choice usually do better than pigs full-fed grain with the supplement hand-fed.

TABLE 5.3
Effect of Feeding Protein at Different Time Intervals (5)

	Corn and supplement fed mixed together	Corn and supplement fed separately		
		Intervals between feeding protein supplement		
		24 hours	36 hours	48 hours
Daily gain (% of initial weight)	1.88	1.81	1.64	1.60
Grams gained/kg of feed	393	383	350	340
Apparently digested nitrogen retained (%)	51.9	51.8	48.3	44.8

V. EXCESS PROTEIN

Excess protein cannot be stored except in limited quantity. Excess proteins are deaminized (the nitrogen removed as ammonia and urea). The remainder of the protein molecule serves as a source of energy or is stored as fat through complex mechanisms in the body. Thus, excess protein is not entirely wasted. However, excess protein should not be fed to supply energy and fat, since it is usually too expensive to do so. Rather, grains and their by-products and other high carbohydrate feeds, which are cheaper sources of energy, should be used for finishing purposes.

VI. AMINO ACID REQUIREMENTS OF THE PIG

The pig requires ten essential amino acids for growth, gestation and lactation. Tables 2.1 and 2.2 show the essential amino acids required by the pig as recommended by the National Research Council Committee on Nutrient Requirements for Swine (3). In all cases, the requirements correspond to the natural isomer (L form) which is the form in which amino acids occur in feeds.

The amino acid requirements of growing swine increase as the levels of dietary protein and caloric (energy) density of the diet increase. The amino acid levels shown in Table 2.1 are adequate to support normal growth and performance, and they apply to diets of the caloric density indicated in Table 2.1. The amino acid requirements for pregnant gilts and sows (Table 2.2) are based on amounts required for satisfactory retention of nitrogen during the late stages of pregnancy and are at least adequate to support development of a normal litter. Optimum levels have not been established. The requirements for lactation have been extrapolated from published requirements for maintenance of adult female swine and from amounts calculated to be required to support production of 13.2 lb of milk daily. These amounts should support satisfactory lactation but are not necessarily minimum requirements (3). Table 5.4 shows differences in amino acid requirements for lactation as proposed by Dr. V. C. Speer based on Iowa data (56). His suggestions should be taken into account as one considers amino acid levels to use.

VII. EFFECT OF AMINO ACID DEFICIENCIES

This will be discussed for each amino acid. With amino acids there are some variations in the requirement figures suggested by different investigators. This is to be expected since many factors can influence amino acid needs. As more research is conducted, a clearer picture will be obtained on the amino acid

TABLE 5.4
Estimating the Essential Amino Acid Requirements for Lactation: Dietary Concentrations

Information source:	1970 Illinois data (55)	1973 NRC data (3)		Iowa estimates (56)		
Sow parity:	Gilt	Gilt	Sow	First litter	Second litter	Third or more litters
Feed intake (lb/day):	8.8	11.0	12.1	9.0	9.9	12
Amino acid	Dietary concentration (%)					
Arginine	0.34	0.34	0.34	0.39	0.43	0.41
Histidine	0.26	0.26	0.26	0.22	0.24	0.23
Isoleucine	0.67	0.67	0.67	0.34	0.37	0.35
Leucine	0.99	0.99	0.99	0.65	0.71	0.68
Lysine	0.81 (0.60)[a]	0.60[a]	0.60[a]	0.59	0.64	0.61
Methionine + cystine	0.36	0.36	0.36	—	—	—
Methionine	—	—	—	0.13	0.14	0.14
Cystine	—	—	—	0.13	0.14	0.13
Phenylalanine + tyrosine	1.00	1.00	1.00	—	—	—
Phenylalanine	—	—	—	0.32	0.35	0.33
Tyrosine	—	—	—	0.35	0.38	0.38
Threonine	0.51	0.51	0.51	0.36	0.38	0.37
Tryptophan	0.13	0.13	0.13	0.10	0.11	0.11
Valine	0.68	0.68	0.68	0.41	0.44	0.43

[a] Baker and Allee lysine value (55) adjusted according to findings of Boomgaardt *et al.* (57).

requirements of the pig. The difference in diet ingredients and amino acid availability accounts for much of the difference obtained in amino acid requirements by various investigators.

A. Lysine

A look at Table 5.2 shows that lysine is apt to be borderline or deficient in swine diets, since corn and the other cereal grains are generally low in lysine. A deficiency of lysine (6) results in reduced appetite, loss of weight, poor feed efficiency, rough, dry hair coat and a general emaciated condition (Fig. 5.1). The lysine requirements of the pig are shown in Tables 2.1 and 2.2. Opaque-2 corn is high in lysine. This decreases the amount of protein and/or lysine supplementation needed in swine diets.

Cornell workers (7) found that weanling pigs required 0.6% L-lysine in the diet when a 10.6% linseed meal protein diet was used, and 1.2% L-lysine when a 22% protein sesame meal diet or a 22% mixture of meat scraps, zein, and wheat

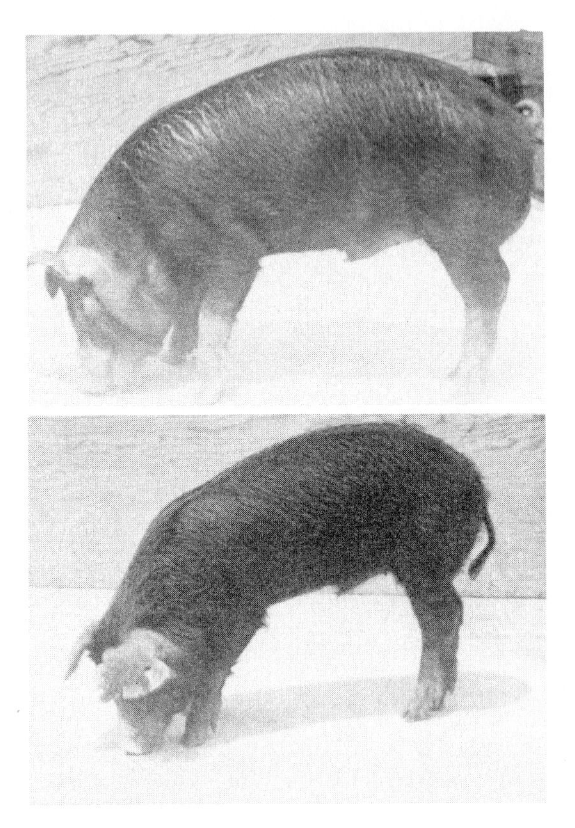

Fig. 5.1. Note effect of lysine in the pig. In 28-day trial the lysine-deficient pig lost 2 lb (bottom photo), whereas the pig receiving lysine gained 25 lb. (Courtesy of W. M. Beeson, Purdue University.)

protein diet was fed. The difference in these requirements is largely eliminated, however, if they are expressed in terms of their proportion to the protein in the diets. The lysine requirements of 0.6 and 1.2% of the diet correspond to 5.7 and 5.5% of the protein in the 10.6 and 22.0% protein diets, respectively.

Missouri (8) workers found that pigs from weaning to about 100 lb liveweight fed diets of corn, soybean meal, tankage, and wheat shorts appeared to have their requirements met by 5.0% L-lysine, expressed as a percent of the protein in the diet. They also found that some pigs on the unsupplemented diets grew at faster rates than their litter mates receiving lysine. This indicates that the lysine requirement is variable between pigs even in the same litter. This means that the variability within litters must be reduced if lysine requirements are to be determined as an exact figure and not as an approximation.

Illinois workers (9) found that pigs from 40 to 100 lb performed satisfactorily on a corn-soybean meal diet containing 14% protein and 0.63% lysine, whereas pigs from 100 to 200 lb needed 12% protein and 0.52% lysine. The Illinois workers also point out that corn is low in lysine and tryptophan. So, the higher the level of corn used in the diet, the lower will be the percent of these two amino acids in the diet. As diets lower in protein are used there will be a higher level of corn used in the diet.

This problem is accentuated when high protein corn is used. The higher the protein level in corn, the lower the amount of soybean meal needed to balance the corn. This means the diet contains a higher percentage of corn and less soybean meal. This, in turn, results in a diet even lower in lysine and tryptophan. The figures in Table 5.5 illustrate this point. Hence swine producers would be wise to feed enough protein and amino acids to include a margin of safety which would allow for the variation which exists in the protein content of corn. This variation ranges from about 6.5 to 11% protein, and even higher, in the United States.

A study in Georgia (50) showed that lysine given via water or feed gave essentially the same results.

A study in England (53) showed that the response to added synthetic lysine was similar on diets in which the source of supplementary protein was either white fish meal or soybean meal.

B. Tryptophan

Tryptophan is another of the essential amino acids which is apt to be low or borderline in certain swine diets, since corn is low in tryptophan. A lack of tryptophan (Fig. 5.2) causes a loss in weight, poor feed consumption, depraved appetite, rough hair coat, and symptoms of inanition in the pig (11,12). Tryptophan addition to the diet of deficient pigs resulted in an immediate response and recovery.

TABLE 5.5
Effect of Corn Protein Level on Lysine and Tryptophan in Diet (10)

	Protein level in corn (%)	
	7.4	8.5
Protein in diet (%)	14	14
Corn protein in total diet (%)	41.0	48.8
Lysine in diet (%)	0.63	0.60
Tryptophan in diet (%)	0.13	0.12

Fig. 5.2. Tryptophan deficiency. In 21-day trial the tryptophan-deficient pig shown in bottom photo lost 8 lb while the tryptophan-supplemented pig gained 25.5 lb. (Courtesy of W. M. Beeson, Purdue University.)

An interrelationship exists between niacin and tryptophan. Tryptophan can be converted to the B vitamin, niacin. But the reverse reaction does not occur. Excess tryptophan can completely replace the need for niacin in the diet, but the efficiency of converting tryptophan to niacin is low. It requires 50 gm of tryptophan to yield 1 gm of niacin. Niacin in excess, however, does not lower the tryptophan requirement. If the tryptophan is at a high enough level in the diet, animals will not develop a niacin deficiency. For example, swine will develop a niacin deficiency on a low protein diet, but not if the diet contains 25 or 26% casein (which is high in tryptophan). Thus, when studying the niacin requirement, one should use diets which are just adequate in tryptophan. When studying the requirements for tryptophan, one should use diets which are adequate in or contain an excess of niacin.

Cornell workers (13) have shown that the hydroxy and keto analogues of tryptophan are utilized to some extent for growth by the pig. This indicates that either the pigs or the microorganisms in the digestive tract or both have the ability to add ammonia to form the amino acid tryptophan.

The requirements for tryptophan are shown in Tables 2.1 and 2.2.

C. Methionine

Methionine is another essential amino acid which might occasionally be borderline in certain swine diets. A lack of methionine results in reduced rate of gain and efficiency of feed utilization (14,15). Methionine can be converted to cystine but cystine cannot be converted to methionine. Cystine can satisfy 50–70% of the total need for methionine + cystine (sulfur amino acids). Methionine can meet the total need for sulfur amino acids in the absence of cystine (3). When an adequate amount of cystine is included in the diet, methionine is no longer needed for cystine synthesis, instead, it is used as methionine to form new tissue and carry out the functions for which methionine is required. Using soybean meal with corn results in diets which are occasionally borderline in methionine. However, there are conflicting experimental results on the value of adding methionine to corn-soybean meal diets for swine. These differences may be due to variation in the methionine content and its availability in soybean meal. It is possible that methionine supplementation may be beneficial with improperly processed soybean meal but of no help with properly prepared meals.

An interrelationship exists between methionine and choline. Methionine can furnish methyl groups for choline synthesis. In a diet which is mildly deficient in both, adding either one will improve growth. Methionine is effective both in correcting a methionine deficiency and in promoting the synthesis of choline. For example, Illinois workers (16) showed no demonstrable dietary choline requirement for the baby pig on a synthetic milk diet containing 1.6% methionine. With 0.8–1.0% methionine in the diet, however, the baby pig required choline at a level of approximately 0.1% of the dry matter of the diet.

Choline is effective only in sparing methionine. Thus, with an adequate level of choline in the diet, methionine is not needed for choline synthesis. There also seems to be an interrelationship between B_{12} and methionine needs of the pig (17,18). More work is still needed on this point, however.

The requirements for methionine are shown in Tables 2.1 and 2.2.

The weanling pig can utilize DL-α-hydroxy-γ-methylmercaptobutyric acid to satisfy at least partially the methionine requirement for normal growth (10). Methionine hydroxy analogue can be used to replace methionine. It has at least 80% the replacement value of methionine.

D. Threonine

Under certain situations, threonine can be a limiting amino acid in the diet (Table 5.2). A deficiency of threonine (Fig. 5.3) decreases feed consumption, rate of gain and efficiency of feed utilization (9,19,20). The requirements for threonine are shown in Tables 2.1 and 2.2.

E. Isoleucine

In some cases, isoleucine may be the third, fourth or fifth limiting amino acid in swine feeds (Table 5.2). Cornell workers (26) showed that a lack of isoleucine in the diet decreased rate of gain, efficiency of feed utilization and nitrogen retention. The requirements for isoleucine are shown in Tables 2.1 and 2.2.

Fig. 5.3. These pigs illustrate the striking growth response from adding threonine to the diet. The pigs received the following levels of L-threonine. Pig 1, 0.36% of dry matter of the diet; Pig 2, 0.54%; Pig 3, 0.73%; and Pig 4, 0.92%. (Courtesy of J. K. Loosli and Cornell University and *Journal of Nutrition.*)

F. Valine

Valine may be the third, fourth or fifth limiting amino acid in some swine feeds (Table 5.2). A deficiency of valine in the diet decreased daily feed consumption, rate of gain and feed efficiency (27,28). The valine requirements of the pig are shown in Tables 2.1 and 2.2.

G. Histidine

Histidine may be the fifth limiting amino acid in meat and bone meal (Table 5.2). A deficiency of histidine in the diet resulted in decreased growth and efficiency of feed utilization (29). When histidine was added to the diet of the deficient pigs, they showed greatly increased appetites and resumed growth almost immediately. The requirements for histidine are shown in Tables 2.1 and 2.2.

H. Phenylalanine

Phenylalanine is usually adequately supplied in practical swine diets. A deficiency of phenylalanine in the pig resulted in decreased growth and efficiency of feed utilization (28). Tyrosine can replace 30% of the phenylalanine requirement of the pig (3,30). Tyrosine is synthesized from phenylalanine. This means that if the diet contains sufficient tyrosine, phenylalanine will not be used up for the synthesis of tyrosine. If the diet is short of tyrosine, then phenylalanine will be used for tyrosine synthesis and it can satisfy the total requirement for phenylalanine plus tyrosine.

The requirement of the pig for phenylalanine is shown in Tables 2.1 and 2.2.

I. Arginine

Arginine is not apt to be deficient in practical swine diets. A deficiency of arginine causes slower growth and lowered efficiency of feed utilization with the pig (28). Arginine can be synthesized at a rate sufficient to permit about 60–75% of normal growth, but a small amount of arginine must be provided from a dietary source to fulfill the total need (3,28). The arginine requirement of the pig is shown in Tables 2.1 and 2.2.

J. Leucine

Leucine is present in sufficient quantity in feeds to meet the requirements of the pig. A deficiency of leucine in the pig causes a decrease of appetite, feed efficiency, and rate of gain (28,31). The leucine requirement of the pig is shown in Tables 2.1 and 2.2. Illinois data (54) have shown that corn starch contains a significant quantity of available leucine. So, anyone using corn starch in a purified diet would have to analyze it for leucine content.

VIII. UREA AS PROTEIN SUBSTITUTE

Minnesota workers (32) found that the addition of 1.5% urea to a low protein (10.6%) diet for weanling pigs had no significant effect on daily feed consumption or rate of gain. The pigs fed urea required 6% more feed per pound of gain. There was no clinical evidence of toxicity at any time in any of the pigs fed urea. This means that cattle feeds (removed from feed troughs and not consumed by the cattle) containing low levels of urea (1.0–1.5%) can be fed to swine without risk of toxicity. Iowa and Canadian studies (33,34) showed that there is a small, but definite, amount of administered urea incorporated into the body protein of the pig. A review of the literature on feeding urea to swine indicates that the effect of urea on daily gain is either neutral or negative. Even when the diet is low in protein, the addition of urea has little, if any, beneficial effect. So, the use of urea in swine diets is not recommended on the basis of present knowledge on the subject.

IX. BALANCE OF AMINO ACIDS

The balance of amino acids in the diet is very important. For example, an Arkansas study (35) showed that pigs benefited from receiving 0.1% level of DL-lysine with a corn-cottonseed meal diet. Levels of lysine greater than 0.1% appeared to depress growth, whereas levels lower than 0.1% failed to support maximum growth. Many other investigators have obtained similar results. This means that if one adds an amino acid to a diet and the pig does not respond, the diet may still lack the amino acid if the correct level was not used. To definitely determine the need for an amino acid, various levels of the amino acid as well as different protein content diets—and in some cases other possible limiting amino acids in various combinations and levels—must be fed to test all possibilities.

In other words, determining the need for amino acids in diets is difficult. Evidently, amino acids need to be fed at the right level and also in the right proportion, and at the right time, with the other essential amino acids for maximum response. Moreover, the requirement for amino acids varies not only with the type of diet used and the relative supply and availability of the amino acids, but also with the dietary supply of other essential substances which can, in case of need, be made from an amino acid. The complexity of the amino acid supplementation problem undoubtedly accounts for the inconsistent results which have been obtained by many workers in supplementing swine diets. Sometimes these workers get beneficial effects and other times they do not.

In the future, supplemental lysine, methionine, tryptophan, and possibly other amino acids, may replace a part of the supplemental protein normally added to cereal grains. But when individual amino acids are used in supplementation, the level of all essential amino acids will have to be checked, or the second, third or fourth limiting amino acid may then manifest its need in the diet (see Table 5.2). Once the amino acid supplementation problem is well understood, it will be possible to use lower protein diets as well as have a better balance of amino acids in the diet. An Ohio study (51) showed that 106 pigs fed added lysine and methionine each returned $2.86 for an investment of $1.20 for the added amino acids.

X. PROTEIN REQUIREMENTS OF THE PIG

Advances in vitamin and mineral nutrition have made it possible to feed less protein in swine diets. Evidently, protein requirements were high in the past because the protein supplements were supplying factors other than amino acids (such as vitamins and minerals).

As more information is obtained on (1) the amino acid content, availability, and variation in feeds; (2) the amino acid requirements of the pig at various stages of its life cycle; (3) the amino acid balance and its interrelationships with other amino acids and nutrients; and (4) the effect of processing on the nutritive value of protein and amino acids, protein requirements of the pig will be lowered even further in the future.

Protein needs are now thought of in terms of supplying adequate amounts and a proper balance of the essential amino acids. For example, Purdue workers (28) found that if the ten essential amino acids were fed at the proper balance and level, a 30-lb weanling pig could grow normally on 7.4% protein equivalent from amino acids and 3.9% protein equivalent from diammonium citrate. This diet had an equivalent of only 11.3% crude protein. The pigs on this diet gained 1.12–1.29 lb daily, which is above the average for pigs of that weight on practical diets.

The Purdue work was conducted with a synthetic diet which is far from a practical diet. However, it can be used as an indication to show that the high level of protein previously required by pigs was due to poor quality protein which was not supplying a proper amount and balance of all amino acids. This work also serves to illustrate the possibilities of reducing the protein requirements of swine diets in the future by proper supplementation with amino acids.

In discussing the protein requirements of the pig, it must be realized that young pigs require more protein in the diet than older animals which are storing less protein and more fat in their bodies. Many feeders overlook this fact and feed pigs the same proportion of protein supplement from weaning time to market. If the diet supplies enough protein for the young pig, there will be considerably more than is needed after the pigs weigh 75 or 100 lb. Similarly, if the diet supplies only enough protein for a 100- or 125-lb pig, then the young, weanling pig will have too little protein. This indicates that complete diets (grain and protein supplement mixed) should have different protein levels and be designed specifically for pigs in different weight groups. Table 5.6 shows the protein requirements recommended by the National Research Council Committee on Nutrient Requirements for Swine (3).

The protein levels recommended by the NRC in Table 5.6 are higher than the levels suggested by several workers. The lower levels of protein suggested, however, contain no margin of safety. The experiments were conducted under excellent conditions of sanitation and management, and the feeds used were of excellent quality, and all diets were fortified with suitable levels of antibiotics, vitamins, and minerals (36). The protein levels were satisfactory under these optimum conditions, but under average farm conditions, management and/or sanitation is not always what it should be. Moreover, the quality of feeds and their proper fortification with vitamins, antibiotics, and minerals may not be adequate. Natural feeds vary considerably in their amino acid content, and needs of pigs also vary depending on many factors. Thus, protein requirements would tend to be higher on the average farm than under the optimum conditions found at

TABLE 5.6
Protein Requirements of Swine (3)

Class of animal	Liveweight range (lb)	Total feed intake (lb)	Crude protein content of diet (%)	Crude protein needed daily (lb)
Growing-finishing	11–22	1.32	22	0.29
	22–44	2.75	18	0.50
	44–77	3.74	16	0.60
	77–132	5.50	14	0.77
	132–220	7.71	13	1.00
Bred gilts	242–352	4.40	14	0.62
Bred sows	352–550	4.40	14	0.62
Lactating gilts	308–440	11.00	15	1.65
Lactating sows	440–550	12.11	15	1.82
Young boars	242–396	5.50	14	0.77
Adult boars	396–550	4.40	14	0.62

an experiment station. This is one reason for the higher figures on protein requirements of the pig recommended by the NRC (3).

The NRC recommendations may be lowered as new developments in vitamins, antibiotics, and minerals become more widely known and applied by those concerned directly and indirectly with swine feeding. Again, it must be remembered that grains will vary in protein and amino acid content depending upon the area of the country in which they are grown, and a safety factor must be provided to take care of this variation. This fact also presents an opportunity for plant breeders and geneticists to pay more attention to the amino acid and protein composition of the new varieties of plants being developed. Enough information is already available to indicate that much progress can be accomplished through breeding and selection for feeds with higher levels and higher quality protein for animal feeding.

XI. ENGERY-PROTEIN RATIO RELATIONSHIPS

The amino acid requirements of growing swine increase as the level of dietary protein and caloric density of the diet increase (3). There is an optimum calorie to protein ratio for each stage of growth (46). The biological value of a protein also depends on the energy level of the diet and on the consumption level of this diet.

XII. PROTEIN LEVEL AND REPRODUCTION

Some investigators have been able to feed sows low protein diets during gestation. As a result, many swine producers thought they no longer needed to supplement their grain diets with a source of protein during gestation. This conclusion is incorrect, and will be discussed briefly here.

Complete deprivation of protein throughout pregnancy may cause the birth of smaller pigs but apparently has little or no effect on number born or on survival. But one should not deprive sows of protein needs, since it may affect the ability of the animal to breed back, and to reproduce at a high level in subsequent litters. A Kentucky study (25) showed that an extended protein deficiency in the sow significantly depresses reproductive efficiency by causing inactive ovaries or lengthening the interval between weaning and return to estrus and lowering ovulation rate. Oklahoma studies (21) showed that protein intake had a more adverse effect on reproduction in gilts before breeding than in early gestation. A Cornell study (24) showed that dietary protein at the time of implantation appears to be an important factor in determining an individual pig's birth weight, but not in determining embryo survival. Another Cornell study showed that in pigs, the dam acts as an efficient "buffer" to at least partially protect the developing fetus against the effects of maternal protein deprivation during the final three-fourths of the gestation period. This would seem to indicate that the sow can take care of the protein needed by the developing young from her own body stores (39). Starting a protein-free diet at day 24–28 of pregnancy in the gilt appears to have less adverse effect than starting it before the gilt is bred. Part of this effect may be due to the interruption of the estrus cycle in the absence of dietary protein which seems to be a more immediate and specific hormone effect (40).

Evidently, the sow can be fed a low protein diet during gestation and litter size will not be affected. But the low protein diet is not adequate for lactation (41–44). It may also have harmful effects on subsequent reproductive performance if the protein (and/or amino acid) deficiency exists for a long enough time (25). Until more is known about this subject, the NRC recommended levels of 14% protein (Table 5.6) should be used in the gestation diets for gilts and sows (3).

XIII. EFFECT OF PROCESSING ON AMINO ACIDS

An excellent review by R. J. Meade (45) gives the following information on the effect of processing on amino acids in feeds:

1. Severe overheating of protein supplemental feeds results in seriously depressed availability of all amino acids. Lysine appears to be more heat sensitive than some of the other essential amino acids.
2. Amino acids are highly available from properly processed soybean meal, cottonseed meal and peanut meal.
3. The amino acid content of fish meals was not found to vary greatly, but availability varied greatly with serious depression of availability if meals were overheated, or scorched, in processing and shipment.
4. Amino acid composition of meat and bone meal will vary greatly depending on the starting material. Availability of amino acids will be depressed if the meals are overheated in processing.
5. There is need for development of a rapid method for determining the availability of amino acids for swine and poultry.

XIV. PROTEIN LEVEL AND CARCASS COMPOSITION

There are many factors which affect the amount of lean and fat deposited in swine carcasses. A brief summary of some of the more important factors follows:

1. If the protein level is low, it will decrease feed intake, rate of gain, feed efficiency, percent lean cuts, loin eye area and will increase the amount of back fat.

2. If the quality of protein is poor in amino acids it will have an effect similar to a low level of protein. So feeding an adequate protein diet with the proper level of amino acids is very important.

3. Different breeds and different strains of pigs will vary in the amount of lean or fat in the carcass. Since carcass quality is highly heritable, rapid progress can be made in breeding and selection for the type of carcass desired.

4. Female market pigs usually have less back fat, larger loin eye areas and a higher percentage of lean cuts than barrows. This indicates they should also be fed about 2% more protein in the diet for maximum carcass leanness.

5. Limiting feed intake to less than a full-feed will increase the lean content in the carcass. The fatter-type pig will respond more to limited feeding than a lean-type pig. Thus the degree of limited feeding used should vary with the type of pig being fed. Dr. A. A. Rerat (48) feels that consideration should be given to increasing the percent of amino acids in the diet when limited feeding is used.

6. The protein level needed for maximum carcass leanness is usually higher than that required for the fastest rate of gain and best feed efficiency. Therefore, a higher level of protein needs to be used for maximum carcass leanness. The level needed will vary with the breeding and carcass quality of the pigs being fed. Dr. D. Lewis of England has done considerable work in this area (47). Table 5.7 illustrates this concept well. It shows that as the dietary protein level is raised, the lean proportion appears to increase steadily whereas the growth rate falls off beyond a certain point (47). There is a top limit, however, beyond which additional protein will not further increase carcass leanness. Under certain conditions a higher level of protein may not increase carcass quality as was recently shown by Dr. R. Braude in England (49).

7. Pigs fed on pasture usually show slightly more lean and less fat than pigs fed in confinement. Some of this could be due to exercise and/or the higher fiber intake which limits feed intake somewhat.

TABLE 5.7
Growth Rate and Lean Production as the Level Is Raised at a Constant Dietary Energy Level (3500 kcal Digestible Energy/kg Feed) (47)

Protein ($N \times 6.25$)	Liveweight gain (gm/day)	Percentage lean (cold weight basis)	Lean (gm/day)
15.0	676	44.7	222
17.5	749	46.6	263
20.0	745	46.8	268
22.5	749	47.6	272
25.0	717	49.0	277
27.5	676	50.0	268

8. Pigs have an optimum slaughter weight for maximum carcass leanness. This can vary with the type and quality of the diet as well as the breeding of the pig. In general, as the pig increases in weight it puts on more fat and less lean tissue. Therefore, pigs should be slaughtered as close as possible to the optimum carcass desired. As energy and other feed shortages become more serious, it becomes increasingly important to avoid feeding beyond the optimum slaughter weight.

All of these factors, plus others, are important in the kind of carcass produced. Ultimately, the price paid for pigs will determine whether the swine producer will market animals with maximum or optimum carcass quality. His program is usually guided by trying to earn the greatest profit possible. Unless the market pays in proportion to carcass merit the swine producer has no real incentive to market a pig with optimum carcass quality.

XV. FREE-CHOICE PROTEIN FEEDING

A study at California (37) showed that given a choice, swine select a diet more adequately balanced in amino acid composition. In another phase of the study (37) pigs were fed a protein-free diet plus diets of high protein concentration to determine if gilts would voluntarily consume and thus dilute a high protein diet with a protein-free diet. They found that the gilts voluntarily diluted the high protein diets with a protein-free diet. They ate more of the protein-free diet as the percentage protein increased in the high protein diets. A Florida study (38) showed that gilts voluntarily consumed more protein than barrows when both had corn and a protein supplement offered free-choice to them. The gilt produces a leaner carcass than the barrow and previous studies have shown that gilts should be fed more protein in the diet than barrows. These studies indicate that the pig has some ability to balance its own protein diet. How accurately the pig can do this is not known, however. But the pig may be able to do a better job than the person who does not understand nutrition and how to properly balance a diet. So, free-choice feeding of a protein supplement still has merit when feeds variable in protein level, or where the protein level is not known, are fed to swine.

XVI. GENERAL INFORMATION

A study in Mexico (52) showed that lysine is the first limiting amino acid in triticale and that methionine is probably second limiting for the growing pig.

REFERENCES

1. Sunde, M. L., *Poult. Sci.* **51,**44 (1972).
2. Harper, A. E., *J. Nutr.* **104,**965 (1974).
3. Cunha, T. J., J. P. Bowland, J. H. Conrad, V. W. Hays, R. J. Meade, and H. S. Teague, *N.A.S.—N.R.C. publi.* (1973).
4. Shelton, D. C., W. M. Beeson, and E. T. Mertz, *Arch. Biochem.* **29,** 446 (1950).
5. Eggert, R. G., M. J. Brinegar, and C. R. Anderson, *J. Nutr.* **50,** 469 (1953).
6. Mertz, E. T., D. C. Shelton, and W. M. Beeson, *J. Anim. Sci.* **8,** 524 (1949).
7. Brinegar, M. J., H. H. Williams, F. H. Ferris, J. K. Loosli, and L. A. Maynard, *J. Nutr.* **42,** 129 (1950).
8. Pfander, W. H., and L. F. Tribble, *J. Anim. Sci.* **14,** 545 (1955).
9. Becker, D. E., J. W. Lassiter, S. W. Terrill, and H. W. Norton, *J. Anim. Sci.* **13,** 611 (1954).
10. Becker, D. E., A. H. Jensen, S. W. Terrill, and H. W. Norton, *J. Anim. Sci.* **14,** 1086 (1955).
11. Beeson, W. M., E. T. Mertz, and D. C. Shelton, *Science* **107,** 599 (1948).
12. Beeson, W. M., E. T. Mertz, and D. C. Shelton, *J. Anim. Sci.* **8,** 532 (1949).
13. Gallo, J. T., and W. G. Pond, *Cornell Univ. Swine Mimeo* No. 67-2 p. 3 (1967).
14. Bell, J. M., H. H. Williams, J. K. Loosli, and L. A. Maynard, *J. Nutr.* **40,** 551 (1950).
15. Shelton, D. C., W. M. Beeson, and E. T. Mertz, *J. Anim. Sci.* **10,** 57 (1951).
16. Nesheim, R. O., and B. C. Johnson, *J. Nutr.* **41,** 149 (1950).
17. Cunha, T. J., H. H. Hopper, J. E. Burnside, A. M. Pearson, R. S. Glasscock, and A. L. Shealy, *Arch. Biochem.* **23,** 510 (1949).
18. Sewell, R. F., T. J. Cunha, C. B. Shawver, W. A. Ney, and H. D. Wallace, *Am. J. Vet. Res.* **13,** 186 (1952).
19. Beeson, W. M., H. D. Jackson, and E. T. Mertz, *J. Anim. Sci.* **12,** 870 (1953).
20. Sewell, R. F., J. K. Loosli, L. A. Maynard, H. H. Williams, and B. E. Sheffy, *J. Nutr.* **49,** 435 (1953).
21. Jones, R. D., and C. V. Maxwell, *J. Anim. Sci.* **37,**283 (1973).
22. Yeo, M. L., and A. G. Chamberlain, *Proc. Nutr. Soc.* **25,**16 (1966).
23. Baker, D. H., N. K. Allen, J. Boomgaardt, G. Graber, and H. W. Norton, *J. Anim. Sci.* **31,**1018 (1970).
24. Strachan, D. N., E. F. Walker, Jr., W. G. Pond, J. R. O'Connor, J. A. Dunn, and R. H. Barnes, *J. Anim. Sci.* **27,**1157 (1968).
25. Svajgr, A. J., V. W. Hays, G. L. Cromwell, and R. H. Dutt, *J. Anim. Sci.* **31,**212 (1970).
26. Brinegar, M. J., J. K. Loosli, L. A. Maynard, and H. H. Williams, *J. Nutr.* **42,** 619 (1950).
27. Jackson, H. D., E. T. Mertz, and W. M. Beeson, *J. Nutr.* **51,** 109 (1953).
28. Mertz, E. T., W. M. Beeson, and H. D. Jackson, *Arch. Biochem. Biophys.* **38,** 121 (1952).
29. Eggert, R. G., L. A. Maynard, B. E. Sheffy, and H. H. Williams, *J. Anim. Sci.* **14,** 556 (1955).
30. Beeson, W. M., E. T. Mertz, and J. N. Henson, *J. Anim. Sci.* **12,** 906 (1953).
31. Mertz, E. T., D. C. DeLong, D. M. Thrasher, and W. M. Beeson, *J. Anim. Sci.* **14,** 1217 (1955).
32. Hanson, L. E., and E. F. Ferrin, *J. Anim. Sci.* **14,** 43 (1955).
33. Liu, C. H., V. W. Hays, H. J. Svec, D. V. Catron, G. C. Ashton, and V. C. Speer, *J. Nutr.* **57,** 241 (1955).
34. Grimson, R. E., J. P. Bowland, and L. P. Mulligan, *Can. J. Anim. Sci.* **51,**103 (1971).
35. Miner, J. J., W. B. Clower, P. R. Noland, and E. L. Stephenson, *J. Anim. Sci.* **14,** 24 (1955).
36. Coalson, J. A., D. W. Huck, and A. J. Clawson, *J. Anim. Sci.* **38,**220 (1974).
37. Jackson, H. M., D. W. Robinson, and F. Khalaf, *J. Anim. Sci.* **31,**204 (1970).

38. Wallace, H. D., A. Z. Palmer, J. W. Carpenter, and G. E. Combs, *J. Anim. Sci.* **36**,203 (1973).
39. Pond, W. G., J. A. Dunn, G. H. Wellington, J. R. Stouffer, and L. D. Van Vleck, *J. Anim. Sci.* **27**,1583 (1968).
40. Pond, W. G., W. C. Wagner, J. A. Dunn, and E. F. Walker, Jr., *J. Nutr.* **94**,309 (1968).
41. Mahan, D. C., and L. Mangan, *J. Anim. Sci.* **39**,185 (1974).
42. Holden, P. J., E. W. Lucas, V. C. Speer, and V. W. Hays, *J. Anim. Sci.* **27**,1587 (1968).
43. Rippel, R. H., O. G. Rasmussen, A. H. Jensen, H. W. Norton, and D. E. Becker, *J. Anim. Sci.* **24**,203 (1965).
44. Baker, D. H., D. E. Becker, A. H. Jensen, and B. G. Harmon, *J. Anim. Sci.* **31**,526 (1970).
45. Meade, R. J., *J. Anim. Sci.* **35**,713 (1972).
46. Rerat, A. A., *Nutr., Proc. Int. Congr., 8th 1969* pp. 759–761 (1971).
47. Lewis, D., *Proc. E.A.A.P. Study Comm., Dublin, Ireland 1907* pp. 1–8 (1968).
48. Rerat, A. A., and J. Loognon, *World Anim. Prod.* **4**,66 (1968).
49. Braude, R., and Z. D. Hosking, *J. Agric. Sci.* **83**,385 (1974).
50. Seerley, R. W., C. E. Meeks, H. C. McCampbell, and R. D. Scarth, *J. Anim. Sci.* **37**,91 (1973).
51. Nelson, M. L., and J. W. Kelley, *J. Anim. Sci.* **35**,1108 (1972).
52. Bravo, F. O., L. Martinez R., and A. Shumada R., *J. Anim. Sci.* **33**,227 (1971).
53. Braude, R., and P. Lerman, *J. Agric. Sci.* **74**,575 (1970).
54. Baker, D. H., N. K. Allen, J. Boomgaardt, G. Graber, and H. W. Norton, *J. Anim. Sci.* **33**,42 (1971).
55. Baker, D. H., and G. L. Allee, *J. Nutr.* **100**,277 (1970).
56. Spear, V. C., *Feedstuffs* **47**,21 (1975).
57. Boomgaardt, J. D., D. H. Baker, A. H. Jensen, and B. G. Harmon, *J. Anim. Sci.* **34**,408 (1972).

6

Carbohydrates and Fiber

I. INTRODUCTION

Carboyhydrates supply the chief form of energy in swine diets. They make up about 75% of the dry matter in most plants on which swine depend primarily for most of their feed supply.

Cellulose, which along with lignin forms most of the skeletal portion of the plant, has low digestibility for the pig. Cellulose and lignin represent most of the fiber in swine feeds. Sugars and starches, however, are readily digested and have a high feeding value for swine.

Newborn pigs cannot use starch and sugar because of low levels of carbohydrate enzymes other than lactase in their intestinal tracts. This was amply demonstrated in 1948 when Bustad tried to raise pigs on a synthetic milk diet with sucrose as the source of carbohydrates (12). But the levels of other carbohydrase enzymes increase rapidly after the first week of life. Under natural conditions, the only carbohydrate ingested by the newborn suckling pig is lactose. This accounts for nature providing for the lactase enzyme early in life.

The use of carbohydrates which are readily hydrolyzed and absorbed is important for the young pig. The cecum and colon in the young pig are occupied by microflora responsible for the fermentation of unabsorbed feed residues. It is reasonable to assume that, if the carbohydrates fed to the pig cannot be digested, this results in a more active microbial fermentation in the cecum and colon causing the pigs to scour and suffer from dehydration and a lack of unabsorbed nutrients.

II. CARBOHYDRATE USE BY THE PIG

Illinois workers (1) found that 1- to 2-day-old pigs react to sucrose feeding with severe diarrhea, rapid weight loss, unthriftiness, thinness, and death. These results are similar to earlier work at Washington State (12). In the Illinois work, death usually followed 4 days of sucrose feeding. If the pigs were 7 days old when started on sucrose however, about 60% of them were able to utilize the sucrose. With pigs started on experiment at 9 weeks of age, sucrose produced equally as satisfactory results as glucose, dextrin, and corn starch. Other studies have shown that the baby pig cannot utilize fructose (6,7,10). Iowa studies (6) showed that the mortality rate of 4-day-old pigs was lower on the fructose and sucrose diets than that of the 2-day-old pigs when started on these same diets. No deaths occurred when these diets were fed to 6- and 7-day-old pigs. Other Iowa studies (11) suggest that fructose plays only a minor role, if any, as an energy source for the fetal and newborn pig. They confirmed the work of others that fructose levels in the blood of the newborn pig are high at birth and decrease rapidly within the first 48 hours after birth. But they suggest the disappearance of fructose is due more to excretion than to utilization as a source of energy.

The newborn pig is not able to utilize fructose and sucrose because enzymatic activity is insufficient. But the ability of the pig to utilize sucrose and fructose increases with age—the reverse is true in the use of lactose by the pig. Newborn pigs can use lactose with excellent results, but their tolerance of lactose diminishes with age. Illinois workers (2) showed that 9-week-old pigs could use 25% lactose in the diet and perform satisfactorily. However, a 50% level of lactose in the diet, a level the newborn pig can do well on, caused a pronounced depressing effect upon feed intake and growth rate. In addition, moderate diarrhea was occasionally observed. Antibiotics tended to depress the severity of the adverse effects (except on efficiency of feed utilization) of the high level of lactose feeding. At 16 weeks of age, the pig was able to tolerate 25% of lactose in the diet without harmful effect.

North Carolina workers (3) compared xylose to glucose in a purified diet for the 2-week-old pig. Xylose is one of the major constituents of wood molasses. The pigs fed xylose exhibited a depression of appetite, consumed less feed, grew at a slower rate, required more feed per pound of gain, had a lower nitrogen retention, and developed cataracts. Moreover, voluntary activity was decreased and the hair coat color was altered from deep red to yellowish-red in pigs fed xylose. These results would indicate that xylose has relatively little value as a source of energy for the growing pig.

Canadian work (13) suggests that the prime factor restricting the digestion of raw starch by newborn pigs is that which is responsible for the initial rupture of the starch granule. It also states that the rate of digestion of soluble starch, maltose and glucose by newborn pigs would appear to be sufficiently high to meet a large proportion of the pigs' energy requirements, although consumption of such quantities of these carbohydrates may result in diarrhea.

A "diabetic-like" syndrome may exist in swine. Of twenty-four animals studied in one experiment, three developed a "diabetic-type" tolerance curve (14). Previously, the Iowa Station (15) had suggested that in swine there are individual differences in glucose tolerance. The young animal in certain species is less susceptible to diabetes. This fact may account for the discrepancy between reports which indicate that swine are apparently good tolerators of glucose (based for the most part on experiments in which young or immature animals were used) and the studies by Bunding *et al.* (14) and others showing that some swine are poor tolerators of glucose.

Studies in England (16) showed that the newborn pig can utilize only lactose and glucose. The ability to assimilate sucrose, maltose and dextrin develops during the first 10 days of life. Pigs from two litters showed decreased utilization of lactose after 2–3 weeks of age, but pigs from a third litter still utilized lactose when 36 days old. So, there undoubtedly are differences in enzyme production and the utilization of various carbohydrates by individual pigs depending on many factors (5).

Illinois workers (4) reported that the intravenous administration of sucrose, fructose, and lactose failed to resuscitate hypoglycemic comatose pigs, and mannose, galactose, and maltose yielded slower responses than glucose in the comatose pig. Evidently, glucose is the most effective—probably the only—sugar in resuscitating fasting newborn pigs in hypoglycemic coma. Other sugars exerting a beneficial effect are probably first converted to glucose before they can act.

Baby pigs like sugar; it is one of the most palatable ingredients used in pig starters (8). As has been shown by many scientists, pigs, as human beings, have a sweet tooth, and sugar is one of the ingredients used to get them to consume more starter feeds.

Georgia studies (9) with 3-week-old pigs fed purified diets showed that those fed either corn starch or dextrin grew at a significantly slower rate than pigs receiving the simpler carbohydrate sources of either glucose or sucrose. The depressed gains appeared to be the result of decreased feed intake since no differences occurred when the pigs were pair-fed the diets. Protein and organic matter digestibility was also significantly lower when dextrin was fed as the only source of carbohydrate as compared with diets containing either glucose, corn starch or a combination of corn starch and sucrose.

Georgia studies (17) indicate that lactose plays a physiologically important role in nutrient absorption and utilization by the very young pig and accounts for at least a part of the improved difference in response to protein from a skim milk source as compared to soybean protein.

It is recognized that fetal liver glycogen in most species of animals is at a maximum shortly before parturition. Studies in Louisiana (18) showed glycogen concentration in pig fetal livers to be more than 50 mg/gm of liver. The higher concentration was observed in the heavier livers. The highest liver concentration of glycogen was about 2 gm per pig at 112 days of the gestation period.

Studies at Hawaii (19,20) showed that vitamin K supplementation was beneficial to pigs fed a high sugar diet. These pigs developed sugar-induced heart lesions and a hemorrhagic syndrome. The use of chlorotetracycline at a level of 200 mg/pound of diet prevented the heart lesions but did not reduce prothrombin time or the hemorrhagic syndrome. Florida studies (21) also showed a need for vitamin K supplementation with pigs fed sugar in the diet. The pigs developed crippling hematomas in the rear legs, fragile capillaries, profuse bleeding into the cranial cavity, anemic tissues, severe subcutaneous hemorrhaging of the rear legs and around the eyes, and some deaths.

A review article by Wisconsin scientists (22) indicated that although carbohydrates serve primarily as a source of calories, the indirect effects of individual carbohydrates may be of considerable nutritional significance. They cited evidence that when a less soluble carbohydrate is substituted in the diet for a more soluble one, the requirement for most members of the B-complex vitamins and for essential amino acids as a percentage of the diet decreased. They state that the effect of complex carbohydrates on lower B vitamin requirements is related to changes in the intestinal microflora. The lower amino acid requirements may be the result of physiological effects on feed intake, digestion or absorption. The effect of different carbohydrate sources on vitamin and amino acid needs of the pig is an area which needs considerable study since very little is known on this subject.

III. FIBER UTILIZATION BY THE PIG

The pig is less able to use crude fiber than other farm livestock. Its digestive tract includes a simple stomach of relatively small capacity in contrast to the complex stomachs of cattle and sheep and the large cecum of horses. Cattle and sheep have a great deal of microbial activity in the rumen. This breaks down a large part of the crude fiber consumed. Thus the pig is limited in the amount of crude fiber it can use in the diet (23,24).

There is no agreement as to how much crude fiber pigs can use (25). This lack of agreement is probably due to differences in source of fiber fed, level of fiber in the diet, level of other nutrients in the diet, plane of nutrition, age and weight of the pigs, character of the nonfibrous part of the diet, as well as to other factors. It has been shown that the heavier the pig the more fiber it can use in the diet (26). The stage in which a forage is harvested also makes a great deal of difference in fiber digestibility for pigs. Moreover, the amount of lignin in a fibrous feed would affect the fiber digestibility (27). Thus the study of fiber digestibility is not simple and a great deal of research remains to be done on this subject. It is assumed that fiber in the diet of pigs facilitates the passage of residues through the digestive tract. Levels between 2 and 5% of fiber in the diet are thought to be adequate for this purpose (41).

Most of the cellulose digestion occurs in the large intestines of the pig (40). Studies by Farrell and Johnson, 1970 (41) showed that some cellulose is digested by the pig. They fed diets containing 8 and 26% cellulose. Measurements of the production rates of volatile fatty acids in the cecum indicated that only 2.7 and 1.9% of the apparent digestible energy of the 26 and 8% cellulose diets respectively came from the volatile fatty acids. So on a percentage basis, the digestibility of cellulose was higher on the 8% cellulose diet. The end products of microbial digestion of the cellulose in the alimentary tract of the pig are mainly the steam-volatile fatty acids which are produced in and absorbed from the cecum and the colon (41,43). Studies at Illinois (39) showed that cellulose added to the diet at levels of up to 60% caused a linear decrease in average daily gain, average noncellulose feed intake and gain per unit of noncellulose feed intake. Studies by Farrell (42) showed that the pig was able to digest 53% of the dry matter on an all alfalfa diet. This is a high digestibility rate when one notes that sheep fed the same alfalfa diet digested 57% of the dry matter in it (42). The digestibility of high quality forages by the pig accounts for the fact that many swine producers make maximum usage of pasture, silage and green-chopped forage in their swine programs in many areas of the world.

IV. FIBER LEVELS FOR GROWING-FINISHING PIGS

There is still disagreement as to what level of fiber should be recommended for growing-finishing pigs (from 35 to 40 lb up) when no effort is made to add extra energy to compensate for the higher fiber level in the diet. Many scientists would agree on a figure of 5–6% fiber as the maximum to include in the diet, but some think a figure of 6–8% may be used. The Wisconsin Station showed that levels of close to 8% fiber can be satisfactorily fed to pigs providing the fibrous feed is finely ground and thoroughly mixed. Coarsely ground roughage in a swine mixture invites the pigs to sort out and waste feed in the trough or self-feeder. This means that fibrous feeds should be finely ground when used in swine feed mixtures. For example, Canadian workers (28) found that pigs gained faster when oats were ground finer. They also found that the fibrous portions, made up largely of hulls, were decidedly unpalatable to the younger pigs and, unless finely ground, were sorted out and refused.

Of considerable importance in determining the fiber level to use is the quality of the fibrous feeds. High quality alfalfa meal and oats can be tolerated at higher levels than corncob meal, poor quality roughages, and other fillers which might be used. This means that any increase in fiber levels allowed in swine diets depends on the quality of fibrous feeds used and the level of energy in the diet.

Higher levels of fiber in swine diets can be used during the latter part of the finishing period. For example, it has been shown (29) that a restriction of the feed intake during the finishing period increases the quality of the hog carcass by reducing fat deposition during that period. Reduction of feed is only one way of causing this reduction in growth rate. On the farm, uniform restriction of feed intake for hand-fed groups of pigs is difficult, and it is impractical if self-feeding is carried out. A reduction in the total energy of the finishing diet by the introduction of crude fiber would seem to be a more practical method.

Canadian workers (30) have shown, for example, that 25% of the barley in the diet could be replaced with 25% of either wheat bran or wild oats for 110 lb pigs. In doing so, they produced a superior carcass without any change in rate of gain, feed intake, or length of feeding period. Replacing the barley with 45% alfalfa or 45% bran, however, caused a decreased rate of gain and increased length of feeding period, but superior bacon carcasses were still produced. These data and other information available in the United States indicate that the use of higher-fiber-content feeds during the latter part of the growing-finishing period is a means of producing a hog with more meat and less fat.

Lard has been a burden on the market for quite a few years and considerable emphasis has been placed on producing a "meat-type" hog. Much can be accomplished in this direction by selection and breeding of a meat-type animal and by feeding practices. In certain areas, grain production is low and large quantities of alfalfa are grown. There is interest in substituting as much alfalfa as possible in the diet to increase the total feed available for swine feeding. This practice would increase swine production in those areas. The Nevada Station (31–33) showed that high levels of alfalfa can be used for growing-finishing pigs. Some of their data are shown in Table 6.1.

Their findings show that alfalfa can be fed to finishing pigs at higher levels than previously thought possible. The alfalfa-fed pigs were consistently leaner but had a lower dressing percentage. The pigs fed high levels of alfalfa produced more ham, loin, and shoulder and had less belly and back fat. Table 6.2 shows the data obtained at the Nevada Station on the carcasses of the pigs fed various levels of alfalfa.

TABLE 6.1
Effect of Level of Alfalfa on Swine Performance (31, 32)

Level of alfalfa in the diet	Rate of gain and feed efficiency		
%	Trial I	Trial II	Trial III
0	—	—	1.82 (306)
10	1.2[a] (451)[b]	1.7 (466)	1.72 (357)
30	1.1 (480)	1.6 (462)	—
35	—	—	1.37 (404)
50	—	1.3 (574)	1.01 (543)
60	0.9 (530)	—	—

[a]Rate of gain.
[b]Feed per 100 lb gain.

TABLE 6.2
The Effect of Alfalfa on the Carcass of the Pig (31)

	Alfalfa (%)			
	0	10	30	50
Back fat (in.)	1.80	1.73	1.31	1.11
Carcass length (in.)	29.6	29.8	30.0	30.3
Dressing percentage	74.6	73.0	69.6	68.1
Percent of carcass				
Loin	14.1	14.7	14.6	15.9
Ham	15.7	16.2	17.7	18.9
Shoulder	14.0	14.2	14.9	15.5
Bacon belly	13.7	14.1	13.3	12.4
Back fat	11.3	10.4	6.9	5.1
Weight in grams				
Stomach	508	504	587	618
Large intestine	1272	1522	1589	1684

The data show that as the pigs consumed higher levels of alfalfa, their digestive tracts enlarged to accommodate the increased fiber and bulk (Fig. 6.1). The Nevada workers (32) found that pelleting the alfalfa-grain mix increased the rate of gain and the efficiency of feed utilization. They also feel that creep-feeding the pigs a diet with 25% alfalfa conditions them so they do better when fed high level alfalfa diets after weaning. In addition to the work at Nevada, other stations have shown that 5–15% high quality alfalfa meal can be used in finishing pig diets.

It might also be desirable to feed a higher level of fiber to pigs which are going to be kept in the herd for breeding purposes. Although these pigs will not gain as fast, the extra fiber may provide a safeguard against overfatness and subsequent weak pasterns or breeding troubles. It has been shown at the Wisconsin (34), Washington (35), and other stations that high quality alfalfa meal contains a factor (or factors) beneficial for reproduction and lactation. The factor(s) is stored for a long period of time. This storage is such that the diet a pig receives during growth will definitely influence the ability of the animal to conceive, reproduce, and lactate many months later. This means that prospective herd replacement gilts should be fed differently from pigs being finished for market—the herd replacements should get more alfalfa meal or pasture, if possible. This will pay dividends later by producing gilts that will settle quickly and farrow larger litters and wean a higher percentage of their pigs.

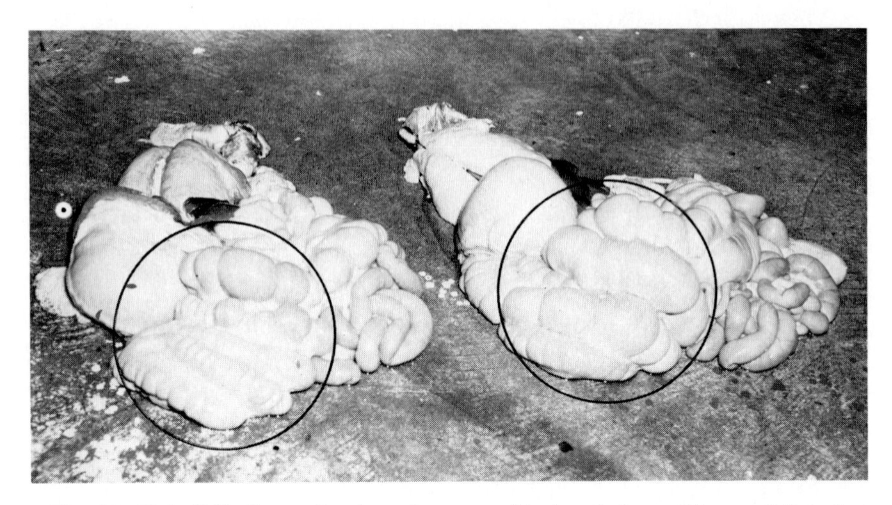

Fig. 6.1. High alfalfa diets produced an enlargement of the intestinal tract. Viscera at left are from a conventionally fed hog. Viscera at right are from an animal fed a 50% alfalfa diet. (Courtesy of J. F. Kidwell and Max C. Fleischmann, College of Agriculture, University of Nevada.)

V. FIBER LEVELS FOR SOWS

Diets for sows can contain more fiber than those for growing-finishing pigs. Sows can digest higher fiber feeds better than young pigs. However, no definite level of fiber has been established as the maximum to use in sow diets. Many scientists agree that sow diets may contain 10–12% fiber. However, they qualify these figures by stating that these higher fiber levels should be allowed only to make use of higher levels of high quality alfalfa meal and oats which are valuable constituents in sow diets. The use of high fiber levels will also tend to prevent sows from becoming too fat, a condition that can lead to poor results in reproduction.

Swine producers are interested in learning more about the use of pasture, silage, and other roughage feeds for sows. Experimental information already available indicates that these feeds can be included in sow diets with good results.

The whole question of what fiber levels to use in swine diets needs some reevaluation in view of the data on alfalfa feeding. Some emphasis should be placed on the production of a meat-type pig and the effect of adding extra energy when high fiber diets are used. A Georgia study (36) showed that crude fiber levels of 3.5, 5.5, 7.5, 9.5, 11.5 and 13.5% in the diet had no effect on rate of gain, efficiency of gain or carcass leaneness if the metabolizable energy in all the diets was kept near equal levels. They concluded that it was the reduced energy intake, and not specifically the crude fiber level, that was responsible for differences in performance and in carcass traits. Energy increases in the diet of about 15% increased gains, reduced percent lean cuts and also increased the back fat. The stress of cold weather reduced gain and feed efficiency. This seasonal effect is in agreement with Iowa studies (37) which showed that pigs required more feed per pound of gain at 0°C than at 25° and 35°C.

Studies in Canada (38) showed that daily nitrogen intake and retention, energy intake and energy gain did not change significantly as the crude fiber levels of the diet increased (levels of 3.1, 6.1, 9.1 and 12.1% crude fiber were used). Carcass measurements were similar for all pigs. The results suggested that deposition of lean and fat do not vary markedly when crude fiber levels in the diet are increased if daily digestible energy intake between groups is kept at a constant level.

Both the Georgia (36) and Canadian studies (38) and those of others indicate that if carcass leanness is to be improved through feeding, the reduction of dietary energy intake is required. Increasing the fiber level, without increasing the energy level, can accomplish this. But fiber itself does not increase carcass leanness; the reduced total daily energy intake does. Even adding sand to the diet has been used effectively in Illinois trials (39) to dilute the energy intake of the pig.

REFERENCES

1. Becker, D. E., D. E. Ullrey, and S. W. Terrill, *Arch. Biochem. Biophys.* **48,** 178 (1954).
2. Becker, D. E., and S. W. Terrill, *Arch. Biochem. Biophys.* **50,** 399 (1954).
3. Wise, M. B., E. R. Barrick, G. H. Wise, and J. C. Osborne, *J. Anim. Sci.* **13,** 365 (1954).
4. Newton, W. C., and J. Sampson, *Cornell Vet.* **41,** 377 (1951).
5. Lewis, C. J., D. V. Catron, V. C. Speer, and G. C. Ashton, *J. Anim. Sci.* **14,** 1214 (1955).
6. Aherne, F. X., V. W. Hays, R. C. Ewan, and V. C. Speer, *J. Anim. Sci.* **29,** 444 (1969).
7. Steele, N. C., L. T. Frobish, L. R. Miller, and E. P. Young, *J. Anim. Sci.* **33,** 983 (1971).
8. Grinstead, L. E., V. C. Speer, and V. W. Hays, *J. Anim. Sci.* **20,** 934 (1961).
9. Sewell, R. F., and C. V. Maxwell, Jr., *J. Anim. Sci.* **25,** 796 (1966).
10. Kidder, D. E., M. J. Manners, M. R. McCrea, and A. D. Osborne, *Br. J. Nutr.* **22,** 501 (1968).
11. Aherne, F. X., V. W. Hays, R. C. Ewan, and V. C. Speer, *J. Anim. Sci.* **29,** 906 (1969).
12. Bustad, L. K., W. E. Ham, and T. J. Cunha, *Arch. Biochem.* **17,** 247 (1948).
13. Cunningham, H. M., *J. Anim. Sci.* **18,** 946 (1959).
14. Bunding, I. M., M. E. Davenport, Jr., and M. A. Schooley, *J. Anim. Sci.* **15,** 234 (1956).
15. Catron, D. V., M. D. Lane, L. Y. Quinn, G. C. Ashton, and H. M. Maddock, *Antibiot. Chemother. (Washington, D.C.)* **3,** 571 (1953).
16. Dollar, A. M., K. G. Mitchell, and J. W. G. Porter, *Proc. Nutr. Soc.* **16,** 12 (1957).
17. Sewell, R. F., and J. P. West, *J. Anim. Sci.* **24,** 239 (1965).
18. Itoh, H., and S. L. Hansard, *J. Anim. Sci.* **24,** 889 (1965).
19. Nakamura, R. M., and C. C. Brooks, *J. Anim. Sci.* **33,** 237 (1971).
20. Brooks, C. C., R. M. Nakamura, and A. Y. Miyahara, *J. Anim. Sci.* **37,** 1344 (1973).
21. Neufville, M. H., H. D. Wallace, and G. E. Combs, *J. Anim. Sci.* **37,** 288 (1973).

22. Harper, A. E., and C. A. Elvehjem, *Agric. Food Chem.* **5,** 754 (1957).
23. Woodman, H. E., and R. E. Evans, *J. Agric. Sci.* **37,** 202 (1947).
24. Woodman, H. E., and R. E. Evans, *J. Agric. Sci.* **37,** 211 (1947).
25. Axelsson, J., and S. Eriksson, *J. Anim. Sci.* **12,** 881 (1953).
26. Teague, H. S., and L. E. Hanson, *J. Anim. Sci.* **13,** 206 (1954).
27. Forbes, R. M., and T. S. Hamilton, *J. Anim. Sci.* **11,** 480 (1952).
28. Crampton, E. W., and J. M. Bell, *J. Anim. Sci.* **5,** 200 (1946).
29. Crampton, E. W., G. C. Ashton, and L. E. Lloyd, *J. Anim. Sci.* **13,** 321 (1954).
30. Crampton, E. W., G. C. Ashton, and L. E. Lloyd, *J. Anim. Sci.* **13,** 327 (1954).
31. Bohman, V. R., personal communication. Nevada Agricultural Experiment station, Reno (1955).
32. McCormick, J. A., and J. F. Kidwell, *Nevada, Agric. Exp. Stn., Circ.* **2** (1953).
33. Bohman, V. R., J. F. Kidwell, and J. A. McCormick, *J. Anim. Sci.* **12,** 876 (1953).
34. Cunha, T. J., O. B. Ross, P. H. Phillips, and G. Bohstedt, *J. Anim. Sci.* **3,** 415 (1944).
35. Cunha, T. J., E. J. Warwick, M. E. Ensminger, and N. K. Hart, *J. Anim. Sci.* **7,** 117 (1948).
36. Baird, D. M., H. C. McCampbell, and J. R. Allison, *J. Anim. Sci.* **31,** 518 (1970).
37. Seymour, E. W., V. C. Speer, V. W. Hays, D. W. Mangold, and T. E. Hazen, *J. Anim. Sci.* **23,** 375 (1964).
38. Bowland, J. P., H. Bickel, H. P. Pfirter, C. P. Wenk, and A. Schurch, *J. Anim. Sci.* **31,** 494 (1970).
39. Baker, D. H., D. E. Becker, B. G. Harmon, and W. F. Nickelson, *J. Anim. Sci.* **24,** 872 (1965).
40. Keys, J. E., Jr., and J. V. DeBarthe, *J. Anim. Sci.* **39,** 53 (1974).
41. Farrell, D. J. and K. A. Johnson, *Anim. Prod.* **14,** 209 (1970).
42. Farrell, D. J., *Anim. Prod.* **16,** 43 (1973).
43. Elsden, S. R., M. W. S. Hitchcock, R. A. Marshal, and A. T. Phillipson, *J. Exp. Biol.* **22,** 191 (1946).

7

Fatty Acids, Fat, and Energy

I. NEED FOR FAT AND FATTY ACIDS

Fat provides energy, adds palatability to certain diets, decreases dustiness of certain feeds, acts as a carrier of nutrients (including the fat-soluble vitamins) and is a source of fatty acids for swine.

The Purdue Station (1) first showed that the pig needs fat in the diet. By using a semipurified diet which contained only 0.06% fat, the following fat deficiency symptoms were obtained in the weanling pig: (a) loss of hair, (b) scaly dandrufflike dermatitis, (c) necrotic areas on the skin around the neck and shoulders, (d) unthrifty appearance, (e) retarded sexual maturity, (f) underdeveloped digestive system, (g) a very small gallbladder, (h) slower growth rate, (i) lower efficiency of feed utilization, and (j) enlarged thyroid glands. Fat deficiency symptoms began to appear after 42 days and were quite severe after 63 days. Adding 1.5% corn oil at that time caused an immediate increase in growth rate and some recovery of other deficiency symptoms. Figure 7.1 shows the deficiency symptoms obtained.

Other scientists have shown that a deficiency of fat has an effect on growth or skin lesions. Figure 7.2 shows an effect on skin lesions from a deficiency of linoleic acid (2). The pigs receiving 0.07% of the calories as linoleic acid had a pronounced scaliness of the skin, first noted after about 13 weeks on the diet (pigs were started on trial weighing 10 lb). The scaliness seemed to be confined to the dorsal surface and was most severe about the shoulders. The hair itself was dry and appeared to stand out from the skin at all angles. There was no loss or discoloration of the hair in Large White X Essex pigs used. The requirement of the pig for linoleic acid was found to vary with age and was maximal up to 12–16 weeks of age at which time the need began to decline. Of the fatty acids from depot fat, 95% consisted of acids containing 16 or 18 carbon atoms, and only 2% had longer chains. Odd numbered fatty acids containing 15 or 17 carbon atoms were detected. This study showed that up to 200 lb liveweight, the pig requires about 1% of its calories as linoleic acid. Studies at Minnesota (3) and Georgia (4) showed the dietary linoleic acid requirement to be near 2% of the calories. In the

Fig. 7.1. Fat deficiency. Note loss of hair and scaly dandrufflike dermatitis, especially on the feet and tail of pig fed a fat-deficient diet. (Courtesy of W. M. Beeson, Purdue University.)

Georgia study (4), however, feeding linoleic acid as 1% of the dietary calories either prevented or remitted the dermal symptoms obtained.

It is evident from these and other studies that the pig needs linoleic acid although there is one university which questions this need (5,15). This may be due to the pig being able to synthesize some of its linoleic acid needs under certain conditions (15). The National Research Council (6) states that a dietary level of 0.03% linoleic acid is sufficient for normal growth but not for normal skin development. The linoleic acid level needed to meet both requirements is between 0.03 and 0.22% of the diet. After the essential fatty acid requirement has been met, further additions of fat increase the energy content of the diet. Linoleic acid is the key essential fatty acid for the pig. Arachadonic acid can be effective in preventing or curing a linoleic acid deficiency. But, arachadonic is synthesized from linoleic acid. So, unless one adds arachadonic acid to the diet, the fatty acid needed in the feeds used in diet formulation is linoleic acid. It would seem that a level of 1.0–1.5% fat in the diet would meet linoleic acid needs. This would indicate that practical swine diets consisting of grain and protein supplements should supply enough fat for the pig. But when diets consist of potatoes, cassava, sugar, starch or other carbohydrate sources (lacking in fat) as the main sources of energy plus a solvent-extracted source of protein (very low in fat), there may be a possibility of fatty acid deficiency occurring. Fatty acid deficiency could also occur in animals weaned at an early age and fed milk substitute diets. It is recommended, therefore, that these diets contain a source of fat and that it be protected against rancidity.

II. ADDING FAT TO DIET

There are advantages to adding fat to the diet. They are as follows: (a) controls dustiness of feeds, (b) improves feed efficiency, (c) increases acceptability and palatability, (d) improves physical appearance, (e) decreases wear on mixing and handling machinery, (f) increases comfort of workers in feed mill due to freedom from dust, (g) increases ease of pelleting, (h) reduces feed wastage in feeding, (i) reduces fire hazard from dustiness, heating, etc., and (j) decreases carotene loss. But, there are also some disadvantages to adding fats to the diet: (a) they are difficult to handle, (b) they require special effort to maintain high quality, and (c)

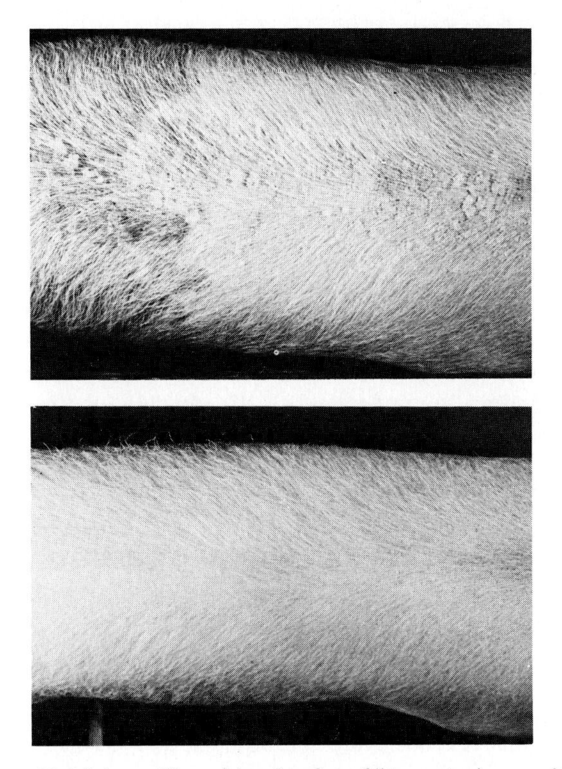

Fig. 7.2. Fatty acid deficiency. View of dorsal surface of litter-mate pigs reared for 18 weeks on a diet containing 0.07% (pig on top) and 0.5% (pig on bottom) of the calories as linoleic acid. (Courtesy of W. M. F. Leat, University of Cambridge.)

they need to be stabilized with a suitable antioxidant to prevent rancidity. Studies indicate that good quality fats are a desirable ingredient in certain diets and an increasing tonnage is being used by the feed industry.

The young pig appears to utilize dietary fat calories about as well as dietary carbohydrate calories (7,13). But this depends on the type of fat used. More study is needed to determine why certain fats are not utilized efficiently. It is apparent that the young pig can utilize certain types of fat as evidenced by the excellent performance on sow's milk which contains about 35% fat on a dry matter basis. Supplementation of weanling pig diets with 2.5–10% fat as tallow, corn oil or lard (and possibly other fats) may improve their performance. A slightly higher level of protein may be needed to compensate for the added calories from the fat. This was demonstrated in a study in England (9) which showed that no improvement in growth rate and feed conversion occurred at levels greater than 18% protein, but there was an increase in carcass quality at the 22% protein level in the diet containing 10% beef tallow. This showed that a higher level of protein may benefit carcass quality even though it does not increase rate of gain and feed efficiency. An antioxidant, such as ethoxyquin, should also be used to protect the fat from rancidity. There is considerable variation in the level of fat which is added to prestarter and starter diets. Most scientists recommend no more than 2.5–5.0% but some recommend up to 10% of the diet. Undoubtedly, the quality of fat, the level of protein in the diet, the remainder of diet ingredients and other factors are involved in determining the level of fat needed to get good results.

With growing-finishing pigs, higher levels of fat (10–20%) have been used throughout the world (8). These studies indicate that the kind and level of dietary fat exerts an influence on the quality of the depot fat by changing its fatty acid composition. The addition of fat at 10–20% levels in the diet has usually increased rate of gain, feed efficiency and back fat thickness. The effect on the carcass, however, must be assessed to ascertain if it is harmful to its selling price and profitability. Usually, carcasses with increased fat content are to be avoided if lean pork is desired. Diets with high levels of fat added to them and fed *ad libitum* to pigs weighing more than 100–125 lb may produce excessively fat bacon and other pork cuts. It is also well established that the iodine number (an indication of the degree of softness in fat) in the body of swine varies with the iodine number of the feed fat. But the extremes exhibited by the fat in the feed are usually not manifested in the body fat. So, the pig does modify the ingested fat somewhat. Only fats which do not produce soft pork should be added to swine diets. Most feed manufacturers use fat at low levels of about 1 or 2% in the diet. Whether adding fats at these levels would affect carcass quality has not been determined. Utilizing fats at low levels would minimize problems, and the right type of fat should not affect carcass quality.

A number of studies (8) have shown that the fatty acid composition of the sow's milk is largely influenced by the dietary fat composition during lactation, more so than in the cow. It has also been shown that the addition of fat to the diet of the lactating sow increases the fat content of milk even when the diet with added fat is isocaloric with the control carbohydrate diet (8). At the same time, the loss of weight by the sow during lactation is diminished by adding fat to the diet. The reason for this may be that the fatty acids in the diet fat are directly used for fat synthesis which is more efficient than the synthesis of fat from diet carbohydrates. A Canadian study (10) showed that the addition of 10% tallow (to supply 20% of the energy in the diets during growth and reproduction) had no significant effect on conception rate, number or weight of pigs born alive or weaned, when fed at an intake to equalize energy consumption with sows full fed on the control diet.

This discussion indicates that considerable information is still needed on adding fats to swine diets. The final solution to the problem of surplus waste fat should not depend on feeding it back to animals, however. Rather, it should lie in breeding and feeding swine that will produce more meat and less fat.

III. DECREASING EXCESS FAT IN CARCASS

The waste fat problem in pork is still a very important one. Excess trimmable fat in pork is undesirable because of its cost and consumer acceptability. Table 7.1 shows that considerable improvement has occurred in decreasing the amount of lard per pig during the last 25 years. Twenty-five years ago the average 100 lb of live slaughter hogs produced 70.1 lb of carcass. This 70.1 lb of carcass yielded 20.7% lard and 79.3% of pork cuts. In 1974 a market hog produced 74.4 lb of carcass per live hundredweight. This 74.4 lb yielded 8.2% lard and 91.8% pork cuts. This is quite a change in the lard produced from pork carcasses. The 243 lb hog in the 1950–1954 period had 35.24 lb of lard whereas the 240 lb pig in 1974 had 14.64 lb of lard. This is a decrease of 20.6 lb of lard and also a corresponding increase of 28.8 lb of pork cuts (Table 7.1) per pig. With 72.8 million pigs slaughtered in 1974, the decrease in lard production per pig amounted to 1,499,680,000 lb less lard than would have been produced 25 years ago. So paying attention and selecting for meatiness has paid off.

This change has been accomplished primarily through breeding and genetic selection for a meatier-type hog. This shift can be carried further to produce still higher ratios of lean to fat in certain pork cuts, especially the loin and the pork shoulder. Carcass quality traits are highly heritable, so increasing carcass quality can be done fairly rapidly. The average estimates of heritability are as follows: carcass length, 56%; carcass back fat thickness, 38%; yield of lean cuts, 29%;

TABLE 7.1
Commercial Hog Slaughter and Pork and Lard Production per 100 Pounds of Liveweight[a]

Year	Hogs slaughtered (millions)	Av. live-weight per hog (lb)	Average yield per 100 lb of live hog			
			Carcass (lb)	Pork (lb)	Lard (lb)	Lard (%)
1950–1954 (Av.)	71	243	70.1	55.6	14.5	20.7
1955–1959 (Av.)	75.6	238	71.1	56.7	14.4	20.2
1960–1964 (Av.)	79.7	239	71.9	59.7	12.2	17.0
1965	71.0	242	72.6	61.8	10.8	15.0
1970	95.4	239	73.2	64.4	8.8	12.0
1972	78.2	240	74.8	68.0	6.8	9.0
1974	72.8	240	74.4	68.3	6.1	8.2

[a]Source: USDA, "Livestock and Meat Statistics" (July, 1973) and "Fats and Oils Situation" (Oct., 1975)

and length of hind leg, 40% (11). The best example of improvement in carcass meatiness is the Denmark Production Testing Program which has resulted in the development of a premium meat-type pig which is difficult to surpass.

Table 1.8 indicates that there is still considerable progress to be made in decreasing the amount of the total separable fat (including lard) in swine carcasses.

In USDA Marketing Bulletin No. 51, it was estimated that 8, 42, 36, 12 and 2% of the barrows and gilts slaughtered in the United States in 1968 were in USDA grades 1, 2, 3, 4 and utility, respectively. It is estimated that in 1975, there were about 10% of the barrows and gilts making USDA grade number 1 in the United States. The data in Table 1.8 show that only the pigs in USDA grade 1 had more separable lean than separable fat in their carcasses. But, even in this grade, the carcass has 38.2% separable fat which is still too high and can be improved in the future.

In grade numbers 2, 3, and 4, all the carcasses had more separable fat than separable lean. This is a very wasteful and expensive carcass to produce, and one the consumer objects to because of the excess fat. Until carcass fat can be decreased to more acceptable levels, the U.S. consumer will not increase its level of pork consumption to an appreciable extent. Therefore, more emphasis needs to be placed on producing meat-type pigs by a much larger number of the swine producers in the United States. Ten percent of the pigs making USDA grade 1 is too low and needs to be greatly increased.

Decreasing excess fat will increase feed efficiency since fat has about 90% dry matter whereas lean tissue has about 30% dry matter—or about one third as much. Therefore, it can be assumed that three times as much dry matter in fat will require about three times as much feed as a pound of lean tissue. The average pig in the United States requires 3.3–3.5 lb of feed per pound of gain from weaning to market weight. Yet the top performing pig in a Swine Evaluation Center in the Midwest recently required only 2.24 lb of feed per pound of gain—or about two thirds as much feed. So, eliminating trimmable fat in the pig will decrease the amount of feed needed and the cost of producing pork.

Many scientists have assumed that a pound of fat requires 2.25 times as much energy as a pound of protein (both on a dry matter basis). But work in Europe by Greta Thorbek and others has shown that protein deposition is inefficient as compared to fat deposition from an energy standpoint. In a review paper (12) she stated that the energy requirement was fairly equal for both with about 12 kcal needed per gram of protein or fat retained in swine. This differs little from the Canadian data (14) which showed these values to be 11.65 ± 4.93 and 16.26 ± 4.41 kcal in trial 1 and 14.62 ± 4.93 and 15.73 ± 2.62 kcal in trial 2, respectively, per gram of protein or fat deposited. In both trials protein deposition required less energy but not as much as one would expect, which verifies the European studies.

IV. DO NOT SACRIFICE CARCASS QUALITY AND REPRODUCTION

A word of caution is needed about programs to develop more of a meat-type pig. Some pigs have been bred to the point where they are too heavy in the ham and loin area and too light in the front end. Some of these animals do not have enough lung, heart, and digestive tract capacity and are encountering problems during the latter part of the growth period and especially during reproduction. Some people refer to them as pigs with "basketball-like hams." These are extreme cases, but it indicates that the problem must be corrected. Dr. H. Clausen of Denmark, who has done such a good job in developing the meat-type Danish Landrace, admits this has happened in Denmark. Now, they are trying to breed away from it. Therefore, as breeding programs are developed for producing a meat-type pig, one must be certain that the resulting animals can still function properly throughout growth and also during reproduction.

Acceptable pork quality levels need to be maintained, or improved, as one selects for a meat-type animal. In the past, carcass quality problems have occurred as selection for extremely heavy muscled pigs took place. Indications are that pale soft exudative pork (PSE) and pork stress syndrome (PSS) may be associated with heavy muscled animals (although not all heavy muscled animals are susceptible). Therefore, swine producers should exert strong selection pressure against PSS and PSE. Moreover, they should make sure their breeding stock comes from PSS and PSE free herds.

The pork stress syndrome (PSS) is a condition that occurs when some hogs are stressed when handled, moved, or regrouped and especially during warm weather. This results in an acute shocklike syndrome, which often causes death. These death losses are most frequent at marketing but also occur at weaning or when pigs are mixed, sorted, or moved. The problem occurs throughout the world. It has been reported in certain strains of almost all pure breeds of swine as well as in crossbred animals. Approximately 70% of PSS swine produce PSE pork.

There are some scientists who are concerned that not only has back fat been bred off many pigs, but also in the process some of the intramuscular fat. This results in some of the pork cuts being dry and less tasty than a good many customers prefer. Thus, it is recommended that selection programs include carcass studies which will lead to adequate intramuscular fat while decreasing the excess back fat.

V. UNSATURATED FAT IN PORK

Unsaturated fat can be produced in pork by feeding unsaturated fats in the diet. For example, the use of peanuts in swine diets will produce an unsaturated fat or soft pork. Studies show that it takes a considerable length of time to harden pigs which have been fed peanuts for any length of time. It may take 3–3.5 times more gain on a hardening diet as compared to the gain on peanuts to produce hard or medium hard hog carcasses. This means that 40 lb of gain on peanuts would probably require 120–140 lb of gain on a hardening diet. A Florida study (16,17) showed that 50 lb pigs fed peanuts to 122, 151 and 165 lb weights, and then fed hardening diets until they weighed approximately 200 lb still produced carcasses which graded soft. Figure 7.3 shows the softness of the carcasses and its effect on the bacon. Peanut feeding tends to yield carcasses with more back fat thickness in relation to carcass length and thus causes lower carcass grades.

Scientists have shown that using 250 ppm copper in the diet will increase the proportion of unsaturated fatty acids in pork fat. This increases the oxidative susceptibility (more prone to rancidity) of the depot fat. A Canadian study (18) showed that supplementation of the diet with vitamin E improved the oxidative stability of the depot fat from copper-fed pigs presumably through its absorbtion and deposition in the depot fat where it exerts its antioxidant property.

There is no doubt that pork with more unsaturated fat can be produced. The question is will the consumer accept it. A Nebraska trial (19) involved adding 20% safflower oil to the diet and comparing it to a control diet of corn-soybean

Fig. 7.3. Note effect of soft pork on bacon which is in proportion to the amount of body weight gain made on peanuts. The pigs in lot V (at far left end) were fed a corn-soybean meal diet. The pigs in lots I, II, III, and IV, made 145, 111, 98, and 71 lb of gain on peanuts, respectively. (Courtesy of A. Z. Palmer, University of Florida.)

meal for feeding pigs. Consumers preferred the sausage and bacon from pigs fed corn-soybean meal diets over those fed safflower oil. But no differences occurred with smoked pork loin or frankfurters from corn-soybean meal diets or diets containing safflower oil. It would appear that until the problem of rancidity in pork products and the problem of consumer acceptance of pork with more unsaturated fat is adequately studied, changes should not be made in adding unsaturated fats in swine diets.

Although much information exists on animal products, saturated fats, and cholesterol and their relation to heart disease, there is still much to be learned on this subject. Some controversy still exists on the effects of dietary changes, such as substituting unsaturated fat for saturated animal fat, and the significant reduction of cardiovascular problems in man. The effects of feeding diets high in unsaturated fatty acids to experimental animals over a long period of time are not entirely clear, but there are indications that serious problems may develop.

VI. LIMITING ENERGY INTAKE

There is no doubt that restricted feed intake will improve carcass quality. There are plenty of data in Europe and the United States to verify this fact. It is important, however, that a premium be paid for improved carcass quality in order to off-set the increased time that pigs are handled on feed due to the lower rate of gain on restricted feeding.

A fat-type pig will respond best to limited feeding and will show the greatest improvement in carcass quality. A meat or lean-type pig will respond least to restricted feeding. Thus, the level of restricted feeding used depends on the type of pig being fed, the diet used, the type of carcass desired and the premium paid for increased carcass quality. Dr. H. Clausen of Denmark has bred and selected some pigs which can be self-fed all the feed they will consume, and their carcass quality is as good as the restricted fed pigs. This indicates that swine producers should breed and select animals that will produce maximum carcass quality without having to restrict their feed intake. This should be the long-term goal since it has already been demonstrated in Denmark and elsewhere in Europe that it can be accomplished. One California Polytechnic University study (20) concluded that restricted time of access to feed for swine may not be as effective in improving carcass quality as restricting the amount of total feed available over a 24-hour period. This is important if one is contemplating restriction of feed intake.

During the last few years, the energy requirements of the pregnant gilt and sow have been markedly reduced. Research has shown that the sows' utilization of energy can be improved by allowing moderate weight gains during gestation and by increasing the total amount of feed given during lactation. However, there has not been general agreement on the influence of energy level on the performance of sows during lactation.

If brood sows are allowed feed *ad libitum,* they consume more energy during gestation than they require for maintenance and for development of the fetuses. Therefore, energy intake should be limited. It not only saves on feed, but it may improve reproduction because sows fed too high an energy level may reduce their reproductive level (21).

In an excellent review by Dr. J. P. Bowland (22) on energetic efficiency of the sow he states that "a system of feeding that attempts to maintain the true weight (total energy content of the body) of sows during lactation and that allows no gain in weight above that accounted for by the fetuses and fetal membranes during gestation would allow the maximum energetic efficiency. For gilts, and possibly for second litter sows, an energy allowance for normal growth would also be required. To lower maintenance requirements, relatively low mature weights would be desirable. Direct energy conversion by the young pig by allowing it to eat as soon as possible should also be encouraged as an aid to total energetic efficiency. To improve over-all efficiency of energy utilization, feed costs should be spread over more young pigs." So increasing litter size is one of the most important goals for the swine producer to accomplish in order to decrease feed costs.

VII. ENERGY VALUES OF FEEDS

The energy portion of swine diets is the largest and most expensive part of it. Energy values are expressed as digestible energy (DE) and metabolizable energy (ME). DE is defined as the gross energy of the feed minus fecal gross energy. ME is defined as the gross energy of the feed minus fecal gross energy minus urinary gross energy. Unfortunately, much of the information available on energy values for feeds used by swine is not in DE or ME but in TDN (total digestible nutrients).

There are advantages in expressing energy requirements of animals in terms of calories. The energy content of feed and excreta can be measured in calories so that DE and ME can be measured directly and easily without the assumptions and approximations used in the calculation of TDN. Because of this, DE and ME are preferred instead of TDN by most people. To change to the calorie system, it is necessary to convert TDN values to DE or ME. Where DE values are computed from TDN, 1 lb of TDN has an average DE value of 2000 kcal. With mixed feeds for swine, ME values approximate 96% of DE values, although individual feed ingredients will vary considerably from this value.

Metabolizable energy is emerging as the preferred method of expressing the energy content of feeds for swine. Metabolizable energy is more accountable for energy losses which occur in the urine and feces. Gas losses remain unaccounted for although by definition they are deducted. But this loss is small and rather constant. Bowland and others (23) reported methane gas production in swine at a level of 1.1% of DE.

Data are accumulating on DE and ME in swine feeds. Much more research is needed, however, to determine the needed ME values of all swine feeds. Some ME values for poultry feeds are similar to swine ME values but some show marked differences. So, it appears that separate ME values will have to be determined for swine (25). Once adequate ME values are obtained, equations can be developed whereby net energy (NE) values for maintenance and for gain can be calculated from ME data. Scientists in Europe, where energy feeds are scarce, are further ahead than the United States in energy studies. A leader in the energy field is Dr. J. Kielanowski of the Animal Physiology and Nutrition Institute at Jablonna, Poland. He has been working in the area of net energy values for swine for many years (24).

REFERENCES

1. Witz, W. M., and W. M. Beeson, *J. Anim. Sci.* **10,** 112 (1951).
2. Leat, W. M. F., *Br. J. Nutr.* **16,** 559 (1962).

3. Hill, E. G., E. L. Warmenen, C. L. Silbernick, and R. T. Holman, *J. Nutr.* **74,** 335 (1961).
4. Sewell, R. F., and L. J. McDowell, *J. Nutr.* **89,** 64 (1966).
5. Babatunde, G. M., W. G. Pond, E. F. Walker, Jr., and P. Chapman, *J. Anim. Sci.* **27,** 1290 (1968).
6. Cunha, T. J., J. P. Bowland, J. H. Conrad, V. W. Hays, R. J. Meade, and H. S. Teague, *N.A.S.—N.R.C.*, publication (1973).
7. Baker, D. H., *Proc. Fla. Swine Day, 1971* pp. 1–3 (1971).
8. Vanschoubroek, F. X., *World Congr. Anim. Nutr., 1966* **1,** 217–262 (1966).
9. Lewis, D., *Feedstuffs* **37,** 18 (1965).
10. Bowland, J. P., *Can. J. Anim. Sci.* **44,** 142 (1964).
11. Cunha, T. J., A. Z. Palmer, R. L. West, and J. W. Carpenter, *Feedstuffs* **47,** 37, 38, and 47 (1975).
12. Thorbek, G., *Nutr., Proc. Int. Congr., 9th,* pp. 37–46 (1973).
13. Allee, G. L., D. H. Baker, and G. A. Leveille, *J. Nutr.* **101,** 1415 (1971).
14. Sharma, V. D., and L. G. Young, *J. Anim. Sci.* **31,** 210 (1970).
15. Kass, M. L., W. G. Pond, and E. F. Walker, Jr., *J. Anim. Sci.* **41,** 804 (1975).
16. Alsmeyer, R. H., M. S. Thesis, University of Florida, Gainesville (1956).
17. Alsmeyer, R. H., A. Z. Palmer, H. D. Wallace, R. L. Shirley, and T. J. Cunha, *J. Anim. Sci.* **14,** 1226 (1955).
18. Amer, M. A., and J. I. Elliott, *J. Anim. Sci.* **37,** 87 (1973).
19. Stelwell, D. E., R. W. Mandigo, and E. R. Peo, Jr., *J. Anim. Sci.* **31,** 191 (1970).
20. Anderson, T. A., H. D. Fausch, and J. Gesler, *Growth* **28,** 213 (1965).
21. Frobisch, L. T., N. C. Steele, and R. J. Davey, *J. Anim. Sci.* **36,** 293 (1973).
22. Bowland, J. P., *J. Anim. Sci.* **26,** 533 (1967).
23. Bowland, J. P., H. Bickel, H. P. Pfirter, C. P. Wenk, and A. Schurch, *J. Anim. Sci.* **31,** 495 (1970).
24. Kielanowski, J., and M. Kotarbinska, *Nutr., Proc. Int. Congr., 8th, 1969* pp. 1–12 (1971).
25. Diggs, B. G., D. E. Becker, A. H. Jensen, and H. W. Norton, *J. Anim. Sci.* **24,** 555 (1965).

8

Water

I. INTRODUCTION

Most people take for granted the water they drink and the oxygen they breathe. However, life would not exist for very long without these two important essentials. A starving animal may lose nearly all of its fat, half of its body protein, and about 40% of its body weight, and still live. But if it loses 10% of its water, disorders will occur, and if it loses 20% of its body water, it will die.

Water is one of the most important nutrients required by swine and yet it is very often neglected. Water affords a medium for the digestion, absorption and transportation of other nutrients throughout the body and for the elimination of waste products. Moreover, it plays a major role in regulation of body temperature. A lack of water results in lowered appetite, lowered efficiency of feed utilization, and impairment of all body processes. Thus, it is apparent that an adequate supply of clean, fresh, palatable, drinking water should be readily available for the animals.

II. WATER REQUIREMENTS

Young pigs and lactating sows have the highest water requirements. Milk has 70–80% water and the small pig's body is about two-thirds water. As the pig grows larger, it requires proportionately less water because it consumes less feed per unit of body weight and the water content of its body is decreasing. Under normal conditions, pigs consume a fairly constant amount of water for each pound of dry feed eaten. If some of the feed is high in water, such as silage, milk by-products, roots, or green forage, the amount of water consumed is correspondingly reduced.

It is difficult to state the exact requirements for water because of the many variables that influence its consumption. Temperature, the kind of diet fed, the level of feed intake, the quality of the water, and the size of the pig are some of the factors affecting water needs. Water is also excreted from the body in the feces, urine, evaporation from the lungs and skin and in the milk. So all of these factors will influence water requirements.

The 1973 National Research Council (NRC) publication (1) states the following on water needs. "Pigs consume an average of 1.9–2.5 lb of water per pound of dry feed. In a high temperature environment, voluntary consumption may be as high as 4.0–4.5 lb of water per pound of dry feed. Pigs 5–8 weeks old may consume as much as 20 lb of water per 100 lb body weight daily, whereas market weight pigs will normally consume about 7 lb of water per 100 lb body weight per day. Lactating sows must have unlimited access to water if they are to milk adequately (since milk has 70–80% water). Suckling pigs will not consume adequate amounts of creep diets unless water is available."

To be on the safe side, free access to water is desirable, regardless of the feeding system used or the stage of the life cycle the pig is in.

III. TOLERANCE LEVEL OF MINERAL SALTS IN WATER

One of the most important criterion in evaluating water quality is the amount of dissolved mineral salts it contains. The total amount of mineral salts in water appears to be much more important than the types of salts it contains. But, there is still a lack of information on the actions and interactions of individual mineral salts in water.

In general, the effect of salts in water is to increase the consumption of water. Since the pig must maintain an osmotic balance, extra salts in the water or the feed are going to require certain adaptive measures. Toxicity due to excess mineral salts in water resembles simple dehydration and will upset the electrolyte balance. For example, toxic symptoms were produced in the pig deprived of salt for several months and then given feed with 6–8% salt all at once and no drinking water (2). But no harmful effects occurred if the pigs had free access to all the water they needed. So, salt is not a problem if plenty of drinking water (without excess salt) is available.

All natural water supplies contain dissolved minerals, but the amounts and relative proportions of the various mineral salts vary considerably. Bicarbonates, sulfates and chlorides are the most common anions, while magnesium, calcium and sodium occur in greater quantities than other cations which may be present.

Canadian workers (3) used 26 lb pigs to conduct studies with pure mineral salt solutions of magnesium sulfate, sodium chloride, and sodium sulfate, or with mixtures of magnesium sulfate and sodium chloride added to city water containing 140–150 ppm of total solids. They found no adverse effect with water containing up to 5000 ppm of any of the mineral salts when performance was compared with that of pigs receiving city water. Based on this study and many others, it appears that under normal conditions of environment and health, the water requirements of all classes of pigs may be met satisfactorily by waters containing various mixtures of salts up to a total of 4500–5000 ppm. There may be some scouring initially, but it does not appear to influence performance of the pig. Levels of 7000 ppm of mineral salt mixtures for growing-finishing pigs appear to have no adverse effect on rate of gain or feed efficiency. But, levels of 10,000–11,000 ppm of mineral salt mixtures do influence performance adversely. Table 8.1 gives the U.S. NRC recommendations on saline water for

TABLE 8.1
A Guide to the Use of Saline Waters for Livestock and Poultry (8)

Total soluble salts content of waters (mg/l)[a]	Comments and discussion
Less than 1,000	These waters have a relatively low level of salinity and should present no serious burden to any class of livestock or poultry.
1,000–2,999	These waters should be satisfactory for all classes of livestock and poultry. They may cause temporary and mild diarrhea in livestock not accustomed to them or watery droppings in poultry (especially at the higher levels), but should not affect their health or performance.
3,000–4,999	These waters should be satisfactory for livestock, although they might very possibly cause temporary diarrhea or be refused at first by animals not accustomed to them. They are poor waters for poultry, often causing watery feces and (at the higher levels of salinity) increased mortality and decreased growth, especially in turkeys.
5,000–6,999	These waters can be used with reasonable safety for dairy and beef cattle, sheep, swine, and horses. It may be well to avoid the use of those approaching the higher levels for pregnant or lactating animals. They are not acceptable waters for poultry, almost always causing some type of problem, especially near the upper limit, where reduced growth and production or increased mortality will probably occur.
7,000–10,000	These waters are unfit for poultry and probably for swine. Considerable risk may exist in using them for pregnant or lactating cows, horses, sheep, the young of these species, or for any animals subjected to heavy heat stress or water loss. In general, their use should be avoided, although older ruminants, horses, and even poultry and swine may subsist on them for long periods of time under conditions of low stress.
More than 10,000	The risks with these highly saline waters are so great that they cannot be recommended for use under any conditions.

[a]Same as parts per million.

farm animals (8). Whenever possible it is best to use water with a low saline level. But sometimes the only water available is high in saline and Table 8.1 can serve as a guide for its use with swine.

There has been some suggestion that water becomes increasingly unpalatable as salinity increases beyond 1000 ppm total solids. Most of the work with pigs, however, would suggest that provision of water with up to 11,000 ppm total mineral salts causes the pigs to drink more water and to increase urinary output.

Under high environmental temperatures, when total water intake is greater, there is a possibility of a problem arising because the total water intake might be so high that the total mineral salt intake cannot be cleared by the kidneys. This should be considered when a safe standard for water is being considered.

Very young pigs are usually less able to adapt to high mineral salt levels than older pigs. There is evidence too that pigs may require a few days to adapt to a new water supply and they may have some incidences of scouring initially.

Water sometimes contains other mineral elements at high enough levels to be toxic. Elements such as chromium, copper, cobalt and zinc seldom cause any problem in farm animals because they do not occur at high levels in soluble form or because they are toxic only in excessive concentrations. Also, these elements do not appear to accumulate in meat or milk to any extent that would constitute a problem for animals drinking water under any but the most unusual conditions. But elements such as lead, mercury and cadmium must be considered actual or potential problems, because they occasionally are found in water at toxic levels.

TABLE 8.2
Safe Level of Mineral Elements in Water

Item	Safe upper limit of concentration (mg/l)[a]
Arsenic	0.2
Cadmium	0.05
Chromium	1.0
Cobalt	1.0
Copper	0.5
Fluoride	2.0
Lead	0.1
Mercury	0.01
Nickel	1.0
Nitrate–N	100.0
Nitrite–N	10.0
Vanadium	0.1
Zinc	25.0

[a]Same as ppm (parts per million).

The levels which the U.S. NRC recommends as being safe in water for a number of mineral elements for livestock are shown in Table 8.2 (8). A safety factor was considered in establishing these levels to take into account the many factors which can influence the tolerance for or toxicity of mineral elements.

A toxic level of a mineral element in water consumed for a short period of time may not have an effect which can be observed. But, consumption of the water for a long time causes serious effects. Different species of animals may react differently to a toxic level of minerals in the water. The young and the healthy animal may not respond in the same manner as unthrifty or mature animals. The rate and level of water intake may also influence the toxicity effects. Different forms of the mineral element may result in different toxicity effects on the animals. Thus, many factors can influence the toxicity of a mineral element in the water. But, the levels shown in Table 8.2 can be used as a guide as to the upper limit allowed in water.

Water sources should be analyzed occasionally to make sure they are safe. It is especially important to check water levels when swine are raised under intensified conditions or near areas where water contamination may occur. In almost all cases there should not be a problem, but checking the water is good insurance.

Iron in excess of 0.3–0.5 ppm will cause water to appear rusty (4). The iron content is occasionally so high that it makes water objectionable because of its taste.

Hydrogen sulfide may be contained in water at a level humans find offensive, but there is no evidence that pigs find this particular taste and odor unacceptable. Hydrogen sulfide makes water corrosive to steel, copper and brass (4).

IV. NITRATES AND NITRITES IN WATER

The pig is not appreciably affected by nitrates in water because it does not have the bacterial flora to readily convert nitrate into nitrite which is harmful. However, in water supplies, some of the nitrate may be converted to nitrite by bacteria present in the water. Under these conditions, the pig may be adversely influenced by the presence of nitrite in the water (5). It has been shown that the conversion of nitrate to nitrite prior to its consumption is necessary to produce toxic symptoms in swine and that a single dose of 0.09 gm of sodium nitrite per 2.2 lb body weight is necessary to cause death (6). South Dakota studies (7) showed that levels of sodium nitrate providing up to 300 ppm of nitrate nitrogen in the drinking water had no adverse effect on weight gain, general unthriftiness, or breeding and reproductive performance in swine. Adding sodium nitrite to drinking water to provide up to 100 ppm nitrite nitrogen gave measurable but small increases in methemoglobin which indicated a harmful effect.

V. EFFECT OF TEMPERATURE ON WATER NEEDS

The pig loses only a very small amount of water by evaporation from the skin under hot weather conditions. There is some water loss from the respiratory tract by breathing, but it is not too much. So, sweating will not be of help to a pig during hot weather. This means it needs to be cooled by shade, mist spray or other means.

As the temperature increases, there is a corresponding increase in water consumption (11). Young pigs consume more water in proportion to their body weight than mature, nonlactating pigs. Lactating sows may need at least 40% more water than a nonlactating sow. It is difficult, therefore, to establish an exact or definite ratio of water intake to dry feed intake even though water intake is closely related to feed intake. English (9) and Nebraska (17) studies have shown that having water available *ad lib* to pigs may be preferable to limiting its availability to two or three times daily. This would especially be true if insufficient water were supplied. In carefully reviewing this subject it is apparent that it is preferable to have water available *ad lib* for pigs. This would take care of temperature difference effects on water consumption as well as the many other factors which affect water needs.

In very cold winter weather, a heating device should be used to keep the water from freezing and thus limiting how much water is available to the pig. Iowa Station (10) tests showed that pigs receiving water warmed to 45°F gained an average of 0.14 lb per day more than pigs on unwarmed water which was frozen part of the time. Warming the water to 55°F showed no advantage over warming to 45°F. Thus, there was no advantage in keeping the temperature of the water higher than necessary to keep it from freezing. It seems logical too that cool water would be desirable for hogs during hot weather. This can be accomplished by maintaining the water supply in the shade to keep it from getting too warm.

VI. EFFECT OF SOURCE OF WATER

Ponds and stagnant waters of any kind are likely to be polluted. Many disease outbreaks and parasite infestations are often traceable to contaminated water. Thus, one needs to exercise caution in watering pigs from ponds. Underground waters are usually free of pollution and are preferred as a source of farm supply. Best results are obtained when the water is piped to well-located and protected troughs or waterers, rather than being carried by hand or hauled. Pigs usually suffer from a lack of water when it is carried or hauled to them because the caretaker may not have or take the time to keep them supplied with water at all times.

A trial conducted at the Florida Station (12) with growing-finishing pigs showed there was not much difference in the performance of the animals fed lake water, well water, distilled water, and water from an unpolluted pond. Thus, if the water is clean and fresh, there probably is not much to choose from in the source of water used for pigs.

VII. USE OF WET FEED

In a review paper (13) the author summarized the information available on adding water to feed. Under certain conditions the use of wet feeding is superior to dry feeding with swine whereas in other conditions it is not. European data (16) show that wet feeding is usually superior to dry feeding when feed intake is restricted. It increases rate of gain and feed efficiency. The best results are obtained with a level of about $2\frac{1}{2}$ parts of water to one part of feed. The same results are obtained if the water is mixed with the meal immediately before feeding as compared to mixing the water with the meal and letting it soak for 24 hours before feeding. It is best to supply water *ad lib* even though pigs are being fed a wet feed.

Data at Ohio State (14,15) have shown that a paste feed increases rate of gain and feed efficiency with the pig. The paste is fed from an automated self-feeding system with a mixture of 1.3–1.5 parts by weight of water to one part of dry feed. This results in a feed with 58–64% moisture. Water needs to be supplied *ad lib* to pigs on paste feed. The paste feed will freeze in cold weather. Therefore, the system would not be suitable in a cold climate without confinement housing and environmental control.

United States data show that a liquid feed has no apparent benefit in rate of gain or feed efficiency when the pigs are full-fed their diets. However, when the feeding rate is restricted, then the liquid fed pigs gain faster with less feed per pound of gain, which is similar to European findings.

REFERENCES

1. Cunha, T. J., J. P. Bowland, J. H. Conrad, V. W. Hays, R. J. Meade, and H. S. Teague. "Nutrient Requirements of Swine," Natl. Res. Counc., Washington, D.C., 1973.
2. Bohstedt, G., and R. H. Grummer, *J. Anim. Sci.* **13,** 933 (1954).
3. Berg, R. T., and J. P. Bowland, *Annu. Feeder's Day Rep., 39th; Univ. Alberta* p. 14 (1960).
4. Matz, S. A., *Avi Publ. Co.,* Westport, Connecticut, 1965.
5. Bowland, J. P., Report. University of Alberta, 1972.
6. Winks, W. R., A. K. Sutherland, and R. M. Salisbury, *Queensl. Dep. Agric. Stock, Div. Anim. Ind. Bull.* **2** (1950).

7. Seerley, R. W., R. J. Emerick, L. B. Embry, and O. E. Olson, *J. Anim. Sci.* **24,** 1014 (1965).
8. Shirley, R. L., C. H. Hill, J. T. Maletic, O. E. Olsen, and W. H. Pfander, "Nutrient and Toxic Substances in Water for Livestock and Poultry," Nat. Res. Counc., Washington, D.C., 1974.
9. Barber, R. S., R. Braude, and K. G. Mitchell, *Anim. Prod.* **5,** 277 (1963).
10. Altman, L. B., Jr., G. C. Ashton, and D. Catron, *Iowa Farm Sci.* **7,** 41 (1952).
11. Mount, L. E., C. W. Holmes, W. H. Close, S. R. Morrison, and I. B. Start, *Anim. Prod.* **13,** 561 (1971).
12. Alsmeyer, R. H., T. J. Cunha, and H. D. Wallace, *Fla., Agric. Exp. Stn., A. H. Mimeo* No. 55-5 pp. 1–4 (1955).
13. Cunha, T. J., *Proc. World Congr. Anim. Feed., 2nd, 1972* pp. 619–626 (1972).
14. Teague, H. S., W. L. Roller, A. P. Grifo, Jr., and R. K. Christenson, *Ohio, Agric. Exp. Stn., Anim. Sci. Mimeo* No. 201 pp. 1–7 (1970).
15. Teague, H. S., W. L. Roller, A. P. Grifo, Jr., and R. K. Christenson, *J. Anim. Sci.* **33,** 329 (1971).
16. Braude, R., *Proc. Easter Sch. Agric. Sci., Univ. Nottingham* **18,** 279–295 (1971).
17. Danielson, D. D., *J. Anim. Sci.* **37,** 242 (1973).

9

Antibiotics and Other Antimicrobial Compounds

I. INTRODUCTION TO ANTIBIOTICS

An antibiotic is a substance that has the capacity of inhibiting the growth of or destroying microorganisms. Hundreds of antibiotics exist, but only a few are beneficial for animal feeding.

The discovery and use of antibiotics in swine diets mark another milestone in the advancement of swine nutrition. Penicillin was discovered in 1942, and became available for public use in 1945. In 1949, it was first reported that products from aureomycin (chlortetracycline) fermentation promoted growth in pigs (1). At that time, it was found that what was then called animal protein factor (APF) supplements contained antibiotics and B_{12}.

The first published scientific paper showing the value of crystalline aureomycin in swine feeding appeared in April, 1950 (2). Since then "antibiotics" has become a common household word for swine producers as well as for feed manufactuers and dealers. Prior to 1950, antibiotics were used for combating and treating pathogenic diseases in animals and human beings. In October of 1950, antibiotics were officially recognized and defined for animal feeding by the American Feed Control officials. Since 1950, a great deal of information has been obtained on antibiotics. More important is the fact that antibiotics are now a part of most feeds sold for pigs.

II. RATIONALE FOR CONTINUED ANTIBIOTIC USE

Antibiotics in feed have been used to help produce well over 100 billion animals in the United States during the last 27 years with no adverse effect on the animals, on the humans feeding the animals or on humans consuming animal products. After 27 years of extensive antibiotic use in animal feeds there are no known cases of injury to humans that can be attributed to or associated with the consumption of meat from animals fed antibiotics (5). There have been occasional reports of traces of antibiotic residues in the tissues of animals produced for meat. When antibiotics are used in feeds as authorized by the U.S. Food and Drug Administration (FDA) and the approved withdrawal periods are followed, the residues in edible tissues are zero or nil and below the safe tolerance level approved by FDA. Moreover, normal methods of cooking should destroy antibiotics if traces are present in the meat.

There are some who are concerned that animals treated with low levels of antibiotics used in the treatment of humans will develop reservoirs of bacteria resistant to the antibiotic used. They are also concerned that the resistance factor may be transferred from the bacteria in the animal to the bacteria in man and thus reduce the effectiveness of that particular antibiotic for clinical use. It is true that resistance appears in bacterial populations exposed to antibiotics. However, the literature on transferability of resistance from animals to man and the documented data on this possible hazard does not support the idea that cross contamination from animals to man has become a problem in maintaining human health. The mere presence of *Escherichia coli* with resistant factors is not in itself a hazard. Both people and animals may have very large numbers of such organisms in their intestines and yet be at least as healthy as similar animals or humans that do not have such organisms in their intestines (3).

Recently, a large amount of research has been conducted showing that the use of antibiotics in feed does not increase the quantity, prevalence or duration of shedding or change the resistance characteristics of salmonella in animals given antibiotics in their feed (3).

On May 1, 1972 the Feed Additives Committee of the American Society of Animal Science recommended to FDA that "antibiotics not be limited exclusively to either animal or human use. This would make them unavailable for use in human or animal epidemic because they were limited to only one area. Any such move in this direction should be based on scientific evidence which does not yet exist." Another recommendation stated "a great deal of published literature supports the prophylactic use of antibiotics in animal production. The published literature based on research does not justify the termination of this feeding practice. Therefore, the American Society of Animal Science recommends that the previous long standing policy of FDA, effective in 1971, with respect to the use of antibiotics in animal feeds be continued unchanged."

The policy cited above of the American Society of Animal Science Feed Additives Committee with regard to antibiotics was still the same in 1977. Thus, the community's most knowledgeable scientists on the use of antibiotics for animal feeding still recommend their continued usage in animal feeding since there is no scientific evidence to indicate otherwise.

III. VALUE OF ANTIBIOTIC USAGE

The United States is gradually changing to more intensified swine production operations. These densely populated and confined type operations involve quite a different biological system from the pasture programs used in past years. The use of antibiotics is more important to avoid catastrophic losses under crowded conditions. It is doubtful that the United States could continue to produce pigs efficiently in these environments without the use of effective antibiotics and chemotherapeutic agents (4).

In the United States, hog farms have been reduced in numbers during the last 10 years from 1,057,000 to 752,000 in 1974 or about a 30% drop compared to an 18% decline in the number of general farms (5). The number of producers marketing 500 hogs or more per year has increased a great deal in the last 5 years. These larger producers are most intensified and thus antibiotics are more important for them. With feed, labor, housing and other costs continuing to increase, it is imperative that the swine producer be able to increase efficiency of operation in the future. The use of antibiotics is one key factor in this regard. Dr. J. L. Krider of Purdue University estimated in 1971 that each dollar spent for antibiotics for animal feed use returned five to eight dollars in increased returns to the producer (6).

At a swine conference in 1971, Dr. J. L. Krider (6) summarized the degree of antibiotic response shown in Table 9.1. The data in Table 9.1 are typical of what one sees in various summaries of antibiotic response. The greatest response occurs early in the life of the pig and declines as the pig approaches market

TABLE 9.1
Effect of Adding Antibiotics to Swine Diets (6)

	Percent increase in	
Weight	Rate of gain	Feed efficiency
Early weaned at 15–20 lb to about 40–50 lb	20–40	7–15
30+ to 100–125 lb	9[c] (1965–1970)	1.4
100 lb to market	10.6[d]	6.1
100–125 lb to market	6.7 (1965–1970)	1.8
Weaning to market[a]	10.7	5.1
Weaning to market[b]	9.9	2.4

[a]Average of 61 comparisons.
[b]Average of 36 comparisons at Purdue University.
[c]Average of 14 comparisons.
[d]Average of 14 comparisons.

weight (Figure 9.1). On farms, where the level of sanitation is not as good as at the University station farms (data shown in Table 9.1), the level of antibiotic response is greater. In one test where a building had been thoroughly cleaned and disinfected, 50 gm of antibiotic per ton of feed improved gains 33% and feed efficiency 11%. In a building which was not thoroughly cleaned, the addition of the antibiotic produced 75% faster gains and 37% better feed efficiency than the controls (6).

Judicious use of effective antibiotics at breeding time will increase conception rate, farrowing rate and litter size. Numerous studies illustrate the beneficial effects of antibiotics at this stage. The conception rate on first service may be improved by 15 to 25% and litter size by an average of 0.5 or more pigs per litter (4). Before and after farrowing is another critical stage of the production cycle in which antibiotics are particularly beneficial (4). Improvements in the general well being and health of the sow aid her in coming to milk quicker and in producing more milk for the young pigs during the first few hours of their life (4).

Creep and starter diets should definitely be fortified with antibiotics. The first few weeks of the pigs' life is the period of greatest response in terms of improved survival rate, increased growth rate and improved feed conversion. Unfortunately, there is no satisfactory method at present for providing antibiotics continuously prior to the time the pig eats substantial quantities of creep or starter feeds. Much of the stunting and mortality that occurs during the first week of life could be overcome if the protection offered by antibiotics could be provided then (4).

Fig. 9.1. Note the difference in size, bloom and smoothness of hair coat of antibiotic-fed pig as compared to control pig on right not fed an antibiotic. (Courtesy of T. J. Cunha and University of Florida.)

Although the experimental data are limited and variable, Dr. J. L. Krider (6) estimated that the proper use of antibiotics for sows and in creep diets could save one more pig per litter. In 1971, Dr. Earl L. Butz (7) estimated that the judicious use of antibiotics in commercial swine production resulted in a feed saving of at least $150 million and increased the production of live pigs worth $65 million for a total saving of $215 million dollars.

After the pigs are 8–10 weeks of age, the percentage response to antibiotics declines. The level of antibiotic needed for maximum rate and efficiency of gain is lower than that needed during early growth and reproduction. But antibiotic use should be continued throughout the growing-finishing period so as not to sacrifice some of the benefits already established. When antibiotics are fed to a weight of 100–120 lb and then dropped from the diet, the rate of gain is reduced. Thus, for continued maximum gain, it is necessary to feed antibiotics from weaning to market (8).

Unfortunately, there is no method for determining in advance whether a given group of apparently healthy looking pigs will benefit from the inclusion of an antibiotic in the diet. Therefore, the continuous use of antibiotics is a good recommendation under intensified swine production conditions.

IV. ANTIBIOTICS CONTINUE TO BENEFIT SWINE

The use of effective antibiotics has produced beneficial effects since 1950. Pigs fed antibiotics with present-day diets grow as rapidly as pigs fed similar diets in past years. The control pigs, however, are usually growing at a faster rate than formerly, which is probably due to the elimination of certain harmful organisms in the premises by the continuous feeding of antibiotics. Since it is not known how long the premises would stay free from harmful organisms after antibiotics are discontinued, it is recommended that antibiotics continue to be used. To date there is no evidence to indicate that the continuous use of antibiotics has caused the development of drug-resistant organisms which depress the performance of swine or other farm animals.

There is increasing evidence to indicate that feeding antibiotics to sows can increase birth weight, survival, and weaning weight of the pigs. There is also evidence that feeding antibiotics at a level of 500–1000 mg per day for a 10–14 day period starting 3–5 days before breeding may increase litter size. This may be due to clearing up some minor infection in the reproductive tract of the sow. No harmful effects have yet been reported from using antibiotics for sows. In some cases, they have helped. So, the use of antibiotics for sows under average farm conditions is worth serious consideration. Under average farm conditions where sanitation is not always good, where the diets are not always well balanced, and where other stress factors may be high, there is a good possibility that antibiotics are beneficial in sow diets. An increasing amount of experimental information has been obtained during the past 5–7 years to indicate this is the case. The higher the stress factors (and the subclinical disease level) on the farm, the greater the benefits of antibiotics. So an evaluation of antibiotic need and its value should take these factors into consideration.

V. RECOMMENDATIONS ON ANTIBIOTIC USE

It is difficult to recommend exact antibiotic levels which will suit all farm conditions. This is because all farms have an "environmental disease level." The greater the stress or disease level on the farm, the greater the response to antibiotics. So, the recommendations are only a guide, and deviations from them need to be made to meet the varying stress factors and/or disease level conditions encountered on the farm.

Canadian workers (12) showed that contamination of the quarters in which the pigs are fed increases their response to aureomycin. For example, animals fed aureomycin in quarters where pigs had been kept before, gained at the rate of 1.29 lb per day as compared to 0.96 lb daily for those fed the same diet without the aureomycin. This was a 34% increase in rate of gain due to aureomycin feeding. There was only a 3% increase in rate of gain, however, for pigs fed aureomycin (1.27 lb daily) as compared to the controls (1.23 lb daily) fed in a new barn where pigs previously had not been kept. Thus, in the old barn where the control pigs were held back in rate of gain due to contamination, there was a considerable response to aureomycin feeding. In the new barn, where there was no contamination and the gain of the control pigs was not retarded, there was only a slight improvement from antibiotic feeding.

Table 9.2 gives the recommended antibiotic levels made by the National Research Council (NRC) (8). Tables 9.3, 9.4 and 9.5 list the recommended antibiotic levels from Kentucky, Michigan State and Iowa. There are some similarities and some differences in the recommended antibiotic levels made by the four sources. This is to be expected since scientists in various areas of the United States have different farm conditions to deal with and hence the variation in their recommendations.

Tables 9.2 to 9.5 can be used as a guide in determining proper antibiotic levels. It is essential that the rules and regulations governing the use of antibiotics be followed. The producer should follow feeding instructions as indicated on the feed tag. Withdrawal requirements must be followed carefully to avoid residues in edible pork tissue. By complying with existing rules and regulations, the producer is assured of continued availability of antibiotics. If not, FDA may take action to correct any use or withdrawal violations. The producer as well as the antibiotic and feed manufacturer, and the regulatory agencies have the responsi-

TABLE 9.2
NRC Recommended Antibiotic Levels (8)

Item	Pig weight (lb)	Antibiotic per ton of feed (gm)
Baby pigs	10– 25	40
Growing pigs	25– 66	10–20
Finishing pigs	66–220	10
Therapeutic level for pigs doing poorly	—	50–100[a]
Supplement to be fed free-choice with grain	—	50–100

[a]If pigs are in very poor condition and will not eat, the antibiotic can be given in the drinking water.

bility of assuring that antibiotics are not misused and that wholesome pork is produced.

VI. INTRODUCTION TO OTHER ANTIMICROBIAL COMPOUNDS

There are many other antimicrobial compounds that inhibit specific harmful microorganisms in swine and improve growth and feed efficiency. They include

TABLE 9.3
Kentucky Recommended Antibiotic Levels (11)

Stage of production	Relative level	Suggested range (gm/ton)[a]
Prebreeding and breeding	Moderate to higher levels	100–200
Gestation	None unless disease and environmental stress exists	0– 20
Farrowing	High levels	100–200
Lactation	Moderate levels	0– 40
Prestarter (early weaned pigs)	High levels	100–250
Starter (12–25 lb in weight)	High levels	100–250
Grower	Moderate levels	50–100
Finisher	Low levels	20– 50

[a]Gram/ton needed varies with type of antibiotic and the health problem involved.

TABLE 9.4
Michigan State Recommended Antibiotic Levels (9)

Diet	Antibiotic per ton on complete feed (gm)
Creep	40
Grower	10–20
Finisher	10
Therapeutic	100–200
Supplement	50–100

the arsenicals, nitrofurans, sulfonamides, and copper sulfate. These bactericidal agents have a specialized role in swine feeding and should be used according to recommendations and experimental data on their value. Withdrawal periods recommended by the manufacturers should be followed. These have previously been approved by FDA.

VII. ARSENICALS

Several arsenicals such as 3-nitro-4-hydroxyphenylarsonic acid (3-nitro) and arsanilic acid have been reported to have growth-promoting properties. Of main concern regarding the use of arsenicals is that they are toxic substances with a fairly narrow range between toxic levels and those needed for optimum growth. Thus, the level of feeding must be carefully controlled. Arsanilic acid and

TABLE 9.5
Iowa State Antibiotic Recommendations (10)

	Weight of pig (lb)	Antibiotics[a] per ton of diet (gm)
Sows, gilts, and boars		
Pregestation, breeding and gestation	—	0–300
Young pigs		
Milk replacer (dry feed)	0 to 12	100–300
Starter	Creep to 40	100–300
Growing-finishing hogs		
Grower	40–120	0–100
Finisher	120–240	0–100

[a]The feed additive may be antibiotics, arsenicals or other chemotherapeutics or combinations.

sodium arsanilate have been used extensively (13,14) for the treatment of swine dysentery (bloody scours).

Only 3-nitro and arsanilic acid are used in swine feeds. They have their greatest impact when a disease situation prevails in the swine herd. Arsanilic acid is used at a level of 90 gm per ton of complete feed and 3-nitro at a level of 22 gm per ton of feed. Both arsenicals should never be used in a single diet at the levels just indicated.

VIII. NITROFURANS

Nitrofurazone and furazolidone are effective against enteric infections (15). Another study showed that furazolidone is effective in increasing rate of gain and feed efficiency (16). Other scientists have shown that nitrofurans benefit enteric infections, rate of gain and feed efficiency.

IX. SULFONAMIDES

Certain sulfonamides are used in swine diets. Almost all of them are used in combination with antibiotics. Some of the sulfonamide-antibiotic combinations are excellent and fill an important need in swine production.

X. COPPER AS AN ANTIMICROBIAL

There is considerable interest in the use of copper at a high level in the diet to give a bacteriostatic effect and therefore increase rate of gain and feed efficiency. In a few cases, however, the use of 250 ppm of copper in the diet has resulted in toxic effects. This may be because a high level of copper increases the requirements for iron and zinc. For example, Irish workers showed that toxic effects from 250 ppm of copper in the diet were counteracted by adding 130 ppm of zinc to the diet. Workers in Scotland showed that high levels of copper increased the iron needs to 150 ppm in the diet of the weanling pig and to 270 ppm with the baby pig.

The kind or level of protein in the diet also affects copper toxicity. For example, pigs fed fish meal are more subject to copper toxic symptoms than those fed soybean meal or dried skim milk. The possibility of copper toxicity also decreases as the protein level increases.

The exact mode of action of copper has not been proved although it seems to act on microorganisms in the intestinal tract. In some cases, its growth response is similar to that obtained with an antibiotic, whereas in others it is not. The information obtained in the United States show that in about 75% of the trials when copper is combined with an antibiotic an additive effect on growth and feed efficiency is obtained. While copper may be beneficial in certain situations, it does not eliminate the use of effective antibiotics.

It has been suggested that a high dietary level of copper, when excreted over a long period of time, may interfere with the bacterial action in the lagoon (9). This should be investigated further since many swine producers use lagoons for manure disposal.

A finely ground powdery form of copper called "windswept copper" should be used since it is excellent for mixing and assures good distribution throughout the feed. Many cases of copper toxicity occur because other forms of copper are used and are not mixed well in the feed.

Since 125–150 ppm of copper in the diet gives almost the same response as 250 ppm of copper, this lower level is recommended for those who wish to use a high level of copper. No toxic effects have been reported with this level of copper if a thorough job of mixing it in the feed is followed. If a level of 250 ppm of copper is fed growing-finishing pigs, one should have 130 ppm of zinc and 150 ppm of iron in the diet to protect against possible copper toxicity. In the baby pig, a level of 270 ppm of iron is needed. An adequate level of protein should also be used and an excess level of calcium should be avoided since it increases zinc needs.

If used properly, copper is another effective antimicrobial compound to be used along with antibiotics, arsenicals, nitrofurans and sulfonamides.

REFERENCES

1. Cunha, T. J., J. E. Burnside, D. M. Buschman, R. S. Glasscock, A. M. Pearson, and A. L. Shealy, *Arch. Biochem.* **23,** 324 (1949).
2. Jukes, T. H., E. L. R. Stokstad, R. R. Taylor, T. J. Cunha, H. M. Edwards, and G. B. Meadows, *Arch. Biochem.* **26,** 324 (1950).
3. Kiser, J., *Am. Soc. Anim. Sci. Meet. 1975* pp. 1–40 (1975).
4. V. W. Hays, *Feed Manage.* **26,** 27 (1975).
5. Gerrits, R. J., *J. Anim. Sci.* **40,** 1255 (1975).
6. Krider, J. L., *Proc. N.C. State Univ. Swine Conf.*, pp. 1–5 (1971).
7. Butz, E. L., *Hog Farm Manage.*, pp. 29–31 (1971).
8. Cunha, T. J., D. E. Becker, J. P. Bowland, J. H. Conrad, V. W. Hays, and W. G. Pond, "Nutrient Requirements of Swine," Nat. Res. Counc., Washington, D.C., 1968.
9. Miller, E. D., E. R. Miller, and D. E. Ullrey, *Mich. Ext. Bull.* **537** (1975).
10. Halden, P., V. C. Speer, E. J. Stevermer, and D. R. Zimmerman, *Iowa, Agric. Exp. Stn.* **PM-489** (1974).
11. Whitaker, M. D., *Ky., Agric. Ext. Serv.* Fact Sheet (1971).
12. Bowland, J. P., *J. Anim. Sci.* **14,** 1243 (1955).

13. Davis, J. W., *J. Am. Vet. Med. Assoc.* **138,** 471 (1961).
14. Smith, I. E., E. M. Kiggins, H. S. Perdue, J. C. Holper, and D. V. Frost, *J. Anim. Sci.* **20,** 768 (1961).
15. Guthrie, J. E., *Proc. Natl. Symp. Nitrofurans Agric., 2nd,* p. 46 (1958).
16. Briggs, J. E., and J. E. Guthrie, *Proc. Natl. Symp. Nitrofurans Agric., 2nd,* p. 58 (1958).

10

Enzymes for Swine

I. INTRODUCTION

Enzymes are needed to break down protein, carbohydrates and fats so the body can utilize them. Enzymes are also involved in activating and hastening chemical reactions in the body which are essential to life. There are hundreds of different enzymes and each one is capable of catalyzing a specific chemical reaction. The newborn pig has the enzymes necessary for the utilization of milk, but it lacks certain enzymes during the first few weeks of life to utilize feeds other than milk. These limitations will be discussed in this chapter.

II. ENZYME LIMITATIONS SHORTLY AFTER BIRTH

Early Iowa Station studies (1–3) showed that baby pigs do not have fully developed enzyme systems for breaking down proteins and carbohydrates. They found that soybean protein diets did not work satisfactorily for pigs from 1 to 5 weeks of age. However, the diets were satisfactory for pigs from 5 to 8 weeks of age. At this point, they surmised that the young pig up to 5 weeks of age might be deficient in proteolytic enzymes for digesting soybean protein. They found that early weaned pigs fed diets high in milk protein made quite satisfactory growth before and after 5 weeks of age. This showed that apparently the baby pig's digestive enzyme system is much more capable of breaking down milk protein than soybean protein during the first 5 weeks of life. Supplementation of the soybean diets with an enzyme (pancreatin or pepsin) improved growth rate (but not feed efficiency), approaching that obtained on a milk protein diet when lactose (milk sugar) was the principal carbohydrate. They confirmed the insufficiency of proteolytic enzymes in the baby pig by assays of the secretory glands of litter-mate pigs at birth and at consecutive weekly intervals until 7 weeks of age. The newborn pig had very little or no pepsin (enzyme) activity as shown by assays of the stomach mucosa. The pepsin activity, however, increased gradually up to 7 weeks of age. Assays of the pancreas indicated adequate trypsin activity from birth, but pancreatic amylase activity was very low in the newborn; however, it rose rapidly with age and reached a maximum at about 4 weeks of age.

Canadian workers (4,5) confirmed the insufficiency of certain enzymes in the baby pig. They (5) found that the amylolytic (enzymes used in carbohydrate breakdown) activity of the pancreas, per unit weight of gland, was negligible in the newborn pig. The amylolytic activity increased rather markedly to 37 days of age. Their data suggested that amylase production undergoes a significant increase at 21 days or about the fourth week of life. They also found that the lipolytic (enzyme used in fat breakdown) activity of the digestive system of the young pig is of a high order at birth and remains high with advancing age.

The Canadian workers (4) found that lactase (enzyme that breaks down lactose) activity is high, and sucrase and maltase (enzymes that break down sucrose and maltose, respectively) activity is low during the first few days of life. Lactase activity reached a peak at about 3 weeks of age. Then there was a rapid decline to minimum levels at 4–5 weeks of age. Sucrase and maltase activity, however, rose steadily from negligible levels at birth to significant levels in 1–2 weeks. These findings would explain why very young pigs are unable to utilize sucrose but can utilize lactose (see Chapter 6, Section II).

These studies indicate that important changes occur in the digestive enzymes as early growth of the pig proceeds. These enzyme changes affect the suitability of various carbohydrates and proteins for the baby pig shortly after birth.

III. CARBOHYDRATE UTILIZATION

The newborn pig can absorb glucose. It cannot digest sucrose or more complex sugars. Lactose (milk sugar) is efficiently utilized by the newborn pig. But lactose utilization decreases as the pig becomes older because the production of the enzyme lactase, which breaks down lactose, decreases.

Studies by Manners and Stevens (6) in 1972 showed no sucrase activity in the intestines of very young pigs, but sucrase production rose steadily from birth to about 30 weeks of age. The sucrase level in 30-week-old pigs was about the same as in adult pigs, but the values for the adult pigs were quite variable. This lack of sucrase in the intestinal mucosa of the newborn pig and the slow rate of increased production of this enzyme account for the danger of death among baby pigs fed a diet rich in sucrose (see Chapter 6, Section II). In other studies (7) they found that sucrose, when present at a level of 15% in the diet, was not always well utilized by 3-week-old pigs. But, 5% sucrose in the diet was utilized fairly efficiently by 2-week-old pigs. Their studies suggest that between 3 and 4 weeks of age there may be a rapid increase in sucrase production which might make a

4-week-old pig much less susceptible than a 3-week-old pig to any detrimental effects of a high sucrose diet.

Manners and Stevens (6) found that lactase levels in the pig fell rapidly from birth to 2 weeks of age and then continued to fall at a slower rate until 8 weeks of age. After 8 weeks of age there was little change in the level of this enzyme in the pig. They found, however, that certain pigs decreased in lactase production to the point of indicating a likely intolerance to high levels of lactose (see Chapter 6, Section II). They suggested that the level of intestinal lactase might be used in selection programs in areas where high levels of whey products (a good source of lactose) are used in swine feeding.

IV. FAT UTILIZATION

The pig is able to hydrolyze and utilize high levels of fat at birth. This is to be expected since milk contains about one-third fat on a dry matter basis. The lypolytic enzyme (used in fat breakdown) activity of the digestive system of the young is high at birth and remains high with advancing age (5).

V. PROTEIN UTILIZATION

The hydrolysis and utilization of milk protein by the newborn pig is very high. But the baby pig does not digest soybean protein as efficiently as milk protein (8,9). This may be due to a lower production of proteinases by the very young pig although this is not definitely established. The secretions of pepsin by the stomach and proteinases by the intestinal mucosa and the pancreas are low at birth and gradually increase to about 8 weeks of age. The difference in the utilization of milk proteins and plant proteins disappears by about 5 weeks of age. Cornell studies (10) indicate that enzymes secreted by the stomach or intestines may be sensitive to dietary protein and this sensitivity may be related to the inferior performance of baby pigs fed soybean protein. Pancreatic chymotrypsinogen, trypsin, amylase and lipase do not appear to be involved. Some other factor or factors may be responsible for decreased soybean protein utilization.

VI. ENZYME SUPPLEMENTATION OF DIETS

The pig is lacking in certain carbohydrate and protein splitting enzymes during the first few weeks of life. Present information indicates that by 5 weeks of age the pig can efficiently utilize feeds other than milk. Pigs have been weaned successfully at 3 weeks of age without enzyme supplementation. To date, information is not available to indicate a need for enzyme supplementation for pigs weaned at 3 weeks. Pigs weaned earlier than 3 weeks of age may be benefited by enzyme supplementation especially when feeds other than milk are used.

More experimental work is needed on enzymes and their possible value in supplementing diets of early weaned pigs. This information will be needed as swine confinement and intensification increases and as earlier weaning is practiced.

More information is also needed on the possible value of enzyme supplementation with growing-finishing pigs. Some trials at Purdue (11), South Dakota (12) and North Dakota (13) have indicated a slight response from enzyme additions with growing-finishing swine. The North Dakota study (13) showed that enzyme supplementation benefited West Coast barley but not North Dakota barley. This is similar to poultry data indicating enzymes may be beneficial with West Coast barley. All of these studies need more verification before enzyme supplementation can be recommended for the growing-finishing pig.

REFERENCES

1. Catron, D. V., personal communications (1955, 1957).
2. Catron, D. V., *Proc., Distill. Feed Conf.* **11,** 27 (1956).
3. Lewis, C. J., D. V. Catron, V. C. Speer, and G. C. Ashton, *J. Anim. Sci.* **14,** 1214 (1955).
4. Bailey, C. B., W. D. Kitts, and A. J. Wood, *Can. J. Agric. Sci.* **36,** 51 (1956).
5. Kitts, W. D., C. B. Bailey, and A. J. Wood, *Can. J. Agric. Sci.* **36,** 45 (1956).
6. Manners, M. J., and J. A. Stevens, *Br. J. Nutr.* **28,** 113 (1972).
7. Kidder, D. E., M. J. Manners, M. R. McCrea, and A. D. Osborne, *Br. J. Nutr.* **22,** 501 (1968).
8. Maner, J. H., W. G. Pond, and J. K. Loosli, *J. Anim. Sci.* **20,** 614 (1961).
9. Hays, V. W., V. C. Speer, P. A. Hartman, and D. V. Catron, *J. Nutr.* **69,** 179 (1959).
10. Pond, W. G., J. T. Snook, D. McNeill, W. I. Snyder, and B. R. Stillings, *J. Anim. Sci.* **33,** 1270 (1971).
11. Conrad, J. H., and W. M. Beeson, *Purdue Agric. Exp. Stn., Mimeo* No. AS-262 (1959).
12. Wahlstrom, R. C., *S.D. Agric. Exp. Stn., Swine Field Day Rep.* pp. 1–6 (1959).
13. Dinusson, W. E., D. Erickson, and D. W. Bolin, *N. D. Agric. Exp. Stn., Mimeo* No AH6-59 (1959).

11

Relative Value of Feeds

I. INTRODUCTION

In compounding diets, one must select feeds that will not only result in a well-balanced diet but also an economical one. Unless the diet is economical, the farmer cannot make a profit. This means one must be acquainted with the relative value of feeds, as well as any limitations they may have in swine diets.

In selecting energy feeds or protein supplements, one must also consider the vitamins and minerals they contain or are lacking in. In selecting feeds, one should consider the purpose for which the feed is to be used; that is, for growth, finishing, gestation, lactation, and whether the feed is to be used in confinement or on pasture. Moreover, the diet must be palatable—the pigs must like it. All these factors must be considered in selecting feeds to be used in various swine diets.

II. DEFICIENCIES IN GRAINS OR ENERGY FEEDS

Grains and energy feeds contain some protein, vitamins, and minerals. However, they are too low in these for good results in swine feeding without supplements. These limitations are discussed in detail in the chapters on proteins, minerals, and vitamins.

Proteins in the cereal grains are poor in quality and lacking in lysine. (See Table 11.6 for the amino acid content of various feeds.) Corn has a lower amount of lysine than the other cereal grains. Corn also has the lowest amount of tryptophan and is decidedly deficient in this amino acid.

Thus, protein supplements used to balance cereal grains must supply not only enough protein but also protein having a good balance of the essential amino acids. Since corn contains the lowest amount of protein of the cereal grains and grain sorghums, more protein supplement is needed to balance a corn diet than one containing wheat, barley, oats, or grain sorghums. This must be considered in deciding which grains to use at given prices.

All energy feeds are lacking in calcium, salt, and vitamin D. Grains are fair sources of phosphorus, but only yellow corn, and its by-products, contain appreciable amounts of carotene (provitamin A). The grains are all good sources of thiamin but are inadequate in riboflavin content. The niacin in cereal grains and in their by-products is in a bound form and thus is largely unavailable to the pig.

If the diet contains a good supply of tryptophan, the pig can synthesize niacin from it and thus reduce the amount of niacin needed in the diet. Since corn is deficient in tryptophan and the other grains are borderline in tryptophan content, it is necessary that diets contain adequate amounts of niacin for the pig. Oats, wheat, rye, and the grain sorghums contain almost twice as much pantothenic acid as corn and barley. Many corn and barley diets are borderline or deficient in pantothenic acid and choline. All grains are lacking in vitamin B_{12}. Since grains form a large part of the diet, many swine diets are borderline or deficient and need supplementation with riboflavin, niacin, pantothenic acid, vitamin B_{12} and choline.

III. RELATIVE VALUE OF GRAIN FEEDS

The cereal grains and their by-products furnish the bulk of the finishing feeds for hogs. A discussion of their value in swine diets follows. Table 11.1 gives the relative feeding value of the grain feeds for swine.

TABLE 11.1
Relative Feeding Value of Grains for Swine

Feed	Pounds/bushel	Bushels/ton	Relative value as compared to corn with corn given a value of 100
			Average value
Corn	56	35.7	100
Wheat	60	33.3	100–105
Barley	48	41.7	90
Triticale	—	—	95
Milo (grain sorghums)	55	36.3	95–97
Oats	32	62.5	80–90[a]
Rye	56	35.7	85–90[b]
Millet	—	—	93

[a]At levels of 10–20% of the diet for young pigs and brood sows, oats are almost as valuable as corn.

[b]Rye, which sometimes contains ergot (a fungus) should not be fed to the swine breding herd or in creep feed to suckling pigs.

A. Corn (Maize)

Corn is the grain feed around which the major portion of swine diets are balanced (Table 11.2). Corn is called maize in most parts of the world. The feeding value of corn is used as the standard with which other cereal grains are compared. White and yellow corn have the same feeding value, provided the diet contains enough carotene or vitamin A, either from pasture or from some other source. Waxy corn has about the same feeding value as yellow dent corn. Corn may be fed shelled, ground, mixed, free-choice or as ear corn. It may be fed dry or as high moisture corn. It seldom pays to grind or soak corn when it is fed free-choice. If the kernels are too hard and dry, however, they should be coarsely ground, especially for pigs under 50 lb in weight. Corn should also be ground if it is to be mixed with other grains and a supplement. The dry matter in soft corn has the same feeding value as the dry matter in sound corn, but the gains on soft corn are usually not so rapid. When beginning to feed soft corn (corn frosted before maturity) or new corn, the change should be made gradually to minimize digestive disturbances. High moisture corn has about the same feeding value, on a dry matter basis, as dry corn. If high moisture corn is fed in a complete diet, the diets should be prepared frequently (every day or two) to prevent spoilage or moldiness. The diets should also be prevented from bridging in the feeders.

B. Opaque-2 Corn (Maize)

The protein and amino acid pattern in corn has been modified genetically. Some corn now being bred is much higher in lysine and tryptophan content. The high lysine corn (called opaque-2 corn) is also higher in total protein than normal corn. Different varieties of corn containing the opaque-2 gene are commercially available, but the yields of these varieties are frequently less than that of regular varieties. They also vary considerably in their lysine level. So, a lysine analysis may be needed to determine its level in the corn. When using mechanical drying, high lysine corn dries more rapidly than normal corn, so its moisture level should be carefully watched, if it is being dried for storage. Some scientists recommend that high lysine corn be ground coarser than normal corn. Some prefer to roll high lysine corn.

Pigs fed high lysine corn need less total protein in the diet. Researchers at the University of Nebraska (1) recommend the following protein levels in the diet (if high lysine corn contains 0.38% lysine or higher on an 86% dry matter basis or 0.44% lysine or higher on a 100% dry matter basis):

1. Reduce growing-finishing diets 2% in crude protein below the protein level being fed in diets containing normal corn.
2. Feed a pregestation and gestation diet containing 12% protein.
3. Feed a lactation diet containing 14% protein.

TABLE 11.2
Suggested Ranges of Commonly Used Ingredients That Have Produced Satisfactory Results at Iowa Station (13)

	Percent of complete diet					Relative value compared to[a]		
Ingredient	Gestation	Lactation	Starter	Growing-finishing	Percent of supplement	Corn	Soybean meal (44%)	Remarks
Alfalfa meal (dehydrated)	0–50	0–10	0	0–5	0–20	75–85	45–50	Low energy, good source of carotene and B vitamins, unpalatable to baby pigs
Alfalfa meal (sun-cured)	0–50	0–10	0	0–5	0–20	60–70	30–40	Same as alfalfa meal (dehydrated)
Animal fat, stabilized	0	0	0–2.5	0	0	210–220	—	High energy, reduces dust
Barley (48 lb/bu)[b]	0–80	0–80	0–25	0–85	—	90–95	—	Corn substitute but lower energy
Beet pulp	0–10	0–10	0	0	0–20	70–80	—	Bulky, high fiber, laxative
Blood meal	0–3	0–3	0	0–3	0–10	—	90–100	Low digestibility, unpalatable, low isoleucine
Corn, yellow (56 lb/bu)[b]	0–80	0–80	0–60	0–85	—	—	—	High energy, palatable, low lysine
Corn and cob meal	0–70	0	0	0	—	80–90	—	Bulk, low energy
Cottonseed meal (solv)	0–5	0–5	0	0–5	0–20	—	85–90	Gossypol toxicity, low lysine
Distillers dried grains with solubles	0–5	0–5	0–5	0–5	0–20	120–125	65–70	B vitamin source, low lysine
Fish meal	0–5	0–5	0–5	0–5	0–20	—	150–180	Excellent amino acid balance
Fish solubles (50% solids)	0–3	0–3	0–3	0–3	0–5	—	60–70	Excellent amino acid balance
Linseed meal	0–5	0–5	0–5	0–5	0–20	—	65–70	Low lysine
Meat and bone meal	0–10	0–5	0–5	0–5	0–30	—	110–125	Low tryptophan and methionine, good source phosphorus

Milo or grain sorghum[b]	0–80	0–80	0–60	0–85	—	95–100	—	Corn substitute, low lysine
Molasses (cane 11.7 lb/gal)	0–5	0–5	0–5	0–5	0–5	55–65	—	Used for energy, better physical appearance and harder pellets
Oats (32 lb/bu)	0–40	0–15	0	0–20	—	80–90	—	Partial substitute for grains, low energy
Oat groats, rolled	—	—	0–20	—	—	110–115	—	Palatable, low lysine
Skim milk, dried	0	0	0–10	0	—	—	95–100	Excellent amino acid balance, palatable
Soybean meal (dehulled)[c]	0–22	0–18	0–22	0–18	0–85	—	110–112	Good amino acid balance when combined with corn, palatable
Soybean meal[c]	0–25	0–20	0–25	0–20	0–85	—	—	Same as soybean meal (dehulled)
Soybeans, whole cooked[c]	0–30	0–25	0–30	0–25	0–85	—	100–115	Similar to soybean meal but higher energy, lower protein
Sugar	0	0	0–5	0	0	80–85	—	High palatability for baby pig
Tankage	0–10	0–5	0	0–5	0–30	—	100–120	Low digestibility, unpalatable
Wheat (60 lb/bu)[b]	0–80	0–80	0–60	0–85	—	100–105	—	Corn substitute, low lysine
Wheat bran	0–30	0–10	0	0	0–20	60–65	35–40	Bulky, high fiber, laxative
Wheat midds	0–30	0–10	0–5	0–10	0–20	90–95	53–57	Partial substitute for grain, low energy
Whey	0–5	0–5	0–20	0–5	0–20	100–110	55–60	Lactose is carbohydrate "of choice" for baby pigs
Yeast, brewers dried	0–3	0–3	0–3	0–3	0–5		100–105	Source of B vitamins

[a]Relative value considers protein content and quality, metabolizable energy, calcium and phosphorus. The cost of an ingredient can be evaluated by comparing its cost with the cost of corn or 44% soybean meal times the particular coefficient for the relative value of the ingredient compared to corn or soybean meal.

[b]Corn, barley, milo or wheat should be the basic energy source with other substitutions made within the ranges suggested.

[c]Soybean meal or whole cooked soybeans should be the basic protein source with other substitutions made within the ranges suggested.

C. Oats

Oats are an excellent feed for young, growing pigs and sows (Tables 11.2, 11.3 and 11.4). They are too high in fiber (10–15%) and too bulky, however, to form a major portion of the diet for young, growing pigs. The fibrous hull makes up about one-third of their weight. If the hulls are removed and the oat groats rolled, the product is an excellent feed, especially in starter diets for baby pigs. For brood sows, oats can be used to replace up to one-half the grain without reducing the efficiency of the diet very much.

Oats provide bulk and fiber which may be helpful for sows especially if they are limited in exercise. Higher levels of oats can be fed during gestation than during lactation when more energy is needed in the diet for milk production. Oats vary in having 80–90% the feeding value of corn, depending on the level fed, the quality of the oats (which varies considerably), and the stage of the life cycle of the pig that is being fed. Oats will have the highest replacement value when fed at lower levels. Good quality, heavy oats contain a smaller percentage of hulls and

TABLE 11.3
Substitution Values for Ingredients Compared to Corn and Soybean Meal—Purdue Station[a] (14)

Ingredient	Maximum level in growing-finishing ration (% diet)	Relative value compared to: Corn (%)	Relative value compared to: Soybean meal (44%) (%)
Alfalfa meal, dehydrated	5	75	45–50
Barley	85	90	—
Blood meal	3	—	90
Cottonseed meal (solv.)	5	—	85–90
Fish meal	5	—	150
Hominy feed	50	95	—
Linseed meal	5	—	65–70
Meat and bone meal	5	—	110–125
Milo (grain sorghum)	85	95	—
Oats	20	80–90	—
Oats, rolled	60	110–115	—
Soybean meal (49%)	—	—	110–112
Soybeans, whole cooked	—	—	110–115
Tankage	5	—	100–120
Wheat	85	100–105	—
Wheat midds	10	90	—
Whey, dried	5	100	55

[a]In Indiana, corn and soybean meal are the most common energy and protein sources, respectively, in swine diets. Other substitutions may be made up to the maximum level indicated. However, total animal protein should not exceed 5% of the diet.

thus can be used at high levels in the diet with better results than light oats. Oats have a lower feeding value on pasture, because the pasture forage is also bulky and fibrous.

Oats should be ground for swine, because it will usually increase their feeding value considerably. A fine or medium grind is better than a coarse grind. Grinding oats usually increases their value by 27–30% for growing-finishing pigs (2). Many swine producers prefer to use rolled or crimped oats. If fed as the major part of the diet, oats tend to produce fat which is a little softer than that produced with corn, but it is not great enough to affect carcass grade. For best results, it is best to mix oats with other grains for swine.

It takes from 155 to 165 lb of whole oats to produce 100 lb of hulled oats. This makes hulled oats rather expensive. This limits their use to starter and creep feeds where a higher price can be justified because the palatability and nutritional value is increased. One hundred pounds of oats with the hulls removed are equal to about 140 lb of corn in feeding value.

Ground oats may be fed to the extent of one third of the grain in the diet without appreciably decreasing the rate of gain of growing-finishing pigs, but feed per pound of grain will increase because of the lower energy value in the diet. The pelleting of swine diets containing oats will improve feed conversion about 10%, but it will not consistently increase rate of gain (3).

Oats are of value in protecting growing pigs against ulcers (Fig. 11.1) which may be encountered with flaked corn (4). The protective effect of the oats is due to an alcohol-soluble fraction in the hulls (5).

D. Wheat

Wheat can have up to 5% more feeding value than corn. Wheat is about equivalent to corn as a source of energy. It is slightly superior in protein quantity and quality. It is generally too high in price, however, to be fed to hogs, since it is produced primarily for human consumption. Low quality wheat not suitable for milling, as well as damaged wheat, is used for swine feeding. Wheat is more palatable than corn for pigs. It has been shown that pigs self-fed wheat and corn separately, free-choice, eat considerably more of the wheat than corn.

Wheat can be fed whole in self-feeders, but many producers prefer to coarsely grind it, especially varieties of wheat which produce the smaller and harder grains. Wheat should always be ground when it is hand-fed, since the pigs are so eager to get their share they fail to chew it properly. Wheat should not be ground too finely, since it will form a pasty mass in the mouth and become less palatable. Studies in England (6) showed that rolled wheat, fed at a level of 35% in the diet, gave better results than ground wheat.

Wheat generally gives excellent results when fed as the only grain to swine. It can replace part or all of the corn without affecting performance. Wheat is

TABLE 11.4
Feed Ingredients Recommended by Michigan Station (15)

	Feed specification code												
	1	2	3	4	5	6	7	8	9	10	11	12	13
										Maximum % of feed in diet			
	Digestible energy (kcal/lb)	Crude protein (%)	Crude fiber (%)	Calcium (%)	Phosphorus (%)	Salt (%)	Lysine (%)	Methionine + cystine (%)	Tryptophan (%)	25–50 (%)	50–75 (%)	75–125 (%)	125–mkt. (%)
Energy Feeds													
Barley	1500	11.0	7.0	0.05	0.30	—	0.35	0.40	0.15	40	60	60	80
Corn	1630	8.8	2.5	0.02	0.26	—	0.25	0.35	0.07	NL[a]	NL	NL	NL
High lysine corn (1)	1630	11.0	2.5	0.02	0.26	—	0.40	0.35	0.12	NL	NL	NL	NL
Sorghum grain	1550	10.0	2.7	0.05	0.27	—	0.20	0.25	0.09	NL	NL	NL	NL
Oats	1320	12.0	12.0	0.08	0.32	—	0.35	0.35	0.15	20	30	30	40
Spelt (Emmer)	1320	12.0	10.0	0.08	0.33	—	0.27	0.35	0.09	20	30	30	40
Rye	1560	11.3	2.0	0.07	0.34	—	0.45	0.32	0.12	20	20	20	25
Molasses, beet	1120	6.0	0.0	0.12	0.03	—	—	—	—	5	5	5	5
Molasses, cane	1150	2.9	0.0	0.81	0.08	—	—	—	—	5	5	5	5
Wheat	1630	11.0	2.3	0.05	0.30	—	0.30	0.35	0.12	NL	NL	NL	NL
Wheat middlings, standard	1450	15.5	7.0	0.06	0.80	—	0.60	0.40	0.20	NL	NL	NL	NL
High lysine corn (2)	1630	9.8	2.5	0.02	0.26	—	0.35	0.35	0.10	NL	NL	NL	NL
Triticale	1550	11.0	2.1	0.06	0.32	—	0.45	0.35	0.11	50	50	50	50
Tallow	3600	—	—	—	—	—	—	—	—	5	4	4	3

Protein Feeds													
Commercial protein sup.										NL	NL	NL	NL
Alfalfa meal, dehydra. 17%	650	17	25.0	1.3	0.24	—	0.72	0.40	0.40	5	5	5	5
Blood meal	1220	80	1.0	0.28	0.22	—	6.0	2.30	1.00	3	3	3	3
Corn gluten meal, 41%	1600	42	4.0	0.16	0.40	—	0.8	1.6	0.20	NL	NL	NL	NL
Corn gluten meal, 60%	1600	60	2.5	0.16	0.50	—	1.0	3.0	0.30	NL	NL	NL	NL
Corn gluten feed, 22%	1100	22	9.0	0.30	0.70	—	0.6	0.9	0.12	NL	NL	NL	NL
Fish meal, menhaden	1400	60	1.0	4.00	3.00	—	4.8	2.2	0.60	NL	NL	NL	NL
Meat and bone meal	1000	50	3.8	8.00	4.00	—	2.6	0.9	0.25	10	6	6	3
Soybean meal, 44%	1500	45	7.0	0.25	0.60	—	2.8	1.3	0.63	NL	NL	NL	NL
Soybean meal, 49%	1550	49	3.0	0.25	0.65	—	3.1	1.4	0.64	NL	NL	NL	NL
Tankage, meat meal	1220	60	2.0	6.00	3.00	—	4.0	1.1	0.60	10	6	6	3
Whey, dried	1550	14	0.0	0.90	0.70	—	0.7	0.4	0.15	10	10	10	10
Whey product, dried	1320	17	0.0	1.50	1.00	—	1.2	1.14	0.36	20	20	20	20
Lysine										NL	NL	NL	NL
Methionine										NL	NL	NL	NL
Tryptophan										NL	NL	NL	NL
Full fat soybeans	1900	37.5	4.5	0.20	0.58	—	2.23	0.90	0.52	NL	NL	NL	NL

[a]NL means there is no upper limit placed on the percentage of the feed that can occur in the least-cost ration. Of course, no feed will be in the diet at 100%; corn, for example, supplies adequate energy to meet the requirements but insufficient protein, calcium, phosphorus and salt.

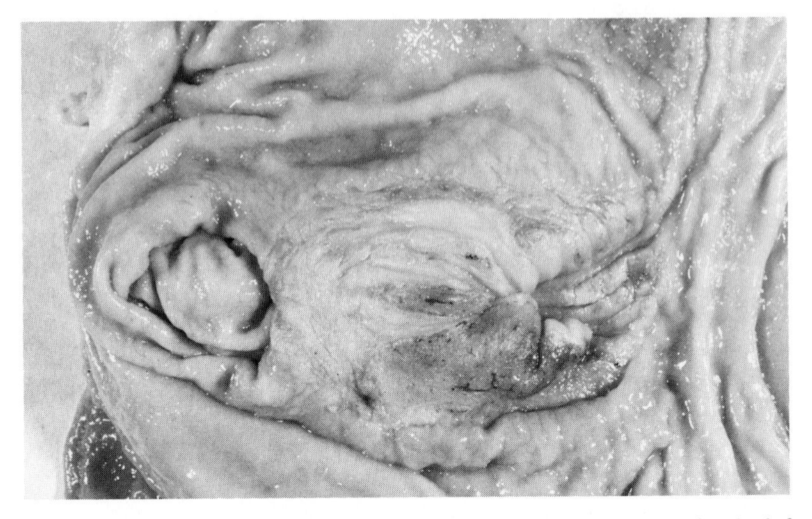

Fig. 11.1. Gastric ulcers (esophageal region) perforating blood vessels and causing death from hemorrhage. (Courtesy of R. H. Grummer, W. G. Hoekstra and T. Kowalczyk, Wisconsin Agric. Exp. Station.)

usually fed in combination with other grains (Table 11.2). Pigs self-fed wheat and a protein supplement free-choice will usually not eat any more of the supplement than they need because the wheat is so palatable.

E. Barley

This is an excellent feed for swine, producing firm pork of high quality (Table 11.2, 11.3 and 11.4). It is high in bulk (6% fiber) and is slightly lower in energy value than corn. It contains more protein than corn, but the amino acid balance is not too good. The feeding value of barley is variable, owing to its fluctuation in weight per bushel. It will average about 90% the feeding value of corn, but certain tests show values somewhat lower than this. Barley is almost equal to corn when used to replace about a third of the grain in the diet. It can, however, replace all the corn in the diet.

Barley should be ground or rolled for swine. It is best to grind barley to a medium degree of fineness. Grinding barley increases its feeding value about 16% (2). Hull-less barley has about the same feeding value as wheat. Oregon studies (7) as well those at Purdue (8) and North Dakota (9) showed that pelleting of barley diets improved weight gain, feed intake, and efficiency of feed utilization. The Purdue and North Dakota studies showed 14% faster gains with 15–17% less feed due to pelleting of barley diets. Barley that is badly infested with scab, a fungus, is unpalatable and produces harmful effects if it comprises more than 5% of the diet. To be on the safe side, it should not be fed to pregnant sows or to very young pigs.

Pigs self-fed barley and a protein supplement, free-choice, will usually consume more of the protein supplement than is needed to balance the diet. This is because barley is less palatable than corn.

F. Rye

Rye is not as palatable as the other grains and for best results should be fed in combination with the more palatable grains (Tables 11.2 and 11.4). It should usually be fed at a level no higher than 10–20% of the grain mixture. Rye gives its best results when fed to finishing pigs on pasture in amounts no greater than 20% of the diet. Tests at Illinois (10) showed that using 30% in the diet reduced gain 8% and increased total feed needs 8%. Information on the value of rye varies considerably. When high quality rye is used and is fed properly, it has about 85–90% the feeding value of corn.

Rye kernels are small and hard and thus should be ground. Rye is frequently contaminated with a fungus called ergot. The ergot makes the rye even more unpalatable. The fungus reduces feed consumption and growth rate when fed to growing-finishing swine. It may also cause abortion and lactation failure and thus should not be fed to pregnant or lactating sows. Ergot should also not be fed in the diets of very young pigs.

G. Grain Sorghums

The grain sorghums, of which there are many, have a feeding value of about 95–97% of corn. Sorghums produce pork equal to that of corn in quality. The sorghum grains are an excellent feed for swine when adequately supplemented and properly fed.

The sorghums should be threshed instead of being fed in the head for swine. The kernel is hard and small and should be coarsely ground. Pelleting of a complete mixed diet with grain sorghum increases rate of gain. Some of the sorghums are not as palatable as others because of the differences in the amount of tannin in the seed and sometimes are slightly less palatable than corn. All the grain sorghums, even the yellow-seeded ones, are deficient in carotene.

Maximum gains and feed efficiency from grain sorghums are obtained when the grain is fed in a complete mixed diet. When a good quality protein supplement is fed with grain sorghums, their feeding value is not much different from corn. But, when a poor quality protein supplement is fed, the grain sorghums have a much lower value than corn. The protein level of grain sorghums varies considerably and ranges from 8 to more than 16% with an average of 11% (11). Lysine is the most limiting amino acid and threonine is the second limiting amino acid of the grain sorghums (11). It is generally recommended that the grain sorghums substitute for corn on a pound for pound basis if their protein level is more than

9%. If the protein level of the grain sorghums is below 9%, the protein level of the diet should be recalculated and additional protein added.

H. Triticale

Triticale is a hybrid obtained by crossing wheat and rye. It has a lower digestible energy value than corn but is higher in crude protein which ranges from 12–15%. It has about 95% the feeding value of corn. Some studies indicate that because of somewhat lower palatability, triticale should not constitute more than 50% of the diet for growing-finishing pigs.

Certain varieties of triticale may become infested with ergot. When this occurs, the grain should not be fed to pregnant sows or very young pigs. Its level in the diet should also be low, but information is lacking on what it should be. The same precautions as discussed with rye (Section III, F) should be followed if the triticale becomes infested with ergot.

I. Rice

Rice kernels are very hard and are enclosed in hard hulls. The rough rice, from which the hull has not been removed, has about the same fiber level as oats. For finishing pigs, the ground rough rice has about 75% the feeding value of corn. For best results, the rough rice should be finely ground and fed at a level of 25–50% of the grain in the diet. Ground rough rice produces good quality, firm pork. Finely ground rough rice is superior to coarsely ground rough rice. Rough rice contains about 8.3% protein and 9.2% fiber.

J. By-Product Feeds from Grains

1. Hominy Feed

Hominy feed is a by-product of the process used for producing corn meal for human use. It consists of a mixture of corn germ, corn bran, and a part of the starchy portion of the kernels. Hominy feed resembles ground corn in composition, and its feeding value approximates 95–100% that of corn for swine. Hominy feed is a satisfactory substitute for corn, but it tends to produce soft pork. The softness of the carcass depends on the level of fat in the hominy feed. Thus, it is recommended that it replace not more than 50% of the corn in finishing diets and preferably be used at levels of 20–25% of the diet. Hominy feed may become rancid if stored too long and at warm temperatures. Thus, it is recommended that hominy feed be used as fresh as possible and be stored in a cool, well ventilated building in order to minimize rancidity problems.

2. Corn Gluten Meal

This feed consists chiefly of the corn gluten which is a by-product of corn starch manufacture. It may include corn solubles and occasionally some corn oil meal. It contains about 42% protein, but it is of poor quality and thus should not be used as the main protein supplement for swine feeding. Corn gluten meal can be used as part (one-fourth to one-half) of the protein supplement and in combination with supplements which supply the amino acids the corn gluten meal lacks. Better results have been obtained when corn gluten meal is fed on pasture rather than in dry lot.

3. Corn Gluten Feed

This consists of corn gluten meal and corn bran and may or may not contain corn solubles. It has about 22% protein, which is of poor quality. Gluten feed is not usually fed to swine, since it has a better feeding value for cattle. It has about 8% fiber, is bulky, and is not too palatable for swine, although they will readily eat diets containing about 15% of corn gluten feed.

4. Wheat Middlings

This feed is made up mostly of fine particles of bran and germ with a little of the wheat red dog and contains about 16% protein. It is a very good hog feed. It has about 90–95% the feeding value of corn, depending on the level used in the diet. Wheat middlings are usually fed at a level of 5–20% in the diet but can be used at higher levels (Tables 11.2, 11.3 and 11.4).

5. Wheat Flour Middlings

This consists of wheat middlings and wheat red dog. It contains about 16% protein but is higher in digestible nutrients than standard middlings. It has a little more feeding value than corn when fed at a level of 20% of the diet for growing-finishing pigs.

6. Wheat Red Dog

This is a feed which consists chiefly of the aleurone layer, with small quantities of flour and fine bran and germ particles. It contains about 18% protein and is a little higher in feed value than corn when limited to 2.5–15.0% of the diet. It is used in diets for young pigs because of its high digestibility and lower fiber content.

7. Wheat Bran

Too bulky for extensive use for growing-finishing pigs, this feed can be used at very low levels of around 5–10% (Table 11.2). It can be used to good advantage at levels of 10–25% of the diet for brood sows, especially during gestation and before and after farrowing. It is valuable in those diets because of its bulk and laxative effect.

8. Wheat Mill Run

This consists of coarse wheat bran, fine particles of wheat bran, wheat shorts, wheat germ, wheat flour and the offal from the "tail of the mill." It has about the same feeding value and can be used in the same manner as wheat bran.

9. Wheat Shorts

Wheat brown shorts, wheat gray shorts, wheat gray middlings and wheat flour middlings have been consolidated into a single product called wheat shorts. Wheat shorts consists of the fine particles of wheat bran, wheat germ, wheat flour and the offal of the "tail of the mill." It has about 87% the value of corn for growing-finishing pigs when fed at levels of 10–20% of the diet.

10. Rice Bran

This feed contains rice bran and hulls removed in milling rice for human food. It has about 13% protein and 13% fat. Because of the high fat content, it often turns rancid on storage. Rice bran has about 90–95% the feeding value of corn if used at a level of not more than 20–30% of the diet. When used at higher levels, its relative feeding value decreases and it tends to produce soft pork. Rice bran should be used fresh. It loses palatability and turns rancid on long storage.

Solvent-extracted rice bran is available for swine feeding. Since the major portion of the oil has been removed, it does not produce soft pork. It also has about the same feeding value as corn when fed at levels no higher than 30% of the diet.

11. Rice Polishings

This feed consists of the finely powdered material obtained in polishing the kernels after the hulls and bran have been removed. It has almost the same protein and fat content as rice bran but only about one-third as much fiber. It tends to become rancid on storage and thus should be used as fresh as possible. It also causes soft pork and so needs to be used at low levels of no more than 10–30% of the grain mixture. Rice polishings have about the same feeding value as corn when limited to no more than 30% of the corn in the diet. If this feed is used at higher levels, its relative feeding value decreases and soft pork will be produced if it makes up more than 30% of the diet. It is apt to cause scouring when fed to young pigs.

12. Brewers' Rice

This consists of the small fragments broken off the kernels during the milling process. The carcasses of pigs fed brewers' rice are hard and firm. Best results are obtained when it is ground and mixed with more palatable feeds. It has about the same feeding value as corn.

13. Rice Meal

This consists of ground brown rice or ground rice after the hull has been removed. It has about the same feeding value as corn and produces hard pork.

IV. SELECTING PROTEIN SUPPLEMENTS

Protein supplements should be selected on the basis of their quantity of protein and balance of amino acids and their value in correcting protein and amino acid deficiencies in the energy feeds used in swine diets. The mineral and vitamin content of the protein supplements should also be evaluated. Table 11.5 lists the composition of the various protein supplements and Table 11.6 the amino acid levels in the protein sources. Consideration also must be given to the cost of the supplement per unit of protein.

Table 11.7 gives an idea as to total feed usage by swine. It also provides information on the kinds of feeds used in swine diets in the United States. The information is for the feeding year 1971–1972, but it applies fairly well to subsequent years. The percentage of various feeds in the diet would vary from the figures shown in Table 11.7 depending on the stage of the life cycle of the pig. They would also vary, depending on the price of feeds and other criteria including the price the producer receives for his product. For example, a lesser percentage of grain was used in 1975 when prices for it were high. Other feeds were substituted for some of the grain in order to decrease diet costs.

V. RELATIVE VALUE OF PROTEIN SUPPLEMENTS

Tables 11.2, 11.3, and 11.4 give information on suggested levels of various protein supplements to use in swine diets. Tables 11.2 and 11.3 present data on their relative nutritional values. These relative values can vary depending on the quality of the individual protein supplements available. Therefore, they are not exact relative values, but should be useful as a guide.

A. Plant Protein Concentrates

More plant protein concentrates than animal protein concentrates are used for swine feeding, thus it is important to know how to use them to supply the proper amount and quality of protein in the diet.

TABLE 11.5
Average Composition of Some Commonly Used Feeds (as Fed Basis)[a]

	Reference number	Dry matter (%)	Energy ME_n (kcal/kg)	Energy Pro (kcal/kg)	Protein (%)	Ether extract	Crude fiber (%)	Calcium (%)	Phosphorus (%)	Potassium (%)	Chlorine (%)	Iron (%)	Magnesium (%)	Manganese (mg/kg)	Sodium (%)
Alfalfa meal															
17% protein	1-00-023	92.	1370	580	17.5	2.0	24.1	1.44	0.22	2.17	0.48	0.048	0.36	30.0	0.12
20% protein	1-00-024	92	1630	850	20.0	3.6	20.2	1.67	0.28	2.21	0.46	0.039	0.36	42.3	0.13
Barley	4-00-549	89	2640	1790	11.6	1.8	5.1	0.24	0.36	0.48	0.15	0.005	0.14	—	0.04
Barley, Pacific Coast	4-07-939	89	2620	1720	9.0	2.0	6.4	0.05	0.32	0.53	0.15	0.011	0.12	16.3	0.02
Blood meal, vat dried	5-00-380	89	2830	2280	70.5	1.6	9.5	0.26	0.25	0.09	0.27	.202	0.16	5.1	0.32
Blood meal, spray or ring dried			3420	2280	85.0	1.0	2.5	0.45	0.37	0.41	0.27	0.30	0.40	6.4	0.33
Bone meal, steamed	6-00-400	97	1090	—	12.6	—	4.8	29.39	12.58	0.09	0.01	0.32	0.11	40.9	0.07
Brewers dried grains	5-02-141	92	2080	1850	25.3	6.2	15.3	0.29	0.52	0.09	0.12	0.025	0.16	37.8	0.15
Buckwheat	4-00-994	88	2660	1800	10.8	2.5	10.5	0.09	0.32	0.40	0.04	—	—	33.8	0.05
Buttermilk	5-01-160	92	2770	1720	31.6	5.0	0.4	1.32	0.93	0.85	0.47	0.001	0.40	3.4	0.73
Casein, dried	5-01-162	90	4130	2500	81.9	0.8	0.2	0.61	1.0	—	—	—	—	4.2	—
Corn, yellow	4-02-931	89	3430	2520	8.8	3.8	2.2	0.02	0.28	0.30	0.04	0.035	0.12	5.0	0.02
Corn and cob meal	4-02-849	85	2770	1980	7.8	3.0	8.7	0.04	0.21	0.45	0.04	0.007	0.13	7.7	0.01
Corn, gluten feed	5-02-903	90	1750	1120	22.0	2.5	8.0	0.4	0.8	0.57	0.22	0.046	0.29	23.8	0.95
gluten meal, 41%	5-02-900	91	2940	1850	41.0	2.5	7.0	0.23	0.55	0.31	0.11	0.040	0.05	8.9	0.07
gluten meal, 60%	5-09-318	90	3720	2820	62.0	2.5	1.3	—	0.50	0.35	0.05	0.040	0.15	4.4	0.02
Cottonseed meal, expeller	5-01-617	93	2320	1520	40.9	3.9	10.8	0.20	1.05	1.19	0.04	0.016	0.52	22.9	0.04
Cottonseed meal, solvent	5-07-872	90	2400	1320	41.4	1.5	13.6	0.15	0.97	1.22	0.03	0.011	0.40	20.0	0.04
Distillers' dried grain w/solubles (corn)	5-02-843	93	2480	1960	27.2	9.0	9.1	0.17	0.72	0.65	0.17	0.028	0.19	23.9	0.48
Distillers' dried solubles (corn)	5-02-844	92	2930	2240	28.5	9.0	4.0	0.35	1.33	1.75	0.26	0.056	0.64	73.7	0.26
Feather meal	5-03-795	93	2360	1320	86.4	3.3	1.0	0.33	0.55	0.31	—	—	0.20	21.0	0.71
Fish meal, anchovy	5-01-985	92	2580	1890	64.2	5.0	1.0	3.73	2.43	0.69	0.29	0.022	0.24	9.5	0.88
Fish meal, herring	5-02-000	93	3190	2050	72.3	10.0	0.7	2.29	1.70	1.09	0.90	0.014	0.15	4.7	0.61
Fish meal, Menhaden	5-02-009	92	2820	1980	60.5	9.4	0.7	5.11	2.88	0.77	0.60	0.044	0.16	33.0	0.41
Fish meal, sardine	5-02-015	92	2880	1980	64.7	5.4	1.0	4.38	2.58	0.25	0.41	0.030	0.10	23.0	0.18
Fish meal, white	5-02-025	95	2570	1815	68.9	3.4	—	5.40	2.60	1.0	—	0.008	—	9.8	1.1
Fish solubles, condensed	5-01-969	51	1460	990	31.5	7.8	0.2	0.30	0.76	1.74	2.65	0.016	0.02	14.4	2.62
Fish solubles, dried	5-01-971	92	2830	1610	63.6	9.3	0.5	1.23	1.63	0.37	—	—	—	50.1	0.37
Hominy	4-02-887	90	2970	1890	10.0	6.9	6.0	0.04	0.50	0.46	0.05	0.007	0.16	14.5	0.10

	Sulfur (%)	Copper (mg/kg)	Selenium (mg/kg)	Zinc (mg/kg)	Biotin (mg/kg)	Choline (mg/kg)	Folacin (mg/kg)	Niacin (mg/kg)	Pantothenic acid (mg/kg)	Pyridoxine (mg/kg)	Riboflavin (mg/kg)	Thiamin (mg/kg)	Vitamin B_{12} (mg/kg)	Vitamin E (mg/kg)
Alfalfa meal														
17% protein	0.17	10.2	0.338	24	0.30	1097	4.2	38	25	6.5	13.6	3.4	0.004	125
20% protein	0.43	11.2	0.288	25	0.33	1171	3.3	40	34	8.0	15.2	5.8	0.004	144
Barley	0.15	10.2	0.10	17	0.15	990	0.7	55	8	3.0	1.8	1.9	—	20
Barley, Pacific Coast	—	7.7	0.102	15	0.15	1034	0.5	48	7	—	1.6	4.0	—	20
Blood meal, vat dried	0.32	9.7	—	—	0.08	695	0.1	29	3	—	2.6	0.4	44.0	—
Blood meal, spray or ring dried	—	8.1	—	306	—	280	—	13	5	4.4	1.3	0.5	—	—
Bone meal, steamed	0.32	8.3	—	425	—	693	—	30	3	—	5.9	0.4	0.069	—
Brewers dried grains	0.31	21.1	—	98	0.96	1723	7.1	29	8	0.65	1.4	0.5	—	25
Buckwheat	—	9.5	—	9	—	440	—	19	12	—	5.5	4.0	—	—
Buttermilk	0.08	—	—	—	0.29	1707	0.4	9	34	2.43	31.3	3.3	0.037	6
Casein, dried	—	4.0	—	—	0.05	205	0.5	1	3	0.4	1.5	0.5	—	—
Corn, yellow	0.08	1.0	—	10	0.06	620	0.4	24	4	7.0	1.0	3.5	—	22
Corn and cob meal	0.18	6.7	0.073	9	0.05	393	0.3	17	4	5.0	0.9	—	—	19
Corn, gluten feed	0.22	47.9	0.1	7	0.33	1518	0.3	66	17	15.0	2.4	2.0	—	15
gluten meal, 41%	0.40	28.3	1.0	—	0.18	926	0.4	50	10	7.9	1.7	0.2	—	20
gluten meal, 60%	—	26.4	1.0	330	0.15	330	0.2	55	3	6.2	2.2	0.3	—	24
Cottonseed meal, expeller	0.40	18.6	—	—	0.60	2753	1.0	38	10	5.3	5.1	6.4	—	39
Cottonseed meal, solvent	—	17.8	—	82	0.55	2933	2.7	40	7	3.0	4.0	3.3	—	—
Distillers' dried grain w/solubles (corn)	0.30	56.6	0.390	80	0.78	2637	0.9	71	11	2.20	8.6	2.9	—	40
Distillers' dried solubles (corn)	0.37	82.7	0.332	85	1.4	4842	1.1	116	21	10	17.0	6.9	—	55
Feather meal	—	—	—	—	0.44	891	—	27	10	—	2.1	0.1	0.078	—
Fish meal, anchovy	0.54	9.3	1.363	103	0.23	4408	0.2	100	15	4.0	7.1	0.1	0.352	4
Fish meal, herring	0.69	5.9	1.930	132	0.31	5306	0.8	93	17	4.0	9.9	0.1	0.403	22
Fish meal, Menhaden	0.45	10.8	2.103	147	0.20	3056	0.6	55	9	4.0	4.9	0.5	0.104	7
Fish meal, sardine	0.30	20.0	1.756	—	0.10	3135	—	70	10	—	6.0	0.3	0.235	—
Fish meal, white	0.5	6.4	—	64	0.12	5180	0.3	49	10	4.1	6.0	2.1	0.081	—
Fish solubles, condensed	0.12	44.9	2.0	38	0.18	3519	—	169	35	12.2	14.6	5.5	0.347	—
Fish solubles, dried	—	—	—	76	0.26	5507	—	271	55	—	7.7	—	0.401	6
Hominy	0.03	13.3	—	3.0	0.13	971	0.3	46	8	11.0	2.2	7.9	—	—

continued

TABLE 11.5 *(continued)*

	Reference number	Dry matter (%)	Energy ME_n (kcal/kg)	Energy Pro (kcal/kg)	Protein (%)	Ether extract	Crude fiber (%)	Calcium (%)	Phosphorus (%)	Potassium (%)	Chlorine (%)	Iron (%)	Magnesium (%)	Manganese (mg/kg)	Sodium (%)
Limestone	6-02-632	98	—	—	—	—	—	36.23	0.02	0.12	0.03	0.341	2.01	247.5	0.06
Liver, meal	5-00-389	92	2860	2400	65.6	15.0	1.4	0.56	1.25	—	—	0.063	—	8.8	—
Meat and bone meal	5-00-388	93	1960	1600	50.4	8.6	2.8	10.1	4.96	1.02	0.74	0.049	1.12	14.2	0.72
Meat meal	5-00-385	92	2000	1670	54.4	7.1	8.7	8.27	4.10	0.6	0.91	0.044	0.58	9.7	1.15
Molasses, beet	4-00-668	79	1990	1560	6.1	—	—	0.13	0.06	4.83	1.30	0.007	0.23	4.7	0.93
Molasses, cane, dried	4-04-696	91	1960	1540	7.8	0.5	3.3	1.10	0.12	2.60	—	0.095	0.33	42.0	0.16
Oats	4-03-309	89	2550	1810	11.4	4.2	10.8	0.06	0.27	0.45	0.11	0.007	0.16	43.2	0.16
Oats, West Coast	4-07-999	91	2610	1760	9.0	—	11.0	0.08	0.30	0.37	0.12	—	—	38.0	—
Oat hulls	1-03-281	92	400	220	4.6	1.4	28.7	0.13	0.10	0.53	0.10	0.010	—	13.6	0.04
Oystershell	6-03-481	95	40	—	0.9	—	—	37.26	0.07	0.09	0.01	0.272	0.28	127.5	0.20
Pea, seed	5-03-600	90	2570	—	23.8	1.3	5.5	0.11	0.42	1.02	0.06	0.005	—	—	0.04
Peanut meal, expeller	5-03-649	90	2500	1870	39.8	7.3	13.0	0.16	0.56	1.13	0.03	—	0.33	25.1	—
Peanut meal, solvent	5-03-650	92	2200	1900	60.7	1.2	11.9	0.20	0.63	1.19	0.03	—	0.04	28.9	0.07
Poultry by-product meal	5-03-798	93	2670	1980	60.0	13.0	2.0	3.0	1.7	0.30	0.54	0.044	0.22	11.0	0.40
Rapeseed meal, solvent	5-03-870	94	2040	—	35.0	8.6	12.4	0.72	1.09	0.8	—	0.018	0.51	61.0	0.5
Rice bran	4-03-928	91	1630	1540	12.9	13.0	11.4	0.07	1.50	1.73	0.07	0.019	0.95	324.5	0.07
Rice, broken	4-03-938	89	2990	2510	8.7	—	9.8	0.08	0.39	—	0.08	—	0.11	18.0	0.07
Rice, polishing	4-03-943	90	3090	2090	12.2	11.0	4.1	0.05	1.31	1.06	0.11	0.016	0.65	—	0.10
Safflower meal		91	1600	1160	42.5	1.3	15.0	0.4	1.3	—	—	—	—	—	—
Sesame meal	5-04-110	93	2210	1720	43.8	8.6	9.7	1.99	1.37	1.20	0.06	—	—	47.9	0.04
Skim milk, dried	5-01-175	93	2520	1670	33.5	0.9	4.9	1.28	1.02	1.59	0.50	0.005	0.11	2.0	0.44
Sorghum, grain (milo)	4-04-444	89	3370	2400	8.9	2.8	2.3	0.03	0.28	0.32	0.09	0.004	0.13	0.1	0.04
Soybeans	5-04-610	90	3300	2170	37.0	18.0	5.5	0.25	0.58	1.61	0.03	0.008	0.28	29.8	0.12
Soybean meal, dehulled	5-04-612	90	2440	1730	48.5	1.0	3.9	0.27	0.62	2.02	0.05	—	—	43.0	0.25
Soybean meal, expeller	5-04-600	90	2430	1720	42.6	4.0	6.2	0.27	0.61	1.83	0.07	0.014	0.26	30.7	0.27
Soybean meal, solvent	5-04-604	89	2230	1570	44.0	0.8	7.3	0.29	0.65	2.00	0.05	0.012	0.27	29.3	0.26
Soybean mill feed	5-04-594	89	720	440	13.3	1.6	33.0	0.37	0.19	1.50	—	—	0.12	28.5	—
Sunflower meal	5-04-739	93	2320	1430	45.4	2.9	12.2	0.37	1.0	1.00	0.10	0.003	0.75	22.9	2.0
Wheat bran	4-05-190	90	1300	1050	15.7	3.0	11.0	0.14	1.15	1.19	0.06	0.017	0.52	113.2	0.05
Wheat, hard	4-05-268	87	2800	2250	14.1	1.9	2.4	0.05	0.37	0.45	0.05	0.005	0.17	31.8	0.04
Wheat, middlings	4-05-205	88	1800	1130	15.5	3.0	3.4	0.12	0.90	0.99	0.03	0.004	0.16	60.1	0.12
Wheat, soft	4-05-284	89	3120	1980	10.2	1.8	2.4	0.05	0.31	0.40	0.08	0.004	0.10	23.8	0.04
Whey, dried	4-01-182	93	1900	1540	13.6	0.8	1.3	0.97	0.76	1.05	0.07	0.013	0.13	6.1	0.48
Whey, low lactose	4-01-186	91	2090	1580	15.5	1.0	0.3	1.95	0.98	3.0	2.10	—	0.25	—	1.50
Yeast, brewers dried	7-05-527	93	1990	1260	44.4	1.0	2.7	0.12	1.40	1.70	0.12	0.012	0.23	5.2	0.07
Yeast, *Torula*	7-05-534	93	2160	1540	47.2	2.5	2.4	0.58	1.67	1.88	0.02	0.009	0.13	12.8	0.01

	Sulfur (%)	Copper (mg/kg)	Selenium (mg/kg)	Zinc (mg/kg)	Biotin (mg/kg)	Choline (mg/kg)	Folacin (mg/kg)	Niacin (mg/kg)	Pantothenic acid (mg/kg)	Pyridoxine (mg/kg)	Riboflavin (mg/kg)	Thiamin (mg/kg)	Vitamin B_{12} (mg/kg)	Vitamin E (mg/kg)
Limestone	0.04	—	—	—	—	—	—	—	—	—	—	—	—	—
Liver, meal	—	88.9	—	—	0.02	11311	5.5	204	29	—	46.3	0.2	0.498	—
Meat and bone meal	0.50	1.5	0.25	93	0.64	1996	0.32	46	4.1	12.8	4.4	0.8	0.070	1.0
Meat meal	0.49	9.8	0.426	103	0.17	2077	0.3	57	5	3.0	5.5	0.2	0.068	1.0
Molasses, beet	0.48	17.7	—	14	—	400	—	42	4	—	2.1	—	—	5.1
Molasses, cane, dried	0.35	0.6	—	30	—	891	—	43	4	—	2.4	—	—	5.4
Oats	0.21	8.3	0.30	1	0.11	946	—	12	—	1.0	1.1	6.0	—	20.0
Oats, West Coast	—	—	0.07	—	0.11	959	0.3	14	13	1.3	1.1	—	—	20.0
Oat hulls	—	3.1	—	0.1	—	284	0.96	7	3	—	1.5	0.6	—	—
Oystershell	—	—	—	—	—	—	—	—	—	—	—	—	—	—
Pea, seed	—	—	—	30	0.18	642	0.4	34	10	1.0	2.3	7.5	—	—
Peanut meal, expeller	0.29	—	—	—	1.76	1655	—	166	47	—	5.2	7.1	—	2.9
Peanut meal, solvent	—	—	—	20	0.39	4396	0.4	170	53	10.0	11.0	—	—	3.0
Poultry by-product meal	0.51	14.0	0.75	120	0.30	5952	1.0	40	12.3	—	11.0	1.0	0.31	2.0
Rapeseed meal, solvent	—	7.0	0.982	44	—	6464	—	153	9	7.0	3.7	1.7	—	19.1
Rice bran	0.18	13.0	—	30	0.42	1135	—	293	23	14.0	2.5	22.5	—	59.8
Rice, broken	0.06	—	—	17	0.08	800	0.2	46	8	—	0.7	—	—	14.5
Rice, polishing	0.17	—	—	—	0.61	1237	—	520	47	—	1.8	19.8	—	90.0
Safflower meal	—	—	—	—	1.40	4130	—	22	40	—	2.4	—	—	0.7
Sesame meal	0.43	—	—	100	0.34	1536	—	30	6	12.5	3.6	2.8	—	—
Skim milk, dried	0.31	11.5	0.12	40	0.33	1393	0.5	11	37	3.9	19.0	3.5	0.037	9.1
Sorghum, grain (milo)	0.16	19.0	—	14	0.18	450	0.2	41	12	3.2	1.1	4.0	—	12.0
Soybeans	0.22	15.8	0.11	16	0.27	2860	4.2	22	11	10.8	2.6	11.0	—	0.9
Soybean meal, dehulled	—	15.0	0.10	45	0.32	2731	3.6	22	15	5.0	2.9	3.2	—	3.3
Soybean meal, expeller	0.33	24.3	0.10	60	0.33	2703	4.4	32	14	—	3.7	3.2	—	6.1
Soybean meal, solvent	0.43	21.5	0.10	27	0.32	2794	1.3	29	16	—	2.9	4.5	—	2.1
Soybean mill feed	0.06	—	—	—	—	640	—	24	13	—	3.5	—	—	—
Sunflower meal	—	3.5	—	—	1.45	2894	—	220	24	16.0	4.7	—	—	11.0
Wheat bran	0.22	14.1	—	133	0.48	797	1.2	186	31	7.0	4.6	8.0	—	13.5
Wheat, hard	0.12	5.8	0.2	31	0.11	1090	0.35	48	9.9	—	1.4	4.5	—	12.6
Wheat, middlings	0.26	6.4	0.8	64	0.37	1439	0.8	98	13	9.0	2.2	16.5	—	40.5
Wheat, soft	0.12	6.9	0.06	28	0.11	1002	0.4	57	11	4.0	1.2	4.3	—	13.2
Whey, dried	1.04	46.0	—	—	0.34	1369	0.8	10	44	4.0	27.1	4.1	0.023	0.2
Whey, low lactose	—	—	—	—	0.64	4392	1.4	18.6	69	3.96	45.8	5.7	0.023	—
Yeast, brewers dried	0.38	32.8	1.0	39	1.05	3984	9.9	448	109	42.8	37.0	91.8	—	—
Yeast, *Torula*	0.34	13.5	1.0	99	1.39	2881	22.4	500	73	—	47.7	6.2	—	—

[a]Data from M. L. Sunde, J. R. Couch, L. S. Jensen, B. E. March, E. C. Naber, L. M. Potter, and P. E. Waibel, "Nutrient Requirements of Poultry," Natl. Res. Counc. Washington, D.C., 1977.

TABLE 11.6
Average Amino Acid Composition of Some Commonly Used Feedstuffs[a,b]

	Reference number	Dry matter (%)	Protein (%)	Arginine (%)	Glycine (%)	Serine (%)	Histidine (%)	Isoleucine (%)	Leucine (%)	Lysine (%)	Methionine (%)	Cystine (%)	Phenylalanine (%)	Tyrosine (%)	Threonine (%)	Tryptophan (%)	Valine (%)
Alfalfa meal																	
17% protein	1-00-023	92	17.5	0.80	0.90	0.77	0.32	0.84	1.26	0.73	0.23	0.20	0.79	0.56	0.70	0.28	0.84
20% protein	1-00-024	92	20.0	0.92	0.97	—	0.34	0.88	1.30	0.87	0.31	0.25	0.85	0.59	0.76	0.33	0.97
Barley	4-00-549	89	11.6	0.59	0.40	0.42	0.29	0.49	0.80	0.40	0.17	0.19	0.64	0.33	0.42	0.14	0.62
Barley, Pacific Coast	4-07-939	89	9.0	0.48	0.36	—	0.21	0.40	0.60	0.29	0.13	0.18	0.48	—	0.30	0.12	0.46
Blood meal	5-00-380	89	70.5	3.21	3.78	4.60	3.96	0.89	10.47	5.88	0.95	1.33	5.41	1.78	3.59	1.02	6.70
Blood meal, spray or ring dried		91	85.0	4.10	4.30	4.60	5.50	1.00	12.70	9.10	3.0	1.50	7.30	3.00	4.90	1.10	9.09
Bone meal, steamed	6-00-400	97	12.6	1.89	2.65	0.48	0.20	0.49	1.03	0.94	0.19	—	0.60	0.05	0.62	0.05	0.76
Brewers dried grains	5-02-141	92	25.3	1.28	1.09	0.80	0.57	1.44	2.48	0.90	0.57	0.39	1.45	1.19	0.98	0.34	1.66
Buckwheat	4-00-994	88	10.8	1.02	—	—	0.26	0.37	0.56	0.61	0.20	0.20	0.44	—	0.46	0.19	0.54
Buttermilk	5-01-160	92	31.6	1.08	0.34	1.39	0.83	2.31	3.10	2.23	0.72	0.40	1.45	0.99	1.47	0.47	2.50
Casein, dried	5-01-162	93	87.2	3.61	1.79	5.81	2.78	4.82	9.00	7.99	2.65	0.21	4.96	5.37	4.29	1.05	6.46
Casein, dried, coprecip.	5-20-837	92	85.0	3.42	1.81	5.52	2.52	4.77	8.62	7.31	2.80	0.15	4.81	5.17	4.00	0.98	5.82
Corn, yellow	4-02-931	89	8.8	0.50	0.37	0.40	0.20	0.37	1.10	0.24	0.20	0.15	0.47	0.45	0.39	0.09	0.52
Corn and cob meal	4-02-849	85	7.8	0.38	0.27	—	0.18	0.35	0.98	0.18	0.14	0.14	0.44	—	0.35	0.07	0.35
Corn, gluten feed	5-02-903	90	22.0	1.01	0.99	0.80	0.71	0.65	1.89	0.63	0.45	0.51	0.77	0.58	0.89	0.10	1.05
gluten meal, 41%	5-02-900	91	40.6	1.38	1.50	1.50	0.98	2.18	7.19	0.78	1.03	0.65	2.67	1.00	1.40	0.21	2.23
gluten meal, 60%	5-09-318	90	62.0	1.93	1.64	3.07	1.22	2.29	10.11	1.00	1.91	1.11	3.77	2.94	1.97	0.25	2.74
Cottonseed meal, expeller	5-01-617	93	40.9	4.26	2.28	—	1.08	1.57	2.47	1.51	0.55	0.59	2.17	0.69	1.38	0.55	1.97
Cottonseed meal, solvent	5-07-872	90	41.4	4.59	1.70	—	1.10	1.33	—	1.71	0.52	0.64	2.22	1.02	1.32	0.47	1.89
Distillers' dried grain w/solubles (corn)	5-02-843	93	27.2	0.98	0.57	1.61	0.66	1.00	2.20	0.75	0.60	0.40	1.20	0.74	0.92	0.19	1.30
Distillers' dried solubles (corn)	5-02-844	93	28.5	1.05	1.10	1.30	0.70	1.25	2.11	0.90	0.50	0.40	1.30	0.95	1.00	0.30	1.39
Feather meal	5-03-795	93	86.4	5.42	6.31	—	0.34	3.26	6.72	1.67	0.42	4.00	3.26	6.31	3.43	0.50	5.57
Fish meal, anchovy	5-01-985	92	64.2	3.66	3.59	2.32	1.53	3.01	4.83	4.90	1.93	0.59	2.70	2.18	2.68	0.74	3.38
Fish meal, herring	5-02-000	93	72.3	4.84	4.61	2.73	1.70	3.22	5.34	5.70	2.10	0.72	2.79	2.27	3.00	0.81	4.38
Fish meal, Menhaden	5-02-009	92	60.5	3.79	4.19	2.25	1.46	2.85	4.50	4.83	1.78	0.56	2.48	1.98	2.50	0.68	3.23
Fish meal, sardines	5-02-015	92	64.7	3.27	4.52	—	1.78	3.38	5.29	5.90	1.98	0.96	2.29	—	2.69	0.68	4.05

Fish solubles, condensed	5-01-969	51	31.5	1.61	3.41	—	1.56	1.06	1.86	1.73	0.50	0.30	0.93	0.40	0.86	0.31	1.16
Fish solubles, dried	5-01-971	92	63.6	2.78	5.89	2.02	2.18	1.95	3.16	3.28	1.00	0.66	1.48	0.78	1.35	0.51	2.22
Gelatin	5-14-503	91	88.0	7.4	20.0	2.80	0.85	1.40	3.10	3.70	0.68	0.09	1.70	0.26	1.30	0.09	1.80
Hominy	4-02-887	90	10.0	0.47	0.40	—	0.20	0.40	0.84	0.40	0.13	0.13	0.35	0.49	0.40	0.10	0.49
Liver, meal	5-00-389	92	65.6	4.14	5.57	2.49	1.47	3.09	5.28	4.80	1.22	0.89	2.89	1.69	2.48	0.59	4.13
Meat and bone meal	5-00-388	93	50.4	3.62	6.79	1.85	1.20	1.40	2.8	2.60	0.65	0.25	1.50	0.76	1.50	0.28	2.00
Meat meal	5-00-385	92	54.4	3.73	6.30	1.60	1.30	1.60	3.32	3.00	0.75	0.66	1.70	0.84	1.74	0.36	2.30
Oats	4-03-309	89	11.4	0.79	0.50	0.40	0.24	0.52	0.89	0.50	0.18	0.22	0.59	0.53	0.43	0.16	0.68
Oats, West Coast	4-07-999	91	9.0	0.60	0.40	0.30	0.10	0.20	0.30	0.40	0.13	0.17	0.20	0.20	0.20	0.12	0.20
Oat Hulls	1-03-281	92	4.6	0.14	0.14	0.14	0.07	0.14	0.25	0.14	0.07	0.06	0.13	0.14	0.13	0.07	0.20
Pea, seed	5-03-600	90	23.8	1.40	1.10	—	0.72	1.10	1.80	1.60	0.31	0.17	1.30	—	0.94	0.24	1.30
Peanut meal, solvent	5-03-650	93	55.0	5.50	2.70	2.22	1.19	2.10	2.99	1.76	0.44	0.76	2.75	2.00	1.45	0.65	1.82
Poultry by-product meal	5-03-798	93	55.0	4.00	5.90	3.68	1.50	2.00	3.70	2.70	1.00	0.69	2.10	0.54	2.00	0.53	2.60
Rapeseed meal	5-03-870	94	35.0	1.93	1.81	1.48	0.87	1.33	2.31	1.75	0.68	0.31	1.41	0.82	1.53	0.45	1.79
Rice bran	4-03-928	91	12.9	0.89	0.80	—	0.33	0.52	0.90	0.59	0.20	0.10	0.58	0.68	0.48	0.15	0.75
Rice polishing	4-03-943	90	12.2	0.78	0.71	—	0.24	0.41	0.80	0.57	0.22	0.10	0.46	0.63	0.40	0.13	0.76
Sesame meal	5-04-220	93	43.8	4.93	4.22	—	1.09	2.12	3.33	1.30	1.20	0.59	2.22	2.00	1.65	0.80	2.41
Skim milk, dried	5-01-175	93	33.5	1.12	0.27	1.59	0.84	2.15	3.23	2.40	0.93	0.44	1.58	1.13	1.60	0.44	2.30
Sorghum, grain (milo)	4-04-444	89	8.9	0.38	0.31	0.53	0.27	0.53	1.42	0.22	0.12	0.15	0.44	0.35	0.27	0.10	0.53
Soybeans, full fat	5-04-610		37.0	2.80	2.00		0.89	2.00	2.80	2.40	0.51	0.64	1.80	1.20	1.50	0.55	1.80
Soybean protein, isolated[c]	5-08-038	93	84.1	6.7	3.3	5.3	2.1	4.6	6.6	5.5	0.81	0.49	4.3	3.1	3.3	0.81	4.4
Soybean meal, dehulled	5-04-612	90	48.5	3.68	2.29	2.89	1.32	2.57	3.82	3.18	0.72	0.73	2.11	2.01	1.91	0.67	2.72
Soybean meal, expeller	5-04-600	90	42.6	3.00	2.38	2.02	1.10	2.81	3.60	2.78	0.67	0.62	2.12	1.40	1.71	0.61	2.21
Soybean meal, solvent	5-04-604	89	44.0	3.28	2.29	2.45	1.15	2.39	3.52	2.93	0.65	0.69	2.27	1.28	1.81	0.62	2.34
Soybean mill feed	5-04-594	89	13.3	0.94	0.40	—	0.18	0.40	0.57	0.48	0.10	0.21	0.37	0.23	0.30	0.10	0.37
Sunflower meal	5-04-739	93	45.4	5.48	2.69	1.75	1.39	2.78	3.88	1.70	0.72	0.71	2.93	1.19	2.13	0.71	3.24
Wheat bran	4-05-190	90	15.7	0.98	0.90	0.90	0.34	0.59	0.91	0.59	0.17	0.25	0.49	0.40	0.42	0.30	0.73
Wheat, hard	4-05-268	87	14.1	0.58	0.72	0.63	0.22	0.58	0.94	0.40	0.19	0.26	0.71	0.60	0.37	0.18	0.63
Wheat, middlings	4-05-205	88	16.0	1.76	0.63	0.75	0.37	0.58	1.07	0.69	0.21	0.32	0.64	0.45	0.49	0.20	0.71
Wheat, soft	4-05-284	89	10.2	0.40	0.49	0.55	0.20	0.42	0.59	0.31	0.15	0.22	0.45	0.39	0.32	0.12	0.44
Whey, dried	4-01-182	93	12.0	0.34	0.30	0.32	0.18	0.82	1.19	0.97	0.19	0.30	0.33	0.25	0.89	0.19	0.68
Whey, low lactose	4-01-186	91	15.5	0.67	1.04	0.76	0.10	0.30	0.20	1.47	0.67	0.57	0.10	0.20	0.50	0.18	0.30
Yeast, brewers dried	7-05-527	93	44.4	2.19	2.09	—	1.07	2.14	3.19	3.23	0.70	0.50	1.81	1.49	2.06	0.49	2.32
Yeast, *Torula*	7-05-534	93	47.2	2.60	2.60	—	1.40	2.90	3.50	3.80	0.80	0.60	3.00	2.10	2.60	0.50	2.90

[a] As fed basis.

[b] Data from M. L. Sunde, J. R. Couch, L. S. Jensen, B. E. March, E. C. Naber, L. M. Potter, and P. E. Waibel, "Nutrient Requirements of Poultry," Natl. Res. Counc., Washington, D.C., 1977.

[c] Soybean protein concentrate (AAFCO).

TABLE 11.7
Feed Units[a] Used by Various Livestock (12) (Feeding Year Beginning 10/1/71)

Feeds[b]	Dairy cattle	Beef cattle: Cattle on feed	Beef cattle: Other beef cattle	Beef cattle: Total	Hogs	Poultry	Total[c]
Corn	13,907	21,071	6,008	27,079	39,392	20,112	100,690
Sorghum grains	901	11,106	1,107	12,213	1,201	3,148	17,463
Other grains[d]	4,869	8,610	1,727	10,337	3,931	4,372	23,509
High protein by-product feeds[e]	4,439	3,517	2,733	6,250	8,170	13,620	32,479
Other by-products[f]	3,559	1,462	1,252	2,714	1,342	2,325	9,940
Hay	16,943	8,930	22,991	31,921	—	—	48,864
Other harvested forage[g]	15,587	5,083	7,169	12,252	—	—	27,839
Pasture and range	16,367	3,010	114,109	117,119	9,515	1,129	144,130
	76,572	62,789	157,096	219,885	63,751	44,706	404,914

Feeds as Percentages of Total Diets for the Different Animal Species

Feeds	Dairy cattle	Beef cattle: Cattle on feed	Beef cattle: Other beef cattle	Beef cattle: Total	Hogs	Poultry	Total
Corn	18.1	33.4	3.8	12.3	62.1	45.0	24.9
Sorghum grains	1.2	17.7	0.7	5.5	1.9	7.0	4.3
Other grains	6.4	13.7	1.1	4.7	6.2	9.9	5.8
High protein by-product feeds	5.7	5.6	1.7	2.9	12.9	30.5	8.0
Other by-products	4.7	2.4	0.8	1.2	2.0	5.1	2.4
Hay	22.1	14.2	14.7	14.5	0.0	0.0	12.1
Other harvested forage	20.4	8.2	4.6	5.6	0.0	0.0	6.9
Pasture and range	21.4	4.8	72.6	53.3	14.9	2.5	35.6
	100.0	100.0	100.0	100.0	100.0	100.0	100.0

[a]A feed unit is the equivalent in feeding value of 1 ton of corn.
[b]Values in thousands of tons.
[c]Horses, mules, sheep and other livestock not included.
[d]Oats, barley, wheat and rye.
[e]Includes oilseed meals, animal protein, grain protein by products.
[f]Includes milling by-products and miscellaneous.
[g]Includes straw, silage and beet pulp.

1. Soybean Meal

This is an excellent source of protein and is the most extensively used protein supplement in swine feeding in the United States. It is becoming available in increasing amounts yearly in other areas of the world. Properly processed soybean meal by the expeller, hydraulic, or solvent method has about the same feeding value. Emphasis should be placed on making certain the meal has been properly heat-treated and processed. Usually methionine supplementation is beneficial with improperly processed soybean meal but of no help with properly cooked meals.

Soybean meal protein is of better quality than the other protein-rich plant protein supplements (Tables 11.2 and 11.3). When soybean meal is self-fed, free-choice, pigs will often eat more than is needed to balance the diet because the meal is extremely palatable. This can be overcome by mixing the soybean meal with less palatable feeds such as with 30–60% oats or 30–60% tankage, meat scraps, or fish meal; 25–35% alfalfa meal; 10–15% minerals; or various combinations of these feeds at somewhat lower levels than if one is used alone.

Soybean meal is excellent as the only protein supplement for swine. Soybean meal and corn provide an excellent diet for swine of all ages if the proper proportions are used to provide a good balance of the amino acids. Studies have shown that methionine is the first limiting amino acid in soybean meal and that both lysine and threonine are next limiting, but which is the second limiting amino acid is not definitely certain. The amino acids which become limiting when soybean meal is used depend on the feeds used, the level of protein in the total diet and the amino acid levels in the other feeds. So if the proper protein level and combinations of feeds are used there is no need to add amino acids to the diet. For example, amino acid supplementation is not beneficial with corn-soybean meal diets if the proper protein levels are used. But, if the total protein level in the diet is reduced, then the limiting amino acid deficiencies can occur.

2. *Cottonseed Meal*

In the United States, cottonseed meal ranks second in tonnage to soybean meal among the plant protein concentrates. It is also available in many areas of the world for swine feeding. Cottonseed meal is used widely in swine diets, but the amount is limited because of the hazard of gossypol poisoning. Gossypol is still a problem in cottonseed meal. Gossypol toxicity is cumulative. It accumulates in the body and death does not occur for a number of weeks. Studies at the Florida Station (16) showed that pigs dying from gossypol toxicity exhibited the following symptoms: excessive quantities of fluid in the pleural and peritoneal cavities; flabby and enlarged hearts; congested and edematous lungs; and a general congestion of other organs such as the liver, spleen, and lymph glands.

The toxicity of gossypol can be prevented by the addition of iron to the diet. Ferrous sulfate is the most effective iron salt to use. The addition of one part of iron (from ferrous sulfate) to one part of gossypol (on a weight ratio) to diets containing more than 0.01% free gossypol is effective in counteracting gossypol toxicity and in increasing gains and feed efficiency. It is postulated that iron and gossypol form a complex in the intestinal tract, thereby preventing the absorption of gossypol. But, even when iron is added, the gains and feed efficiency are not equal to that obtained with a corn-soybean meal diet. But when only 50% of the supplemental protein is cottonseed meal and the other 50% is soybean meal, the performance of growing-finishing swine is equal to that of swine receiving soybean meal as the only supplemental protein (2,16). A Florida study (16) showed that low-gossypol cottonseed meal compared well to soybean meal with sows fed on pasture. So, low-gossypol cottonseed meal can be effectively used to substitute for one half the soybean meal in the diet of growing-finishing pigs and sows. It is recommended, however, that the amount of free gossypol be no higher than 0.01% in the total diet.

Cottonseed meals differ considerably in their nutritional value depending on the processing methods used. Those with low free gossypol and high protein quality are the most desirable. Cottonseed meal is a poor source of lysine and contains from one-half to two-thirds that of soybean meal (Table 11.6). Moreover, during processing, the application of excessive heat may render 10–35% of the lysine unavailable (11). Since the cereal grains are also low in lysine, it is very important that either crystalline L-lysine or feeds rich in lysine (soybean meal, fish meal and others) be combined with cottonseed meal in diets for swine.

Plant geneticists are developing cotton varieties which produce cottonseed very low in gossypol-containing glands. Some of this cottonseed meal has already been fed to swine (17). It contains only 0.009% free gossypol. It is hoped that these studies will eventually result in eliminating the gossypol problem in cottonseed meal for swine feeding.

3. Peanut Meal

Peanut meal is the product which remains after the extraction of oil from peanuts. It is also called groundnut meal in some foreign countries. Peanut meal, which contains 5–7% fat, tends to become rancid when held too long in warm and moist climates. Thus, it should not be stored for much longer than 5 or 6 weeks in the summer or 8 to 12 weeks during the winter. Peanut meal is very low in lysine (Table 11.6), and should be added to feeds high in this amino acid such as was suggested for cottonseed meel.

Peanut meal is so palatable that pigs will eat more than they need to balance their diet when it is self-fed, free-choice, with corn. This can be alleviated by mixing peanut meal with less palatable feeds such as was suggested for soybean meal. Florida studies (18) showed peanut meal to be as valuable as soybean meal when fed to weanling pigs on a corn diet supplemented with vitamins and aureomycin. However, it is recommended that peanut meal be fed in combination with soybean meal so that each supplies about 50% of the supplemental protein.

The protein quality of peanut meal varies with the type and level of heat it receives during processing and the quantity of hulls in the meal. In addition to lysine, peanut meal is low in methionine.

Peanuts are subject to the growth of certain molds. *Aspergillus flavus* can occur in peanuts and peanut meal. The toxic factors produced are called aflatoxins; aflatoxin B is the main one found. These aflatoxins cause a loss of appetite, reduce growth rate, and reduce liver vitamin A levels (19). A rapid qualitative test for aflatoxins has been published (20). A negative test showing no aflatoxin is necessary for the meal to be safe.

4. *Linseed Meal*

Linseed meal is the product which remains after the extraction of oil from flaxseed. Linseed meal is low in lysine and feeds high in this amino acid should be used with linseed meal such as was recommended for cottonseed meal.

Pigs fed linseed meal gain at a slower rate than those fed tankage. The difference in rate of gain is less when the pigs are fed on pasture. The difference is larger, however, the younger the pigs are when started on experiment. Linseed meal does not give as good results as soybean meal. This would indicate that linseed meal should be used at the lowest level with young pigs, at higher levels with older pigs, and at the highest level with finishing pigs on pasture. Linseed meal gives best results when fed at a level of 5–20% of the protein supplement used in balancing a diet (Table 11.2). Younger pigs should be fed the lower level and larger pigs on pasture the higher level. Linseed meal has been shown to be more satisfactory as a protein supplement when used with wheat or barley diets as compared to corn diets. Wheat and barley are higher in lysine and tryptophan than corn (Table 11.6), containing almost twice as much.

5. *Rapeseed Meal*

Rapeseed meal is a by-product of oil extraction from rapeseed. The rapeseed yields about 40% oil and 50% meal. Rapeseed contains undesirable substances which limit its use in swine diets. These undesirable substances are related to its thioglucoside content which yields isothiocyanates and oxazolidinethione on enzymatic hydrolysis with myrosinase. These two compounds, especially oxazolidinethione, exert a goiterogenic effect on the pig. If fed at high levels, rate of gain, feed efficiency and reproduction is interfered with.

Rapeseed contains less lysine than soybean meal but more methionine. It is less palatable than soybean meal. It is suggested that not more than 3% of rapeseed meal be used in sow diets fed during gestation and lactation. For very young pigs, no more than 4–5% rapeseed meal should be used in the diet. For larger growing-finishing pigs (110–200 lb) levels of 5–8% of rapeseed meal can be fed (11,21). These recommendations are based on rapeseed meals that are not excessivly high in thioglucoside level. There are many rapeseed varieties, and they vary in thioglucoside content. Therefore, the use of rapeseed meal in the diet should be guided by the variety used and the experimental results obtained with it. Studies in Canada (22) indicate that new rapeseed varieties low in thioglucoside activity may be fed at higher levels to swine than currently recommended. For example, in one study (23) rapeseed meal of *Brassica campestris* origin was fed at a level of 6% in the diet and was not detrimental to sow performance when fed during growth, throughout growth and reproduction, or when introduced into the diet at the time of breeding. An Australian study (24) showed that rapeseed meal can replace up to 50% of meat meal protein as the sole protein supplement to a wheat-based diet for growing-finishing pigs. So when new rapeseed varieties with low thioglucoside levels become commercially available, they may be fed at higher levels in swine diets.

6. *Sesame Meal*

Sesame meal is the product which remains after the oil is extracted from sesame seed. Sesame meal is a poor source of lysine but is an excellent source of methionine, the level being twice as high as in soybean meal (Table 11.6). It should be blended with soybean meal or fish meal, which are good sources of lysine when used to supplement cereal-based diets. Studies in Latin America showed that sesame meal can replace one half or all of the soybean meal in a corn-soybean meal diet containing 4% fish meal, and it can be substituted for soybean meal in quantities as high as 10% of the diet in corn and soybean meal diets (11,25). In another study (11,25), blends of 50–50 sesame meal and cottonseed meal were fed with cereal grains and 5–6% fish meal and produced satisfactory gains and feed utilization (11,26).

7. *Sunflower Meal*

It is produced from the seed of the sunflower plant. Sunflower meal is low in lysine, but it has twice as much methionine as soybean meal. The meals with the highest protein levels are those which have the largest quantity of seed hulls removed. The 45.5% protein sunflower meal has 12.2% fiber (Table 11.5). A recent Georgia study (27) showed that sunflower seed meal (decorticated) replaced 25% of the soybean meal with equal gains, but feed required per pound of gain was higher. Based on this trial and previous data, sunflower seed meal should replace no more than 20% of the soybean meal in the diet (28–30). Preferably, it should be fed to animals that weigh over 75–100 lb. Lysine supplementation benefits sunflower meal diets (27).

8. *Safflower Meal*

This meal will have about 40–44% protein if decorticated, and lower protein levels of 18–22% if not decorticated. The 18–22% protein safflower meal contains about 60% hulls which limits its energy value and utilization in swine diets (31). Even the decorticated safflower meal contains about 14% fiber. Safflower meal is low in lysine and methionine. Studies in Mexico (32–34) showed that safflower meal should not supply more than 12.5% of the supplemented protein in the diet. At higher levels the rate of gain and feed efficiency was decreased. An Australian study (35) showed that decorticated safflower meal was unsatisfactory as the sole supplementary protein source for sorghum and wheat-based diets for growing-finishing pigs. They recommended that if safflower meal is to be used in swine diets, it should be used in conjunction with other lysine-rich protein concentrates and restricted to pigs with liveweights exceeding 100 lb.

9. Coconut Meal

The coconut meal is the residual product left after extracting oil from the dried meat of the coconut. It is also known as copra meal. Even though it is low in protein, it is an important feed in many tropical areas of the world where other feeds are not available. It contains a little over 20% protein, about 6% fat and 10% fiber. Hawaiian workers (37) reported a 50.7% protein digestibility in coconut meal whereas Philippine workers (36) obtained 73% protein digestibility. The variation in protein digestibility may be due to differences in processing methods and the temperature used. Hawaiian studies (38) showed that 10% coconut meal in the diet was fine, 20% caused only a minor reduction in performance, but 40% in the diet reduced gains about 40% and decreased feed efficiency about 30%. The addition of lysine was not beneficial with 20 and 40% coconut meal in the diet. It could be, however, that some factor, or factors, other than adquate protein or lysine is responsible for the growth depressing effect of coconut meal. It is suggested that no more than 15–20% of coconut meal be added to diets for growing-finishing pigs if optimum performance is to be expected.

B. Animal Protein Concentrates

High-quality animal protein concentrates are good sources of amino acids (Table 11.6). They are also good sources of vitamins and minerals (Table 11.5) and an unidentified factor or factors.

1. Meat Meal (or Meat Scrap), Meat and Bone Meal (or Meat and Bone Scraps), Meat Meal Tankage, and Rendering Plant Tankage

Meat meal and meat and bone meal are the most used animal products. Meat meal tankage is used when a higher protein product is desired. The difference is that blood meal has been added to make tankage, whereas meat meal contains no added blood.

If meat meal or meat meal tankage contains more than 4.4% phosphorus it must be designated meat and bone meal or meat and bone meal tankage, respectively. ''Stick,'' which is the evaporated residue of the tank water, is added back to wet-rendered tankage. Meat scraps is the dry-rendered product without added ''stick'' or blood, and usually contains 55% protein. Meat and bone scraps are meat scraps with added bone, and usually contain 45–50% protein. Rendering-plant tankage is a product from small plants that render scrap meat and bones from butcher shops and carcasses of dead animals. It varies considerably in composition, and may contain 35–45% protein and a variable amount of bone.

There is considerable variation in the nutritional value of these meat by-products, depending on how much meat and internal organs they contain. Years ago these meat by-products contained more meat and internal organs; as a result, they were a superior nutritional product than many now being produced.

Many experiment stations have not obtained as good results as might be expected when certain of these meat by-products are used in various trials. This has resulted in most stations recommending that these meat by-products be used at levels no higher than 3–10% of the diet (Tables 11.2, 11.3 and 11.4). Thus, one should pay attention to the quality of these meat by-products and obtain them from firms which are producing high quality products. Use of these animal protein sources should depend on their local availability and price. Their use may reduce growth rate and feed efficiency under certain conditions. But, availability of feeds and economic returns may still justify their use.

2. *Fish Meals*

Fish meals are by-products of the fisheries industry and consist of dried, ground, undecomposed whole fish or fish cuttings or both. Commonly used fish meals are menhaden, sardine, herring, anchovy, salmon, pilchard, tuna, redfish, and white fish meal. They range in protein content from 55 to 72%. Properly processed fish meal is an excellent protein supplement for swine. If fish meal contains too large an amount of bone and fish heads, its feeding value decreases because of the lower nutritive value of bone as compared to the flesh of the fish.

Fish meals vary in quality and composition depending on the quality of fish material used and method and temperature used in processing. High quality fish meals are usually superior to meat meal and meat and bone meal for swine feeding. Good results are obtained by feeding high quality fish meal as the only protein supplement to grain for pigs and brood sows. Because of the usual higher cost of fish meal, however, it is recommended that fish meal be fed at levels of 2–5% in the total diet. When good quality fish meal is properly used, it does not produce a fishy flavor in pork.

There are some indications that high quality fish meal may supply some unidentified factor or factors of value in swine nutrition. But there are also some trials which do not verify this. This matter should be resolved in the future as more information is obtained with swine in complete confinement away from pasture and dirt which could also be sources of the unidentified factor or factors.

3. *Milk Products*

Milk products are excellent feeds. The value of most milk products in the human diet makes them too valuable to be used in swine feeding. But a certain amount of dried skim milk is being used in pig starter diets, where it has real value. Dried whey is also being used in many kinds of swine diets.

Dried skim milk is too expensive to be used in large amounts in protein supplements for swine. It is an excellent feed for pig starters, where it is being used extensively. It is an excellent source of protein and B vitamins and also contains some unidentified factor or factors. It is also a source of lactose, which has been shown to be the best sugar for baby pigs. Dried skim milk is very palatable and will be eaten in excess if self-fed. It is a source of higher quality protein than tankage or fish meal.

Dried whey contains 65–70% of lactose. It is an excellent source of B vitamins and is even higher in riboflavin than skim milk. It is used in many diets at levels of 2–3% and sometimes at higher levels. There are limitations on how high it can be used in the diet as indicated in Tables 11.2, 11.3 and 11.4 and discussed in Chapter 6, Section II and Chapter 10, Section III.

Liquid skim milk and buttermilk are excellent supplements for farm grains for pigs of all ages. Both have approximately the same feeding value for swine. The small supply available and its bulk, which increases transportation costs, account for the small amount of these two liquid products being used in swine feeding.

Condensed or semisolid buttermilk is an excellent supplement to farm grains for swine. It is more valuable for young pigs than for heavier animals.

In a number of countries there is a considerable amount of liquid whey fed to pigs. In Europe, whey is fed *ad lib* to growing-finishing pigs plus 2.5 lb of concentrate feed (containing about 17% protein) per pig daily (39). Water is not provided for pigs fed liquid whey. Heating whey to 40°C (as compared to cool whey at 15°C) increased rate of gain 11% at 16°C air temperature and 5% at 22°C air temperature (40). The economics of feeding liquid whey and the concentrate mixture will depend on their relative prices.

VI. OTHER FEEDS

There are many other feeds which are used in swine feeding. They will be discussed briefly. Those wanting more detail on their use should check with their local area Experiment Stations for more detailed information on the kind of product available, how it is processed, and how best to feed it to swine under the conditions and feeds available in the area. Some of these feeds are used a great deal whereas others are used very little.

1. Buckwheat

It contains about 14% protein and 9–11% fiber. It also contains a photosensitizing agent which sometimes causes skin eruptions in white pigs exposed to sunlight. It should be ground and used at levels of one-fourth to one-third of the diet because of its high fiber content. It is best to use it primarily for growing-finishing pigs after they weigh about 100 lb.

2. *Oat Mill Feed*

This is a bulky feed high in fiber and low in total digestible nutrients. If used in swine diets, it should be fed at levels of 5 or 10% and no higher than 15% of the diet. At the 15% level it would have one-fourth to one-third the feeding value of corn for finishing pigs. It can be used to add bulk to a diet for sows during the gestation period.

3. *Oat Hulls*

The oat hulls have very little feeding value for swine. They are high in fiber and low in total digestible nutrients. They can be used to add bulk to a diet used for sows during the gestation period.

4. *Cane Molasses*

Sometimes called blackstrap molasses, this is an excellent feed for swine when used properly. In processing a ton of sugar cane, about 220 lb of refined sugar and 55–110 lb of cane molasses (11) are obtained. The results obtained with cane molasses have varied considerably, depending on the level used, the diet fed, and the size of the pigs when started on the molasses diet. Molasses has its highest value when used at a low level of 2.5–10% of the diet. The higher the level used, the lower becomes the relative feeding value of molasses as compared to grain. When the availability and price of molasses warrant it, higher levels of molasses can be used in swine diets.

Heavier pigs can use higher levels of molasses more effectively than smaller animals. The level used will vary depending on the diet fed, method of feeding, price of molasses and grain, and other factors. Table 11.8 gives recommended levels of cane molasses to use where the availability and price of molasses warrant using high levels in the diet.

The sugars present in molasses are well utilized by the pig. But the young pig, until it is 3–4 weeks of age, does not produce enough sucrase, the enzyme which breaks down sucrose (see Chapter 6, Section II and Chapter 10, Section III), consequently, for best results, cane molasses should not be used in baby pig diets until they are at least 4 weeks old.

There is considerable variation in the chemical composition of cane molasses depending on its variety, quality, age, soil type it is grown on, level of fertilization, method of processing and other factors. This accounts for some of the variability in the results obtained in different feeding trials with cane molasses. Most investigators have reported on the laxative effect or "loose-feces" when molasses is fed at high levels. But even though the feces are loose, the pigs can

TABLE 11.8
Level of Cane Molasses to Use in Diet for Growing-Finishing Pigs

	Level of molasses in diet (%)	
Weight of pig (lb)	Range	Probable safe level
40–100	10–15	10
100–150	20–30	20
150–225	40–50	40

still do well. The levels of molasses shown in Table 11.8 will cause some loose feces. So, the main concern should be (1) whether the pigs do well on high levels of molasses and (2) whether it is economical to feed high levels of cane molasses. The major problem in using high levels of molasses will be to devise proper feeding and management practices to facilitate using the liquid product. There is considerable variation in the individual mineral levels in cane molasses (11). This could be involved in the laxative effect of the molasses although there are studies which raise some doubt about this (11). Some of the laxative effect of molasses can be taken care of by adding fibrous products to the molasses. This was helpful in Hawaiian studies (41) in which 13% cane bagasse was added to a diet containing 50% molasses, but this reduces the total energy level in the diet. This can be remedied by adding 20% crude sugar (42) or 10–20% fat to the diet (41,43). The economics and feasibility of adding crude sugar and fat to the high molasses diets would have to be considered, however. Under certain situations it could be advantageous to do so.

5. *Beet Molasses*

It has an even greater laxative effect than cane molasses. It is apt to cause scours in pigs unless they are fed limited amounts and are started on it rather gradually. Utah studies (44) showed that young pigs weighing 25–50 lb developed a staggering, wobbly gait, performed poorly, and often died when fed a level of 15% beet molasses. Hogs weighing over 100 lb at the start, however, were fed as much as 40% beet molasses without any detrimental effects. Newborn pigs, however, were affected with an incoordinated gait, and death losses were high if their dams had been fed a 40% beet molasses diet during gestation.

The addition of 5% of brewers' yeast or 5% of freshly cut alfalfa (on a dry matter basis) prevented the symptoms in the growing pigs as well as in newborn pigs. If the alfalfa was allowed to dry out in the sun for a day, it lost whatever the factor was that prevented the symptoms. Beet molasses can be used at approximately the same levels as recommended for cane molasses (Table 11.8). However, the pigs should be carefully observed to ascertain that they are doing well; if not, the level of molasses fed should be decreased until they are.

6. *Citrus Molasses*

It has a bitter taste which makes it very unpalatable for the pig. Florida studies (45) showed that it takes the pig 3–7 days to get used to the taste of citrus molasses in the feed mixture. After that, the pig will consume the mixed diet containing citrus molasses. Because of the bitter taste, it has been found necessary to mix citrus molasses with other feeds for pigs to consume it readily. Very little difference was found in the feeding value between citrus and cane molasses for growing-finishing pigs. Citrus molasses can be fed at approximately the same levels recommended for cane molasses (Table 11.8). However, a certain amount of extra effort will be required to get the pigs started consuming the feed mixture because of the bitter taste of citrus molasses.

7. *Citrus Pulp*

This is a bulky feed. Two Florida trials have shown 20% of citrus pulp in the diet to be unsatisfactory for growing-finishing pigs. In one of these trials, dried grapefruit pulp replaced corn at a level of 5, 10, and 20%. Gains and feed requirements were similar when 5% of grapefruit pulp was substituted for corn, although care had to be taken to avoid digestive disturbances. Feeding more than 5% of grapefruit pulp, however, caused frequent digestive disturbances, decreased the rate of gain, and increased feed requirements (46).

8. *Citrus Meal*

Citrus meal is the finer particles of dried citrus pulp. Levels of 0.5, 2.0, and 5.0% of citrus meal were incorporated into a corn-soybean diet and fed to weanling pigs in dry lot. The results obtained indicated that as much as 5% citrus meal can be satisfactorily fed to young pigs (47).

9. *Citrus Seed Meal*

This meal is a by-product of citrus seed from which the oil has been extracted. It contains approximately 35% protein and is a valuable cattle feed.

A Florida trial (48) showed the citrus seed meal to be harmful to the pig when fed at levels of 10 and 25% of the diet. It caused rough haircoats and unthrifty pigs. Citrus seed meal is unpalatable and caused a markedly reduced feed intake by the pig. It also caused a decreased rate of gain and increased feed requirement per pound of gain. It is thought that the toxic factor is the bitter principle in citrus seed called limonin.

Thus, citrus seed meal is a good feed for cattle but is harmful for the pig and should not be used in swine diets.

10. Cull Citrus Fruit

This fruit can be eaten by pigs. They will eat tangerines and oranges in preference to grapefruit when these are fed free-choice. They consume the juice, seeds, and rag of the fresh citrus (46). Cull tangerines were studied in a later Florida trial (49). Pigs fed cull tangerines alone plus minerals made practically no gain in weight, although they increased in body size at the expense of condition. Good results were obtained when the cull tangerines were self-fed, free-choice, with a trio mixture (peanut meal, meat scraps, and alfalfa meal) and minerals.

11. Dried Bakery Product

It is produced from reclaimed bakery products that have been blended to give a feed containing about 9.5% protein and 13% fat. Studies were conducted at Florida with early weaned 2-week-old pigs fed a starter diet with 0, 5, 10, 15 and 20% of dried bakery product substituted for equal amounts of dried skim milk (49). Although the feed required per pound of gain was slightly increased with increasing levels of dried bakery product, the daily gains were about the same and feed cost data favored the use of dried bakery product. Other studies at Virginia, Georgia and Nebraska showed that dried bakery product can be used advantageously in pig starter diets with a decrease in cost of the feed mixture. A Florida study (50) showed that dried bakery product can be substituted for at least 30% of the corn in the diet of the pig with excellent results. A Virginia study (51) involved using 0, 12, 24, and 36% dried bakery product in the diet of 33.9–117 lb pigs to a market weight of 210 lb. Feed efficiency was improved at all levels of dried bakery product. There was a trend for improved daily gain with the 12 and 24% levels of dried bakery product. Feed costs were less for all levels of dried bakery product used. All of those studies indicate that properly processed dried bakery product is an excellent feed for swine. It would have its maximum value in starter diets, but it can also be used at levels up to 36% in growing-finishing diets if the relative feed costs favor its use.

12. Sugar

Sugar is an excellent source of energy in swine diets, except for the baby pig which, until it is 3–4 weeks of age, does not produce enough sucrase for sucrose breakdown. Studies at Florida (56) showed that ''C'' sugar (the lowest grade of sugar from the standpoint of polarization and freedom from impurities) could be used in creep diets for pigs at levels of 10–40%. In another Florida study (55) in which growing-finishing pigs were fed 10–40% ''C'' sugar, similar results were obtained as with a control diet in which sugar replaced corn. Studies by Maner (11,52) showed that refined and crude sugar replaced corn at levels of 15–60% in the diet with comparable results. There was a slight increase in rate of gain and feed efficiency with the sugar diets. Studies in Hawaii (53) showed that pigs fed

45–63% sugar (crude or refined) developed heart lesions and a hemorrhagic syndrome which was prevented by vitamin K. The need for vitamin K was verified in Florida studies (54). These studies indicate that vitamin K supplementation needs to be considered when sugar is used to an appreciable extent in swine diets.

13. Potatoes

Potatoes have about 2.5–3.0% protein and 22–23% dry matter. Only cull or surplus potatoes are fed to livestock in the United States. Raw potatoes produce poor results when fed to swine and thus need to be cooked. Cooking helps to reduce the effects of solanin (an organic compound in potatoes) and changes the starch into a form more readily digested by the pig. It is best to add salt to the water in which the potatotes are cooked to increase their palatability. This water should be discarded, as it is not palatable.

Potatoes should be used at a level of 3–4 parts of potatoes to one part of grain in a well-balanced diet. At this level, 350–450 lb of potatoes will be equal to 100 lb of corn in feeding value. Dehydrated potatoes are equal to corn in feeding value when limited to one third of the diet for pigs on pasture.

European studies (11,57) showed that feeding 2.5 lb daily of a 19% protein concentrate and cooked potatoes *ad lib* gave similar results to an all-concentrate diet. With this program, the daily consumption of cooked potatoes increased with increasing pig weight. Pigs fed raw potatoes in a similar program required 40 days longer to reach market weight and gained 31% more slowly (57).

14. Sweet Potatoes

These potatoes are grown chiefly for human food, but many of the culls are fed to swine. They contain about 32% dry matter and 1.5% protein. Hogging-off sweet potatoes gives good results, but the cost of production makes this an uneconomical practice. Cooking sweet potatoes improves their feeding value. Better results are obtained when they are fed to older pigs, since they are somewhat bulky for small pigs.

Best results are obtained when they replace 40–50% of the grain in a well-balanced diet. When fed properly, it takes 3–4 lb of sweet potatoes to replace a pound of corn in a well-balanced diet. Dehydrated sweet potatoes have approximately 80–90% the feeding value of corn when used at a level of one-third to two-thirds of a well-balanced diet.

15. Chufas

Chufas produce a tuber which remains in the ground and is hogged-off in some sections of the South. Chufas are low in protein and thus should be supplemented with a good protein supplement and minerals. They produce soft pork which can be hardened by feeding a hardening diet for an adequate period of time.

16. Peanuts

Peanuts are a palatable feed for swine. Peanuts lack calcium and salt but will give good results if these and other minerals are supplied. Peanuts without the hulls contain about 30% protein and 48% fat. Pigs fed peanuts produce soft pork. Studies conducted to date show that it takes a considerable length of time to harden the fat in pigs which have been fed peanuts for any length of time. It may take 3–3.5 times or more gain on a hardening diet compared to the gain on peanuts to produce hard or medium-hard hog carcasses. This means that 40 lb of gain on peanuts would probably require 120–140 lb of gain on a hardening diet.

A Florida study (58,59) showed that 50 lb pigs fed peanuts to 122, 151, and 165 lb weights and then fed hardening diets until they weighed approximately 200 lb still produced carcasses which graded soft. These groups had 111.3 lb of "soft" gain and 27.7 lb of "hard" gain in group I, 98.3 lb of "soft" and 51.5 lb of "hard" in group II, and 71.5 lb of "hard" and 101.0 lb of "soft" gain in group III. Peanut feeding tended to yield carcasses with more back fat thickness in relation to carcass length and thus caused lower carcass grades. More studies are needed on this problem especially to determine whether any special feed ingredients might be used to harden the pork of peanut-fed pigs. To eliminate soft pork the pigs should not be fed peanuts after they weigh 75–85 lb.

17. Soybeans

Soybeans are usually more profitable when harvested rather than when hogged-down. Soybeans produce soft pork (if they are fed as a major part of the diet) and also the raw soybean is lower in feeding value than well-cooked soybeans or soybean meal. Raw soybeans contain several inhibitors which are destroyed on heating, thus soybeans must be heated before the pig can utilize them efficiently. Raw soybeans do not give good results when used as the only protein supplement for brood sows or for young pigs. They give better results with well-grown pigs and are more satisfactory on pasture than in dry lot.

If pigs are used to hog-down a field of raw soybeans, they will make good gains if they are fed a mineral mixture and a limited amount of corn as well. The carcasses, however, will be soft. They can be hardened by a period of feeding on a hardening diet. This period will be less than with peanuts, but will vary, depending on the weight of the pigs when started on soybeans, the length of time they are fed, and the other feeds used.

Brood sows may be fed three-fourths to one pound of soybeans as a protein supplement per head daily in otherwise balanced diets during gestation (2). South Dakota (60) studies showed that growing-finishing swine could be fed 22% whole cooked soybeans in a corn diet with a similar rate of gain and no adverse effect on consumer acceptance of the pork. The pork fat was noticeably softer, however, and was also higher in linoleic acid. So whole soybeans can be used in swine feeding if economic considerations warrant it. For best results, however, a cooker should be used and the intake of soybeans limited in order to avoid soft pork.

18. Sunflower Seed Meal

Sunflower seed meal contains about 41–47% protein. It has not given good results when used as the only protein supplement in swine diets. It is deficient in lysine unless properly processed. High temperature processing reduces lysine and other amino acids. Sunflower seed meal can be used as 20–30% of the protein supplement for growing-finishing pigs. Preferably, it should be fed to animals that weigh 75–100 lb (28–30). It gives best results when combined with soybean meal, fish meal, and other feeds rich in lysine. Recently, there has been considerable improvement in the quality of sunflower seed meal as processing methods have lowered its fiber and oil content and decreased the temperatures required in processing.

19. Cull Peas

Cull peas contain about 26% protein and can be used as a grain or protein substitute (61,62). Diets of wheat and barley—supplemented wholly or in part with cull peas and fed to growing-finishing pigs on sudan grass pasture—produced as rapid and efficient gains as diets supplemented with soybean meal or meat meal (62).

Cull peas compare very favorably with soybean meal and meat meal as a protein supplement for growing-finishing pigs fed in dry lot. The diets used contained wheat, or a mixture of wheat and barley, minerals, and either 5 or 15% alfalfa meal. The pigs fed 15% alfalfa meal during growth performed better during conception, gestation, and lactation. The information obtained indicated that the diet fed during growth has a very important bearing on results obtained later during conception, reproduction, and lactation (63). Cull peas were shown to be a satisfactory protein supplement for sows when fed with wheat-barley diets containing 20% alfalfa meal (64).

20. Cull Beans

Cull beans are frequently fed to swine. They should be cooked, since this improves their feeding value and palatability for swine. A good quality protein supplement should be fed along with the cooked beans. They can replace from one-fourth to two-thirds of the corn in finishing diets and will most likely give their best results with 75–100 lb pigs being finished for market.

21. Velvet Beans

Velvet beans contain dihydroxphenylalanine, which is closely related to adrenaline, and which is toxic for the pig. Velvet beans also cause diarrhea and vomiting. The toxicity of the beans is partly reduced by cooking, which makes them more digestible and palatable. Even then, however, they are not a satisfactory hog feed.

Velvet beans can be used with fair results if they do not form over one-fourth of the diet and if a high quality protein supplement is fed also. Velvet bean pasture is not recommended for swine. Pigs will not do well if they follow cattle fed heavily on velvet beans unless they also get considerable other feed as part of their diet. If used, velvet beans should be fed in small amounts and only to heavier pigs, and not to brood sows.

22. *Fish Solubles*

Fish solubles are excellent sources of B-complex vitamins. They are used at levels of 1–3% in swine diets primarily as a source of vitamins as well as for some unidentified factor or factors.

23. *Crab Meal*

This is unpalatable to pigs and has not been used successfully in swine feeding (65).

24. *Shrimp Meal*

Shrimp meal contains about 40–45% protein and is about equal to tankage in feeding value (66). It is the dried waste of the shrimp industry and can vary in composition depending on how much whole shrimp it contains.

25. *Shark Meal*

Shark meal contains about 78% protein and is a satisfactory protein supplement for growing-finishing pigs. It did not produce any off or fishy flavor in the lean or fat of the animals when fed at levels of 5.9–13.98% of the diet (67).

26. *Blood Meal*

Blood meal contains 70–85% protein. It is less digestible and of poorer quality than good quality tankage for pigs. Blood meal is low in isoleucine. It is best to use blood meal at levels of 1–4% of the diet and in combination with high quality protein supplements. It should not be used in baby pig diets. A number of studies have shown blood meal can be used satisfactorily in swine diets if properly supplemented (68–70).

27. *Cheese Rind*

This contains about 60% protein and has about the same feeding value as tankage for the pig (71).

28. Liver Meal

Liver meal is an excellent source of B-complex vitamins and contains an unidentified factor or factors. It is used at very low levels as a source of these nutrients.

29. Feather Meal

It contains 85–87% protein. It needs to be hydrolyzed in order to be used efficiently in swine feeding. Feather meal should be used at levels of 2.5–5.0% of the diet for swine for best results (72–74). When used at higher levels, growth rate and nitrogen retention decrease.

30. Poultry By-Product Meal

It contains about 60% protein. Florida studies by Dr. George Combs showed that levels of 8.5–16.0% poultry by-product meal could be used successfully in swine growing-finishing diets. It is best to limit this meal to levels of 5–8% in the diet since variability can occur in the quality of the product available.

31. Incubator Eggs

Occasionally incubator eggs which have not hatched are available for swine feeding. A number of studies (75,76) have shown that hatchery eggs can be fed to growing-finishing swine with good results. The eggs were comparable to tankage in feeding value. A possible biotin deficiency (77) could occur, however, if the eggs have not been heated long enough to denature the avidin in the egg white. The raw egg white protein (avidin) ties up biotin and causes a deficiency of this vitamin (77). Heating denatures the avidin and prevents biotin deficiency. So, heating the eggs may be necessary to insure best results.

32. Distillers' Dried Solubles

These are most valuable as a source of B-complex vitamins in swine diets. The protein of distillers' solubles is deficient in lysine and tryptophan as is corn. However, it can be used to replace some of the protein supplement in a trio mix or mixed supplement. It is less satisfactory as a substitute for alfalfa meal in a trio mix. It also has been of less value when added to a diet containing high quality alfalfa meal. The more alfalfa in the diet and the higher its quality, the less beneficial is the effect of adding distillers' solubles to the diet (78,79). They are used at levels of 2.5–5.0% in the diet.

33. Distillers' Dried Grains with Solubles

They vary considerably in composition because of the various types of distillers' dried grains available. They are usually fed at levels of 2.5–5% in the diet. They are an excellent source of B-complex vitamins but are low in lysine.

34. Yeast

Yeast, when irradiated, is used as a source of vitamin D. Dried brewers' yeast and *Torula* yeast are excellent sources of B-complex vitamins. Usually yeast is fed at levels of 2–5% of the diet for supplying B-complex vitamins. Yeast contains about 50% protein, but it is too expensive to be used as a source of protein in swine diets (80).

35. Alfalfa Meal

Alfalfa meal is the most widely used of the legume meals for swine feeding. It is an excellent source of carotene and the known B-complex vitamins. It also contains an unidentified factor or factors which is of value for pigs and sows fed under dry lot conditions. California studies showed that alfalfa has an unidentified factor or factors advantageous to the guinea pig (86).

Prospective herd replacement pigs fed in dry lot should have 5–10% alfalfa meal in the diet during the growing-finishing period. Growing-finishing pigs which are going to be slaughtered can be fed 2.5–5% dehydrated alfalfa meal in the diet. With some diets, this level can be increased up to 10% without affecting rate of gain or efficiency of feed utilization.

Levels of 15% or more alfalfa meal in the diet usually tend to slow rate of gain and increase feed requirement per pound of gain. Levels in excess of 10% may be used, however, if rate of gain is not an important consideration and if the cost of gain is reduced. Moreover, experimental work has shown that pigs fed 5–10% alfalfa meal during growth store some factor or factors which later made them conceive faster and reproduce and lactate more satisfactorily than sows which had been fed less than 5% of alfalfa during growth (63,81,85).

During gestation and lactation, sows can be fed from 8–20% or more of alfalfa meal in the diet. The level to use will depend on the other feeds in the diet and their nutritional adequacy. Many experiments have shown that using higher levels of alfalfa in sow diets was of considerable benefit (63,81–85,87). In 1975, the results of three studies showed that the gestation diet containing the highest level of alfalfa hay produced the greatest percentage of gilts farrowing in each study (87).

The use of high quality dehydrated alfalfa meal with swine fed in dry lot is a means of substituting for pasture. It is the closest substitute available for pasture and thus insures pigs of most of the nutritional value of pasture.

36. Other Hays

There is not much information available on the feeding value of other hays for swine feeding. High quality soybean, lespedeza, kudza, red clover, cowpea, sudan grass, bermuda grass, and other hays may be used in swine feeding. These hays would be worth somewhat less than alfalfa in swine diets but can be satisfactory under many conditions.

37. Silage

Very little information is available on feeding silage for growing-finishing pigs. Some experimental information is available on feeding silage to sows. Silage will undoubtedly be used in sow diets in the future. As more information is obtained, the role of silage in sow diets will become clearer. Recommendations on feeding silage to sows is given in Chapter 14.

38. Tung Nut Meal

This is unpalatable to pigs. They refused to consume mixtures containing 30–50% of detoxified tung nut meal in combination with soybean meal. When the tung nut meal was reduced to 5, 10, or 15% of the protein mixture, limited consumption occurred. Consumption, however, appeared to be less than required for optimum gains (88).

39. Fresh Avocado Pulp

Fresh avocado pulp, in which the fruit was halved and the seeds removed, was fed to 60 lb growing-finishing pigs. Good gains were obtained. The pigs like the pulp very well and consumed an average of 7.15 lb (wastage included) of it per day in addition to 2.48 lb of the control concentrate diet. It took 9.64 lb of the avocado pulp to replace one pound of the control corn-soybean meal diet. Avocado pulp fed alone gave poor results. The avocado seed was extremely unpalatable to the pig. A pig offered only avocado seeds refused to eat them for a period of one week even though no other feed was available (89).

40. Chickpea

It is also known as garbanzo and is used for human consumption. It contains 18–26% protein and is low in methionine. The chickpea has a feeding value similar to that of a mixture of corn and soybean meal. It probably should be fed at a level no higher than one-half the corn and soybean meal in the diet for growing-finishing swine. The chickpea can be fed to swine without cooking or heating.

41. Cowpeas

They have about 24% protein and are low in methionine. The nutritional value of cowpeas is greatly increased by cooking. Cooked cowpeas probably should be limited to a level no higher than one-half in a corn soybean meal diet for growing-finishing pigs.

42. Field Bean

It is also called a horse bean and contains 25–30% protein which is low in methionine. It can be used to replace 50% of the supplemental protein for growing-finishing pigs. It has been fed at a 10% level in the diet of sows with no adverse effects on the litters produced (90).

43. Algae

Algae have potential as a microbial protein source for animal feeding. But there are problems to be solved regarding its production and harvesting. Algae have a bitter taste which decreases palatability of the diet, thereby reducing feed intake. Protein availability is also low because the algal cell wall is difficult to break down. California studies showed that the protein of algae was only 54% digestible for the pig (91). California workers (91,92) studied algae in swine feeding and recommend that it be fed at a level no greater than 10% of the diet for growing-finishing swine. If algae are to be used in significant quantities in swine feeding in the future, the problems of bitter taste and low protein availability as well as production and harvesting methods must be resolved.

44. Seaweed

It is apparent that seaweed is not a good source of either energy or protein. If it is used in swine diets, it should be as a source of minerals and especially trace minerals.

45. Bananas

The most comprehensive work on bananas for swine feeding has been conducted by Dr. Jerome Maner and Associates at CIAT (95). Reject and other waste bananas are sometimes available for swine feeding. Whole bananas (including the peel) contain about 20% dry matter and 1% each of protein, ash, and fiber. Bananas can be fed fresh or as a dried meal (93). Pigs will voluntarily consume approximately 50% as much of green bananas as compared to ripe bananas (93). The green banana is bitter to the taste as compared to ripened bananas which are higher in moisture and sweeter. Cooking the green bananas increases both their consumption and pig performance but not to the level of ripe bananas (93). The green banana has an astringent taste which, in part, is due to tannins. These free tannins become bound during ripening and then have no effect on palatability of the banans (93). During ripening, which takes 10–11 days, there is a decline in starch content and a corresponding increase in sugars (mainly sucrose) (93). Pigs fed whole, ripe bananas will first consume the banana pulp leaving much of the peel. But, if the total daily quantity of bananas fed is limited, the pig will consume both pulp and peel (93).

The high moisture and low protein content of bananas means that a supplementary source of protein, energy, as well as minerals and vitamins is needed. A 30% protein supplement is better than a 40% protein supplement with ripened bananas (94). The improvement in gain and feed conversion with the 30% protein supplement was assumed to be due to the increased daily consumption of metabolizable energy. The use of a 20% protein supplement was no better than the 30% protein supplement. Growing-finishing pigs will average consuming 17.6–19.4 lb per day of ripened bananas during the entire period. Pigs weighing 55–66 lb will consume 11–13 lb of bananas per day and 22–24 lb as they approach market weight.

Fresh bananas can be used efficiently for the gestating sow. The sows were fed 9.9 lb of bananas and 600 gm of a 40% protein supplement per day from breeding until the seventy-sixth day of gestation. The diet was then changed to 13.2 lb of bananas plus 800 gm of the 40% protein supplement per day until the one hundred tenth day of gestation. The sows fed bananas performed as well as those fed a well-balanced diet with no bananas. During lactation, however, the level of bananas fed must be limited since the sow cannot consume enough bananas to meet her energy needs. Moreover, sows develop a diarrhea when they consume 30–33 lb of bananas daily. Therefore, banana feeding should be limited to the extent that the lactating sow obtains enough energy to satisfy about 11 lb of dry matter intake in the daily diet.

46. Plantains

The plantain is similar to a banana in appearance, but it is used for cooking and not eaten fresh. It contains more dry matter but less sugar than the banana. This may be due to the plantain requiring 6–8 days longer than the banana to ripen. The plantain should be fed to swine in the same manner as the banana. The results obtained, however, will be slightly inferior to the performance of the growing-finishing pigs fed ripe bananas. No differences were obtained with sows fed the plantain or the bananas however (93–95).

47. Dried Bananas and Plantains

Both can be dried, but it is difficult to dry the ripened product. Therefore, the dried meal is usually prepared from either green bananas or green plantains. Both meals contain about 4.3% protein, but the plantain meal has 6.2% fiber versus 3.0% for the banana meal. Both banana and plantain meals are used in complete diets to replace part of the grain. These meals should be limited to 30–50% of the diet for growing-finishing pigs and for sows during gestation and lactation. The exact level to use will depend on the remainder of the diet and how well the pigs perform with the different levels used (93–95).

48. Garbage

Garbage varies considerably in nutritional value depending on its source. A major problem in feeding garbage involves disease and sanitation. Garbage can result in spreading hog cholera and other diseases. Therefore, it needs to be properly cooked to destroy all the disease organisms present. Garbage usually contains 70–85% or more water. The high moisture content means the young pig has a difficult time consuming enough total dry matter to do well. Therefore, most garbage feeders start pigs weighing from 100 to 125 lb on garbage feeding. New Jersey studies showed that supplementing cooked garbage with a 15–17.5% protein supplement improved rate of gain and efficiency of garbage utilization (96). Levels of 0.75–2.7 lb of the protein supplement were used in the New Jersey studies. The level of protein supplement to use would depend on the quality of the garbage and the weight of the pigs. Therefore, protein, mineral and vitamin supplementation would vary depending on the conditions encountered. Mortality is often high in garbage feeding. The use of an antibiotic and other antimicrobial compounds would be desirable with garbage-fed swine. Garbage feeding can result in higher fat deposition and a higher level of soft fat. This will be largely dependent, however, on the kind of garbage being fed.

49. Coffee Grounds

Kansas data showed that coffee grounds cannot be used efficiently at more than 10% as a grain replacement in pig diets (97). But even at a 10% level in the diet, growth rate and feed efficiency decreased. Pigs fed higher levels decreased in performance and those fed 40% coffee grounds actually lost weight.

50. Dried Apple Pomace

Canadian studies showed that up to 30% dried apple pomace could replace barley and oats in the diet of growing-finishing pigs with good results. Levels of 40% pomace decreased rate or gain and feed efficiency (98).

51. Cassava

Cassava is a carbohydrate feed source which has considerable potential in the tropical areas. It is also known as yucca, manioc, tapioca or manioca. It is used for human consumption, but its use for animal feeding is only in the early stages of development. According to Pond and Maner (11), cassava has about 35% dry matter, 1.25% protein, 1.45% fiber, 1.43% ash, 0.29% fat, 0.12% calcium, 0.16% phosphorus, 0.06% sodium, 0.37% magnesium, and 0.86% potassium. There are many varieties of cassava. Some will contain different levels of nutrients then those included herein. But in general, cassava use will necessitate proper supplementation with protein, minerals and vitamins. Both methionine and cystine are the first limiting amino acids (11).

The roots and leaves of the cassava contain hydrocyanic acid (HCN) or prussic acid. Free HCN, which is toxic, is formed when normal plant growth is retarded or when the leaves and roots are cut or bruised. Acute HCN toxicity is very harmful. But small quantities of HCN are not sufficient to cause death. They do, however, affect the general health of the animal and reduce its efficiency of using a number of nutrients. Oven-drying, with little moisture present, eliminates the free HCN. Cooking of the cassava in water destroys the HCN. Chopping, mixing, and sun-drying is an effective and practical method of liberating HCN. These methods not only need to liberate the HCN but also destroy the enzymes which liberate HCN from the glucosides present in cassava.

Cassava in its fresh state cannot be stored for more than 3–5 days without fermentation and decreased feeding value occurrring. The use of refrigeration will increase storage time. Freezing of the roots will result in preservation for almost a year. The roots can also be stored for about a year if they are dried first or are ensiled.

Many feeding trials have been conducted with cassava for the growing-finishing pig. Most of these studies were conducted in Latin America. Pond and Maner (11) reviewed these studies in detail. Essentially they show that cassava, properly processed to eliminate toxicity from HCN, is an excellent source of energy for the pig. The energy digestibility is similar to that of the grains used for swine feeding. It appears that up to 40% cassava meal can be used with good results for the growing-finishing pig. Higher levels of cassava meal in the diet reduce growth and feed efficiency and may require methionine and possibly other nutrient supplementation for best results. Some of the reduced performance may be due to reduced feed intake because of the slightly bitter taste of low levels of HCN which may be in cassava meal. More studies are needed on cassava as a feed for sows to determine their effect on reproduction and lactation. But, preliminary studies by Dr. J. H. Maner (11) indicate that cassava was satisfactory for sows although litter size was somewhat less. The consumption of fresh cassava was also low in lactating sows.

REFERENCES

1. Gilster, K. E., W. T. Ahlschwede, E. R. Peo, Jr., M. Danielson, R. D. Fritschen, and B. D. Moser, *Nebr., Agric. Exp. Stn.* **EC74-210** (1974).
2. Krider, J. L., and W. E. Carroll, "Swine Production," 4th ed. McGraw-Hill, New York, 1971.
3. Meade, R. J., W. R. Dukelow, and R. S. Grant, *J. Anim. Sci.* **25,** 58 (1966).
4. Maxson, D. W., G. R. Stanley, T. W. Perry, R. A. Pickett, and T. M. Curtin, *J. Anim. Sci.* **27,** 1006 (1968).
5. Seether, K. A., T. S. Miya, T. W. Perry, and P. N. Boehm, *J. Anim. Sci.* **32,** 1160 (1971).
6. Braude, R., J. Townsend, G. Harrington, and J. G. Rowell, *J. Agric. Sci.* **57,** 257 (1961).
7. Larson, L. M., and J. E. Oldfield, *J. Anim. Sci.* **19,** 601 (1960).
8. Conrad, J. H., *Feedstuffs* **30,** 38 (1958).
9. Dinusson, W. E., *Feedstuffs* **32,** 28 (1960).
10. Jensen, A. H., W. F. Nickelson, B. G. Harmon, and D. E. Becker, *Ill., Agric. Exp. Stn.* **AS-633c** (1966).
11. Pond, W. G., and J. H. Maner, "Swine Production in Temperate and Tropical Environments." Freeman, San Francisco, California, 1974.
12. Bundie, E., *Feedstuffs* **46,** 16 (1974).

13. Holden, P., V. C. Speer, E. J. Stevermer, and D. R. Zimmerman, *Iowa, Agric. Exp. Stn., Rev.* **Pm-489** (1974).
14. Foster, J. R., V. B. Mayrose, T. R. Cline, H. W. Jones, M. P. Plumlee, D. M. Forsyth, and W. H. Blake, *Purdue Agric. Exp. Stn., Swine Day Rep.* (1975).
15. Miller, E. C., E. R. Miller, and D. E. Ullrey, *Mich. Ext. Bull.* **537** (1975).
16. Wallace, H. D., T. J. Cunha, and G. E. Combs, *Fla. Agric. Exp. Stn., Bull.* **566** (1955).
17. Noland, P. R., M. Funderburg, J. Atterberry, and K. W. Scott, *J. Anim. Sci.* **27,** 1319 (1968).
18. Burnside, J. E., T. J. Cunha, A. M. Pearson, R. S. Glasscock, and A. L. Shealy, *Arch. Biochem.* **23,** 328 (1949).
19. Barber, R. S., R. Braude, K. G. Mitchell, J. D. J. Harding, G. Lewis, and R. M. Loosmore, *Br. J. Nutr.* **22,** 535 (1968).
20. Knake, R. P., C. S. Rao, and C. W. Deyoe, *Feedstuffs,* **44,** 32 (1972).
21. Saben, H., and J. P. Bowland, *Feedstuffs* **44,** 22 (1972).
22. Bowland, J. P., *Can. J. Anim. Sci.* **55,** 409 (1975).
23. Bowland, J. P., and R. T. Hardin, *Can. J. Anim. Sci.* **53,** 355 (1973).
24. Taverner, M. R., and P. D. Mullaney, *Aust. J. Exp. Agric. Anim. Husb.* **13,** 375 (1973).
25. Maner, J. H., and J. T. Gallo, *Congr. Nac. Ind. Porcina, 1st,* pp. 1–3 Bogota, Columbia (1963).
26. Hervas E., J. Viteri, and J. H. Maner, INIAP Quito, Ecuador, unpublished data (1965).
27. Seerley, R. W., D. Burdick, W. C. Russom, R. S. Lowrey, H. C. McCampbell, and H. E. Amos, *J. Anim. Sci.* **38,** 947 (1974).
28. Krider, J. L., D. E. Becker, W. E. Carroll, and H. D. Wallace, *J. Anim. Sci.* **6,** 401 (1947).
29. Wallace, H. D., M. Milicevic, and D. Kropf, *Fla., Agric. Exp. Stn., A. H. Mimeo* No. 4 (1953).
30. Pearson, A. M., H. D. Wallace, and J. F. Hentges, Jr., *Fla., Agric. Exp. Stn., Bull.* **533** (1954).
31. Wallace, H. D., Florida Agricultural Experiment Station, Gainesville, personal communication (1955).
32. Shimada, A. S., and A. Aguilera, *Tec. Pecv. Mex.* **7,** 6 (1966).
33. Shimada, A. S., and S. Brambila, *Tec. Pecv. Mex.* **8,** 30 (1966).
34. Bravo, F. O., and E. Ceballo, *Tec. Pecv. Mex.* **11,** 38 (1969).
35. Williams, K. C., and L. J. Daniels, *Aust. J. Exp. Agric. Anim. Husb.* **13,** 48 (1973).
36. Loosli, J. K., J. O. Pena, L. A. Ynalves, and V. Villegos, *Philipp. Agric.* **38,** 191 (1954).
37. Creswell, D. C., and C. C. Brooks, *J. Anim. Sci.* **33,** 366 (1971).
38. Creswell, D. C., and C. C. Brooks, *J. Anim. Sci.* **33,** 370 (1971).
39. Braude, R., University of Reading, England, personal communication (1971).
40. Holmes, C. W., *Anim. Prod.* **13,** 1 (1971).
41. Brooks, C. C., and I. I. Iwanaga, *J. Anim. Sci.* **26,** 741 (1967).
42. Preston, T. R., and M. B. Willis, *Feedstuffs* **42,** 20 (1970).
43. Brooks, C. C., *J. Anim. Sci.* **34,** 217 (1972).
44. Rasmussen, R. A., H. H. Smith, R. W. Philips, and T. J. Cunha, *Utah, Agric. Exp. Stn., Bull.* **302** (1942).
45. Cunha, T. J., A. M. Pearson, R. S. Glasscock, D. M. Buschman, and S. J. Folks, *Fla., Agric. Exp. Stn., Circ.* **S-10** (1950).
46. Kirk, W. G., and R. M. Crown, *Fla., Agric. Exp. Stn., Bull.* **428** (1947).
47. Wallace, H. D., G. E. Combs, and T. J. Cunha, *Fla., Agric. Exp. Stn., A. H. Mimeo Ser.* No. 2 (1953).
48. Glasscock, R. S., T. J. Cunha, A. M. Pearson, J. E. Pace, and D. M. Buschman, *Fla., Agric. Exp. Stn., Circ.* **S-12** (1950).
49. Combs, G. E., H. D. Wallace, and T. H. Berry, *Fla., Agric. Exp. Stn., A. H. Mimeo Rep.* No. AN65-1 (1964).

50. Wallace, H. D., *Feedstuffs* **37,** 52 (1965).
51. Kornegay, E. T., *Feedstuffs* **46,** 23 (1974).
52. Maner, J. H., H. Obando, R. Portela, and J. Gallo, *J. Anim. Sci.* **29,** 139 (1969).
53. Brooks, C. C., R. M. Nakamura, and A. Y. Miyahara, *J. Anim. Sci.* **37,** 1344 (1973).
54. Neufville, M. H., H. D. Wallace, and G. E. Combs, *J. Anim. Sci.* **37,** 288 (1973).
55. Combs, G. E., H. D. Wallace, J. W. Carpenter, A. Z. Palmer, and R. H. Alsmeyer, *J. Anim. Sci.* **18,** 1405 (1959).
56. Combs, G. E., C. E. Haines, and H. D. Wallace, *Fla., Agric. Exp. Stn., A. H. Mimeo Ser.* No. 55-15 (1956).
57. Braude, R., and K. G. Mitchell, *Agriculture (London)* **57,** 501 (1951).
58. Alsmeyer, R. H., M.S. Thesis, University of Florida, Gainesville (1956).
59. Alsmeyer, R. H., A. Z. Palmer, H. D. Wallace, R. L. Shirley, and T. J. Cunha, *J. Anim. Sci.* **14,** 1226 (1955).
60. Wahlstrom, R. C., G. W. Libal, and R. J. Berns, *J. Anim. Sci.* **32,** 891 (1971).
61. Lehrer, W. P., Jr., W. M. Beeson, and A. Wilson, *Idaho, Agric. Exp. Stn., Circ.* **108** (1946).
62. Warwick, E. J., T. J. Cunha, and M. E. Ensminger, *Wash., Agric. Exp. Stn., Bull.* **500** (1948).
63. Cunha, T. J., E. J. Warwick, M. E. Ensminger, and N. K. Hart, *J. Anim. Sci.* **7,** 117 (1948).
64. Colby, R. W., T. J. Cunha, and M. E. Ensminger, *Wash., Agric. Exp. Stn., Circ.* **153** (1951).
65. Kyzer, E. D., and J. E. Pace, *S.C., Agric. Exp. Stn., Rep.* p. 90 (1942).
66. Bray, C. I., J. B. Francioni, Jr., and E. M. Gregory, *La., Agric. Exp. Stn., Bull.* **228** (1932).
67. Marshall, S. P., and G. K. Davis, *J. Anim. Sci.* **5,** 211 (1946).
68. Hillier, J. C., R. MacVicar, S. A. Ewing, and W. Nickelson, *Okla., Agric. Exp. Stn., Mis. Publ.* **MP-43** (1956).
69. Meade, R. S., and W. S. Teter, *J. Anim. Sci.* **16,** 892 (1957).
70. Becker, D. E., I. D. Smith, S. W. Terrill, A. H. Jensen, and H. W. Norton, *J. Anim. Sci.* **22,** 1093 (1963).
71. Bohstedt, G., and J. M. Fargo, *Wis., Agric. Exp. Stn., Bull.* **430** (1935).
72. Combs, G. E., W. L. Alsmeyer, and H. D. Wallace, *J. Anim. Sci.* **17,** 468 (1958).
73. Barber, R. G., R. Braude, A. G. Chamberlain, Z. D. Hosking, and K. G. Mitchell, *Anim. Prod.* **7,** 103 (1965).
74. Hall, O. G., *J. Anim. Sci.* **16,** 1076 (1957).
75. Willman, J. P., C. M. McCay, O. N. Salmon, and J. L. Krider, *J. Anim. Sci.* **1,** 38 (1942).
76. Tomhave, A. E., and E. Hoffman, *J. Anim. Sci.* **4,** 247 (1945).
77. Cunha, T. J., D. C. Lindley, and M. E. Ensminger, *J. Anim. Sci.* **5,** 219 (1946).
78. Hanson, L. E., M. L. Baker, G. N. Baker, and M. G. A. Rumery, *Nebr., Agric. Exp. Stn., Bull.* **415** (1952).
79. Krider, J. L., and W. E. Carroll, *J. Anim. Sci.* **3,** 107 (1944).
80. Prouty, C. C., *Wash., Agric. Exp. Stn., Bull.* **484** (1947).
81. Cunha, T. J., O. B. Ross, P. H. Phillips, and G. Bohstedt, *J. Anim. Sci.* **3,** 415 (1944).
82. Fairbanks, B. W., J. L. Krider, and W. E. Carroll, *J. Anim. Sci.* **4,** 410 (1945).
83. Krider, J. L., B. W. Fairbanks, R. F. Van Poucke, D. E. Becker, and W. E. Carroll, *J. Anim. Sci.* **5,** 256 (1946).
84. Ross, O. B., P. H. Phillips, G. Bohstedt, and T. J. Cunha, *J. Anim. Sci.* **3,** 406 (1944).
85. Seerley, R. W., and R. C. Wahlstrom, *J. Anim. Sci.* **24,** 448 (1965).
86. Lakhanpal, R. K., J. R. Davis, J. T. Typpo, and G. M. Briggs, *J. Nutr.* **89,** 341 (1966).
87. Danielson, D. M., and J. J. Noonan, *J. Anim. Sci.* **41,** 94 (1975).
88. Wallace, H. D., and L. Gillespie, *Fla., Agric. Exp. Stn., A. H. Mimeo Ser.* No. 55-3 (1955).
89. Gillespie, L., and H. D. Wallace, *Fla., Agric. Exp. Stn., A. H. Mimeo Ser.* No. 55-8 (1955).
90. Clark, H. E., *Proc. Nutr. Soc.* **29,** 64 (1970).

91. Hintz, H. F., H. Heitman, Jr., W. C. Weir, D. T. Torell, and J. H. Meyer, *J. Anim. Sci.* **25,** 675 (1966).
92. Hintz, H. F., and H. Heitman, Jr., *Anim. Prod.* **9,** 135 (1967).
93. Clavijo, H., and J. H. Maner, *CIAT Ser. EE* No. 6, pp. 1–20 (1974).
94. Calles, A., H. Clavijo, E. Hervas, and J. Maner, *J. Anim. Sci.* **31,** 197 (1970).
95. Maner, J., CIAT, Cali, Columbia, personal communication (1974).
96. Kornegay, E. T., G. W. Vander Noot, K. M. Barth, G. Graber, W. S. McGrath, R. L. Gilbreath, and F. J. Biek, *Rutgers Univ., Agric. Exp. Stn., Bull.* **829** (1970).
97. Balogun, T. F., and B. A. Koch, *Trop. Agric. (Trinidad)* **52,** 243 (1975).
98. Bawden, D. M., and J. C. Berry, *Can. J. Anim. Sci.* **39,** 26 (1959).

12

Feeding the Baby Pig

I. INTRODUCTION

Feeding of pigs has advanced to the point where the growth period is divided into several segments, and the pig is fed according to the stage of the growth cycle it is in. The younger the pig, the more critical the period becomes. Consequently, higher quality and more highly fortified diets are needed. This chapter will discuss feeding the baby to weaning age. Chapter 13 will deal with feeding the growing-finishing pig from weaning to market weight.

The trend is toward early weaning of pigs at 3–5 weeks of age so sows can be rebred and maximize the number of litters obtained yearly. Thus, prestarter diets and starter diets are used, depending on age of weaning, milking ability of sows, litter size and other factors.

II. SOW'S DIET AFFECTS PIG'S GROWTH

The diet which the sow receives during her growing period, as well as during gestation, will definitely affect the ability of the baby pigs to survive and grow after weaning (1–3). This means that gilts kept for reproduction must be fed well-balanced diets during their growing period to develop the normal, reproductive tract needed for successful production of large litters. A high quality diet must be fed to the sow during gestation to ensure proper growth and development of the fetus. One of the biggest mistakes swine producers make is to start feeding a sow a well-balanced diet only 3 or 4 weeks before farrowing. They have heard that the developing fetuses grow most in this period (which is true). As a result, they feed the sow a well-balanced diet only then. Until that time, they may feed the sow a poor quality diet in an effort to save on feed costs. As a result, the sow does not obtain the nutrients needed and so farrows many weak or dead pigs and ends up weaning small, poor-looking litters. The answer to this problem is never to deprive the sow of needed protein, minerals, and vitamins. Any lack of these nutrients in the diet means a loss of pigs, and these lost pigs are never marketed.

Presently, 35–45% of the ovulated eggs fail to be represented by live pigs at birth and 20–25% of the live pigs at birth die before weaning (19). This is a tremendous loss which must be eliminated.

The information in Table 12.1 from the Iowa Station shows that the heavier the pig is at birth, the heavier it will be at weaning. On the average, a difference of one pound in weight at birth was accompanied by 7.78 lb difference at weaning (4). This is in good agreement with studies at the Missouri Station (5). This information shows the value of properly feeding the sow during gestation so that she will farrow heavy, thrifty pigs. Such pigs not only will have a better chance to survive but also will be heavier at weaning.

The pigs which are heavier at weaning also tend to do better afterwards. A comprehensive study at the Minnesota Station (6) involving 745 gilt litters, composed of 5562 pigs born alive, of which 3918 pigs survived to weaning also showed that the average birth weight had the greatest effect, of the factors studied, upon both survival and total weaning weight of the litters.

III. EARLY WEANING OF PIGS

An early attempt (7) to feed a synthetic milk and thus wean pigs right after birth was unsuccessful because sucrose was fed as the source of carbohydrate. Later, Illinois workers showed that the newborn pig cannot utilize sucrose (8–10); this partly accounted for the early failure. Shortly after the development of

TABLE 12.1
Average Birth and Weekly Weights of Suckling Pigs Classified according to Birth Weight at the Iowa Station[a] (4)

	Weight range (lb)									
	<1.76	1.76–2.00	2.01–2.25	2.26–2.50	2.51–2.75	2.76–3.00	3.01–3.25	3.26–3.50	3.51–3.75	>3.75
Birth	1.5	1.9	2.2	2.4	2.6	2.9	3.1	3.4	3.6	3.9
1 week	3.1	3.5	4.0	4.5	4.6	5.1	5.6	5.9	6.3	6.6
2 weeks	5.0	5.7	6.4	7.2	7.3	7.9	8.6	8.9	9.6	10.0
3 weeks	6.8	7.7	9.1	10.0	10.0	10.8	11.8	12.1	12.8	13.6
4 weeks	8.5	9.7	11.6	12.6	12.6	13.5	14.6	15.0	16.1	17.2
5 weeks	10.2	11.6	13.6	14.9	15.1	15.9	17.5	18.2	19.2	20.9
6 weeks	11.7	13.7	16.0	17.7	18.0	19.0	21.0	22.1	22.9	25.1
7 weeks	14.0	16.5	19.1	21.4	21.8	23.0	25.5	27.1	27.5	30.6
8 weeks	17.1	20.0	23.3	26.2	26.5	28.1	31.2	32.8	33.5	37.3

[a]Based on data obtained from 1296 Duroc pigs from 191 litters. Pigs which died prior to weaning were not included in the study.

synthetic milks in 1951, pig hatcheries sprang up in many sections of the country. These pig hatchers were establishments which specialized in producing and selling weanling pigs throughout the year for growing and finishing on the farms of purchasers. The hatcheries ranged in scale of operation from 40 or 50 sows on up to 400 or more.

The pig hatchery, however, was characterized by many failures. Disease, lack of sanitation, and poor management were the chief causes of the failures. Synthetic milks were not widely accepted. Liquid milk often spilled over and soaked the litter on the floor which chilled the pigs, caused scouring, etc. Also, keeping the equipment clean and sanitary provided many problems; many cases of scours and deaths resulted from unsanitary equipment. Souring of the milk in the feeders was particularly troublesome when the pigs did not promptly eat their feed. As a result, only a few of the hatcheries, those with qualified personnel and with good management procedures, were able to raise pigs successfully under this system.

Before long, the emphasis was shifted to dry meal formulations and away from liquid milk. In a few years, dry prestarter and starter feeds were developed to substitute for sow's milk and thus allow early weaning of pigs.

Table 12.2 shows information obtained at the Wisconsin Station (11) on the constituents of colostrum and milk from sows milked at the time of parturition, the third day, the end of the first week, and each subsequent week throughout an 8-week lactation period. The third to fourth week seems to be the period when there is a change in some of the constituents of sow's milk. This is also the period of time when the sow's milk secretion starts to decrease. The information in Table 12.2 can serve as a guide to those studying milk replacement feeds for early weaned pigs.

IV. HOW EARLY SHOULD PIGS BE WEANED?

The trend in swine production is toward early weaning of pigs; this practice is becoming increasingly popular. Many swine producers are already weaning their pigs at 3 weeks of age with the use of high quality starter feeds. The majority who wean early, however, are weaning at 4–5 weeks of age.

Farmers with only average management ability and facilities should wait until the pigs are about 5–6 weeks old before weaning. The earlier the swine producer weans pigs, the more know-how, facilities, and well-trained personnel will be needed to operate successfully. The swine producer who lacks these tools and skills should enter the program slowly, and gradually lessen the time pigs are kept on the sow in line with the success experienced.

A great deal of research has been conducted on early weaning. Pigs have been successfully weaned at birth when special techniques were used. Research in early weaning will undoubtedly make it possible to continuously lower the time

TABLE 12.2
Lactational Trend of the Milk Constituents Studied with Sows at Wisconsin Station (11)

Constituent	Lot	Total no. of samples	Period in lactation										Av. 1–8 weeks
			1 day[a]	3 days	1 week	2 weeks	3 weeks	4 weeks	5 weeks	6 weeks	7 weeks	8 weeks	
Total solids	Pasture	66	22.81	22.41	20.63	20.11	18.98	19.39	17.56	19.55	19.23	20.34	19.47
(%)	Dry lot	67	22.81	26.23	21.93	20.67	19.82	20.49	20.46	20.74	20.18	20.44	20.69
Solids-not-fat	Pasture	66	15.92	12.83	12.48	13.04	12.71	12.99	12.44	13.98	13.66	14.32	13.16
(%)	Dry lot	67	17.21	12.81	13.69	12.71	12.43	13.14	13.57	13.65	14.24	13.89	13.38
Protein (%)	Pasture	39	11.25	8.95	7.10	6.90	6.14	7.45	6.10	6.67	8.19	8.30	7.09
	Dry lot	32	14.29	8.23	7.61	6.83	6.16	8.17	7.18	8.23	7.85	7.37	7.42
Lactose (%)	Pasture	39	2.89	4.99	4.35	5.33	6.12	3.93	5.58	6.06	4.55	4.77	5.18
	Dry lot	32	3.42	3.56	5.43	4.95	4.79	4.23	5.79	5.32	4.60	5.54	5.08
Ash (%)	Pasture	67	0.72	0.83	0.86	0.79	0.85	1.03	1.04	1.08	1.09	1.30	0.99
	Dry lot	67	0.73	0.94	0.83	0.79	0.90	0.90	0.99	1.08	1.14	1.22	0.98

[a]Colostrum milk sampled at time of parturition.

the pig will need to stay with the sow. Eventually, in specialized farm units, pigs will be weaned shortly after birth and the sow will serve primarily as an incubator for them (see Chapter 1, Section V).

V. ADVANTAGES AND DISADVANTAGES OF EARLY WEANING

At present, among the advantages of early weaning (if done properly) are the following:

1. Heavier and more uniform pigs at 8 weeks of age as well as fewer runts. This has been verified in Purdue research (12) and in other studies. This is because the sow reaches maximum milk flow about 3 weeks after farrowing. This peak may occur later in certain sows, but the trend is usually downward by the fourth week. Early weaning will mean the baby pig can obtain more to eat than if it continues to nurse the sow to 8 weeks of age. But prestarter and starter feeds must be highly palatable in order to induce the pigs to eat heavily. Moreover, they need to be fortified with the needed protein, minerals, and vitamins.

2. More pigs raised per litter through reduction in losses and saving of surplus pigs. Most pig losses still occur shortly after birth and during the first few days. Early weaning will not prevent these losses. Only proper feeding of the sow during growth and gestation, together with good management and disease control, will lessen these small pig losses. Early weaning, however, will save pigs from being laid on or crippled by the sow. It will also save orphan pigs that die when sows die or go dry or that become runts when sows have more pigs than they can adequately nurse.

3. Complete control of the nutrients in the prestarter and starter feeds, thus fulfilling the optimum nutritional requirements of the young pig. If these feeds are made palatable enough, the pigs will eat enough and will get all the carbohydrates, fat, protein, minerals, vitamins, and antibiotics (or other feed additives) needed for maximum growth and development.

4. Better disease control. Early-weaned pigs have fewer diseases and parasites which are transmitted from the sow to her pigs. Moreover, early-weaned pigs stay healthier and have less anemia and scours. Sow's milk is low in iron and copper, but as much of these two minerals as needed can be added to the milk substitutes. More diseases, however, can result if good management is not followed with early weaned pigs.

5. Saving on the sow's feed. Against this saving, however, one must figure the cost of the prestarter and starter feeds the young pigs eat.

6. Sows lose less weight during lactation. This means the sow can be sold soon after farrowing if she is no longer needed or if the market price is favorable then.

7. Sows can be rebred earlier if one wants to produce more litters per year.

8. Encourages spread of the farrowing season throughout the year. This gives the advantage of more uniform marketing and, as a result, will tend to eliminate some of the highs and lows which occur in hog marketing and prices during the year.

Some of the limitations of early weaning are as follows:

1. Requires fortified, well-balanced, and highly palatable prestarter and starter feeds. Just any feed mixture will not do a successful job of raising early-weaned pigs.

2. Requires excellent management and know-how. Once the pigs are taken away from the sow, they cannot be neglected. The sow can cover up poor management practices, inadequate diets, and failure to feed and water regularly to a certain extent. She cannot do this, however, once the pigs are taken from her care. So, unless one can do as good a job as the sow, it is best to leave the pigs with her.

3. Requires better-than-average equipment and excellent sanitation practices. These facilities and practices, however, are usually no better than those already used by most good swine producers.

Undoubtedly, there are other advantages and limitations to early weaning. The foregoing, however, are among the most important and should be carefully considered by those who plan to enter the program.

VI. PRESTARTER FEEDS

At the present time, many producers have eliminated prestarter feeds and use only starter feeds. This has occurred because very few swine producers are weaning at less than 3 weeks of age. The starter feeds do an adequate job with 3-week-old pigs. Moreover, most pigs will not consume much creep feed (such as a prestarter or starter feed) before they are 3 weeks of age. This is especially true if the sow is a good milker and if she does not have an extremely large litter. The prestarter feed is meant to be a milk replacer and to be used for pigs weaned before 3 weeks of age. It is usually fed in limited amounts and until the pigs weigh 12–15 lb. Then the pigs are switched to a starter diet. But, some swine producers prefer to use prestarter diets for a while even though they wean at 3–5 weeks of age. This is especially the case when their program is benefited by the use of a highly fortified, high quality starter diet.

Table 12.3 gives an example of a milk replacer or prestarter diet recommended by Iowa workers (13). A study of the diet shows it is highly fortified with protein, minerals, vitamins, antibiotics, and highly palatable energy feeds. It includes trace minerals to prevent anemia. High levels of antibiotics or other feed additives are used to prevent scouring and to promote fast growth. This feed is

TABLE 12.3
Milk Replacers—Iowa Station (13)[a,b]

Percent protein	Ingredient	Diet[c] 1	2	3	4
8.9	Ground yellow corn	881	652	551	530
48.5	Solvent soybean meal	455	500	600	500
15	Rolled oat groats	—	—	—	200
33	Dried skim milk	200	400	200	200
12	Dried whey	200	200	400	400
31	Fish solubles	—	50	50	—
	Sugar	200	100	100	100
	Stabilized animal fat	—	50	50	20
	Calcium carbonate (38% Ca)	14	10	10	10
	Dicalcium phosphate (26% Ca, 18.5% P)	22	10	10	10
	Salt	5	5	5	5
	Trace mineral premix	3	3	3	3
	Vitamin premix	20	20	20	20
	DL-Methionine	—	—	1	2
	Feed additives (gm/ton)[d]	100–300	100–300	100–300	100–300
		2000	2000	2000	2000
	Calculated analysis				
	Protein (%)	19.45	23.61	23.48	21.69
	Calcium (%)	0.98	0.72	0.69	0.69
	Phosphorus (%)	0.81	0.62	0.61	0.61
	Lysine (%)	1.23	1.58	1.56	1.40
	Methionine (%)	0.33	0.43	0.44	0.45
	Cystine (%)	0.37	0.35	0.36	0.34
	Tryptophan (%)	0.24	0.30	0.30	0.28
	Metabolizable energy (kcal/lb)	1380	1429	1418	1405

[a]For baby pigs before 3 weeks of age.

[b]The milk replacer diet is normally fed in only limited amounts. It should be used for pigs weaned prior to 3 weeks of age until they reach approximately 12 lb. Then they can be switched to a starter diet. It is a good diet to feed orphan pigs when the sow dies, extreme disease outbreak (TGE) occurs, or the sow fails to produce milk.

[c]Values in pounds per ton.

[d]The feed additive may be part of the vitamin premix, or if a separate premix, it should replace an equal amount of corn.

used at one of the most critical periods in the life cycle of the pig, and this is the reason for such a high quality diet.

Table 12.4 gives another example of a prestarter diet. This one is recommended by the University of Kentucky (14). The diets shown in Table 12.3 and 12.4 are typical of the prestarter diets being fed. There are minor variations in the

TABLE 12.4
Prestarter Diets—University of Kentucky (14)[a]

Ingredient	Diet[b] 1	2	3	4
Ground yellow corn	533	561	327	355
Dried skim milk	800	800	400	400
Solvent soybean meal (50%)	282	302	484	504
Fish solubles	50	50	50	50
Dried whey (high lactose)	—	—	400	400
Sugar (cane or beet)	200	200	200	200
Distillers' dried solubles	50	—	50	—
Stabilized fat	50	50	50	50
Calcium carbonate	8	8	10	10
Dicalcium phosphate	—	2	2	4
Iodized salt	5	5	5	5
Trace mineral mix	2	2	2	2
Vitamin premix	20	20	20	20
Antibiotic mix[c]	+	+	+	+
	2000	2000	2000	2000
Calculated analysis				
Protein (%)	24.0	24.0	24.0	23.9
Calcium (%)	0.69	0.71	0.70	0.72
Phosphorus (%)	0.60	0.60	0.60	0.60

[a]For pigs weaned earlier than 3 weeks of age.
[b]Values in pounds per ton.
[c]Antibiotics should be added to provide 100–250 gm/ton.

feeds used, level of nutrient supplementation, and the feed additives included (as well as their quantity in the diet), but the two examples given are a guide as to what these diets consist of.

VII. STARTER DIETS

Starter diets are used to a large extent by swine producers. The starter or creep feed, as some are called, are supplied at least by the second week after birth. They are used to encourage dry-feed consumption as a means of supplementing the sow's milk. This is especially important if the sow is a poor milker or if the sow has a larger litter than her milk supply can properly support.

Starter feeds can be used as: (a) a follow-up feed to the prestarter diet for early weaned pigs, or (b) as a creep feed for pigs which are to be weaned at 3–5 weeks

TABLE 12.5
Pig Starter Diets—Iowa Station (13)[a]

Percent protein	Ingredient	Diet[b] 1	2	3	4	5
8.9	Ground yellow corn	1091	1087	1216	966	887
48.5	Solvent soybean meal	440	—	450	450	—
44	Solvent soybean meal	—	550	—	—	—
37	Soybeans, whole cooked	—	—	—	—	750
33	Dried skim milk	—	—	50	50	—
12	Dried whey	400	300	200	300	300
31	Fish solubles	—	—	—	50	—
	Sugar	—	—	—	100	—
	Stabilized animal fat	—	—	20	20	—
	Calcium carbonate (38% Ca)	15	10	10	10	10
	Dicalcium phosphate (26% Ca, 18.5% P)	20	25	25	25	25
	Iodized salt	5	5	5	5	5
	Trace mineral premix	2	3	3	3	3
	Vitamin premix	25	20	20	20	20
	DL-Methionine[c]	2	—	1	1	—
	Feed additives (gm/ton)[d]	100–300	100–300	100–300	100–300	100–300
		2000	2000	2000	2000	2000
	Calcualted analysis					
	Protein (%)	18.1	18.62	18.28	18.66	19.51
	Calcium (%)	0.7	0.72	0.69	0.73	0.75
	Phosphorus (%)	0.6	0.63	0.62	0.64	0.66
	Lysine (%)	1.06	1.11	1.06	1.12	1.16
	Methionine (%)	0.38	0.30	0.35	0.36	0.30
	Cystine (%)	0.30	0.31	0.30	0.30	0.31
	Tryptophan (%)	0.22	0.23	0.22	0.22	0.26
	Metabolizable energy (kcal/lb)	1350	1309	1375	1359	1422

[a]The pig starter diet should be used as a creep diet before weaning and fed after weaning until the pigs reach approximately 40 lb. Then the pigs can be switched to a grower-finisher diet.

[b]Values in pounds per ton.

[c]Diets 2 and 5 are borderline in methionine content. The addition of 1 lb of DL-methionine to these diets may improve performance and feed efficiency slightly, although pigs will perform very satisfactorily on these diets without the DL-methionine addition.

[d]The feed additive may be part of the vitamin premix, or if a separate premix, it should replace an equal amount of corn.

of age. After about a week or so, as soon as baby pigs learn to eat the prestarter, they can be switched to a good starter diet. The starter feed does not have to be as highly fortified as the prestarter and thus will be less expensive. Pelleting of starter feeds will reduce feed wastage. Swine producers who wish to leave their pigs on the sows and wean at a later age may get their pigs started eating with either the prestarter or starter diet. If pigs eat a good starter diet early and in large enough quantities to keep them gaining well, it is possible to wean them at 3–5 weeks of age with good results. These pigs get the starter self-fed free-choice in a creep.

Table 12.5 gives examples of pig starter diets from the Iowa Station (13). The University of Kentucky recommends the starter diets shown in Table 12.6 (14). Table 12.7 gives University of Nebraska recommendations of prestarter, starter, and starter grower diets (20). Most university studies recommend starter diets.

TABLE 12.6
Starter Diets—University of Kentucky (14)[a]

	Diet[b]			
Ingredient	1	2	3	4
Ground yellow corn	873	948	981	1035
Solvent soybean meal(50%)	516	490	556	504
Dried skim milk	—	100	—	100
Fish solubles	50	50	—	—
Distillers' dried solubles	50	—	—	—
Dried whey (high lactose)	300	200	300	200
Stabilized fat	50	50	—	—
Sugar (cane or beet)	100	100	100	100
Calcium carbonate	15	14	15	14
Dicalcium phosphate	19	21	21	20
Iodized salt	5	5	5	5
Trace mineral premix	2	2	2	2
Vitamin premix	20	20	20	20
Antibiotics[c]	+	+	+	+
	2000	2000	2000	2000
Calculated analysis				
Protein (%)	20.0	20.0	20.0	20.0
Calcium (%)	0.70	0.70	0.71	0.69
Phosphorus (%)	0.60	0.60	0.60	0.60

[a]For creep diets and for weaned pigs weighing 12–30 lb.
[b]Values in pounds per ton.
[c]Antibiotics should be added to provide 100–250 gm/ton.

TABLE 12.7
Diets for Various Classes of Swine—University of Nebraska (20)

	Percent protein[a]			
	22	18	18	16
Ingredients	Prestarter	Simple starter	Starter–grower	Starter–grower
Sugar (beet or cane)	15.00	10.00	—	—
Ground yellow corn	19.35	35.45	61.48	69.10
Ground oats	5.00	10.00	—	—
Ground wheat	5.00	—	—	—
44% soybean meal	4.00	24.20	23.00	18.70
Dried skim milk	40.00	—	—	—
Dried whey	—	10.00	5.00	2.50
Dried fish solubles	5.00	2.50	2.50	2.50
Dried brewer's yeast	1.00	1.00	1.00	—
Lard or fat (stabilized)	2.50	2.50	2.50	2.50
Dicalcium phosphate	0.10	1.10	1.17	1.50
Monosodium phosphate	0.40	—	—	—
Ground limestone	—	0.60	0.70	0.60
Trace mineral mix[b]	0.15	0.15	0.15	0.10
Salt (iodized)[b]	0.50	0.50	0.50	0.50
Vitamin-antibiotic mix	2.00[c]	2.00[d]	2.00[d]	2.00[e]
	100.00	100.00	100.00	100.00

[a]All diets are calculated to contain 0.7% calcium and 0.6% phosphorus.

[b]The trace mix and/or iodized salt should supply: 90 gm zinc; 0.15–0.20 gm iodine; 90 gm iron; 9 gm copper; 25 gm manganese and 10 lb salt/ton of feed.

[c]Added at the following rate per pound of diet: vitamin A, 2000 IU; vitamin D_2, 180 IU; vitamin B_{12}; 20μg; riboflavin, 1 mg; calcium pantothenate, 3.0 mg; choline chloride, 80.0 mg; thiamin, 2.0 mg; niacin, 6.0 mg; pyridoxine, 2.0 mg; vitamin K (MSB), 1.0 mg; and antibiotics; 50–125 mg.

[d]Added at the following rate per pound of diet: vitamin A, 2000 IU; vitamin D_2, 180 IU; vitamin B_{12}, 20 μg; riboflavin, 1.5 mg; niacin 10.0 mg; calcium pantothenate, 2.0 mg; choline chloride, 100.0 mg; MSB, 1.0 mg and antibiotics, 50–125 mg.

[e]Added at the following rate per pound of diet: vitamin A, 1500 IU; vitamin D_2, 180 IU; vitamin B_{12}, 7.5 μg; riboflavin, 1.5 mg; niacin, 4.0 mg; calcium pantothenate, 4.5 mg; choline chloride, 50.0 mg; MSB, 1.0 mg and antibiotics, 25–125 mg.

Note: MSB (Menadione Sodium Bisulfite, or equivalent) may be used as a source of Vitamin K.

They all show some similarity to those in Tables 12.5, 12.6 and 12.7. Some diets are more simplified than others. Others will vary in the nutrient and level of supplementation. There will also be some difference in feed additives used and in level of supplementation provided. All of the starter diets, however, are high quality feeds which are palatable and nutritious for the very young pig.

VIII. INCREASING STARTER DIET PALATABILITY

Iowa workers (15) found sugar to be one of the most palatable ingredients to include in pig prestarter and starter diets. Early Iowa work showed that sugar-coating the starter feed caused the greatest consumption of it. Later experiments there, however, showed that the advantage of sugar-coating was not so evident. Just to have the sugar in the starter feed was the important thing. Illinois workers (16) found that suckling pigs preferred hulled oats to all other feeds studied. They also ate mixtures containing molasses in greater quantities than the same feeds with no molasses added (Table 12.8). Florida workers (17) found that 10% cane sugar and 10% stabilized beef tallow increased the palatability the most of the creep diets they studied.

Table 12.9 shows the results obtained in a second Florida trial where various combinations of sugar, lard, and beef tallow were studied. In this trial, a diet

TABLE 12.8
Choice of Feeds by Suckling Pigs at Illinois Station (16)

Feeds offered	Feed consumed (lb)	Percentage of total eaten[a]
Hulled oats	158.0	43.8
Rolled oats 75%, molasses concentrate[b] 25%	73.5	20.4
Pig starter diet fed in Lot 3, Experiment 2		
Pellets	52.0	14.4
Meal	31.0	8.6
Dry skim milk 75%, molasses concentrate[b] 25%	25.0	6.9
Shelled corn	4.0	1.1
Rolled oats	3.0	0.8
Dry skim milk	3.0	0.8
Meat scraps (55% crude protein)	3.0	0.8
Dry synthetic milk	2.0	0.6
Pig supplement fed in Lot 1, Expts. 1 and 2	2.0	0.6
High efficiency broiler diet		
Meal	1.0	0.3
Pellets	1.0	0.3
Ground yellow corn	1.0	0.3
Solvent soybean oil meal (44% crude protein)	1.0	0.3
Solvent soybean oil meal (50% crude protein)	0.0	0.0
	360.5	100.0

[a]Percentage of total eaten when all these various feeds were placed before pigs to choose from.

[b]Pure cane molasses blended with sugar cane pulp and dried.

TABLE 12.9
Effect of Feeds on Palatability of Creep Diets at Florida Station (17)

Percent added to diet[a]	Percent of total feed[b]
5% Stabilized lard + 5% cane sugar	30.3
10% Stabilized lard	24.8
10% Cane sugar	14.7
5% Stabilized beef tallow + 5% cane sugar	12.5
5% Stabilized beef tallow + 5% stabilized lard	11.5
10% Stabilized beef tallow	6.3

[a]All diets were pelleted. The diet consisted of ground yellow corn, ground rolled oats, soybean meal, dried skim milk, bone meal, limestone, trace mineralized salt, plus fortification with aureomycin, riboflavin, niacin, pantothenic acid, choline, B_{12}, and vitamin A.

[b]Percent of total feed consumed by pigs when all these various feeds were placed before them to choose from.

containing a combination of cane sugar and lard was more acceptable to pigs than a diet containing only cane sugar. When fed singly, the order of preference was for lard, cane sugar, and tallow. In both experiments, the pigs at first showed a preference for the diets containing waste fat. As the pigs grew older, however, they consumed an ever-increasing amount of the sugar diets.

Table 12.10 shows data obtained in Illinois studies (18) which indicated that pigs ate more starter feed as the sugar content was increased on up to 20% of the

TABLE 12.10
Palatability of Starter Diets for Suckling Pigs (18)

	Average percent of total starter consumed[a]	
Diet	Meal	Pellets
20% Cane sugar	38	37
15% Cane sugar	20	16
High level of dried skim milk	17	29
10% Cane sugar	13	9
5% Cane sugar	5	3
0.05% Saccharin	4	4
0% Cane sugar	2	1

[a]Average percent of total starter consumed when all feeds were placed before pigs to choose. The diet consisted of rolled oats, ground yellow corn, soybean meal, dried skim milk, condensed fish solubles, crude corn oil, plus vitamin and mineral supplements.

TABLE 12.11
Recommended Nutrient Allowances for Swine—University of Kentucky (14)[a]

Stage of production	Weight of pig (lb)	Percent of diet Protein	Percent of diet Ca	Percent of diet P	Units/lb of diet vitamin A[b]	Units/lb of diet vitamin D	μg/lb of diet vitamin B_{12}	mg/lb of diet Riboflavin	mg/lb of diet Pantothenic acid	mg/lb of diet Niacin	mg/lb of diet Choline	gm/ton antibiotic[c]
Boars and gilts												
Developer	200 to —	15	0.80	0.60	2500	300	10	2.5	8	15	400	—
Sows												
Pregestation and gestation	— —	15	0.80	0.60	2500	300	10	2.5	8	15	400	—[d]
Lactation	— —	15	0.80	0.60	2500	300	10	2.5	8	15	400	50–100[e]
Young pigs												
Prestarter	— to 12	22–24	0.70	0.60	2000	400	10	4.0	10	25	500	100–250
Starter or creep	12 to 30	18–20	0.70	0.60	2000	400	10	4.0	10	25	500	100–250
Growing-finishing pigs												
Grower	30 to 75	16–18	0.65	0.55	1500	300	8	3.0	8	20	400	50–100
Developer	75 to 125	14–16	0.60	0.50	1000	200	5	2.0	5	15	300	20–50
Finisher[f]	125 to 220	12–14	0.60	0.50	1000	200	5	2.0	5	15	300	20–50

[a]The nutrient allowances are suggested for maximum performance, not as minimum requirements. They are based on research work with natural feedstuffs and have been found to give satisfactory results.

[b]About 3.0 times as many units of provitamin A (carotene) are needed as compared with true A.

[c]Arsenicals, nitrofurans or other chemotherapeutics may be used instead of or in combination with antibiotics. Levels and combinations used and stage of production for which they are used must comply with current Food and Drug Administration regulations.

[d]Under high disease level conditions, antibiotics may be beneficial in diet used just prior to breeding (1.0 gm/sow/day).

[e]With low disease level conditions, antibiotics may not be needed during lactation.

[f]Improvements in efficiency and carcass grade may justify the higher levels of protein during the growing-finishing periods.

diet. They also found that dried skim milk is a highly palatable feed in a pig starter, but that saccharin was not of much value in increasing the palatability of starter feeds.

Prestarter and starter diets are fed as complete diets. In most cases, they are fed as pelleted diets. Swine producers buy commercial prestarter and starter feeds.

IX. LEVELS OF NUTRIENT SUPPLEMENTATION TO USE

Most University swine scientists vary in the level of supplementation which they recommend. Table 12.11 gives the level of nutrients recommended by the University of Kentucky (14). Table 12.12 gives the University of Kentucky recommended allowances for amino acids (14). Table 12.13 gives the Kentucky

TABLE 12.12
Amino Acid Allowances for Growing Swine—University of Kentucky (14)

	Stage of production				
	Prestarter	Starter	Grower	Developer	Finisher
Weight range (lb):	0–12	12–30	30–60	60–125	125–220
Protein level[a]:	24	20	17	15	13
Amino Acid					
Arginine	0.43[b]	0.36	0.29	0.25	0.22
Histidine	0.36	0.30	0.24	0.21	0.18
Isoleucine	0.84	0.70	0.56	0.49	0.42
Leucine	1.10	0.92	0.74	0.64	0.55
Lysine	1.20	1.00	0.80	0.70	0.60
Methionine + cystine[c]	0.79	0.66	0.53	0.46	0.40
Phenylalanine	0.60	0.50	0.40	0.35	0.30
Threonine	0.84	0.70	0.56	0.49	0.42
Tryptophan	0.19	0.16	0.13	0.11	0.10
Valine	0.74	0.62	0.50	0.43	0.37

[a]Approximate protein levels needed to meet minimum levels of all essential amino acids when corn and soybean meal are used as major ingredients. Lysine and methionine are usually the limiting amino acids in diets composed of natural feedstuffs.

[b]Percent of total diet.

[c]Approximately ½ as methionine if adequate cystine is present.

TABLE 12.13
Trace Mineral Premix—University of Kentucky (14)

Mineral	Concentration in premix (%)	1.0 lb/ton[a] supplies to diet (ppm)
Copper	1.0	5.0
Iodine[b]	0.03	1.5
Iron	10.0	50.0
Manganese	6.0	30.0
Zinc	10.0	50.0

[a]Use 1.0 lb/ton in developer and finisher diets, 1.5 lb/ton in grower diets, and 2.0 lb/ton in baby pig, gestation and lactation diets.

[b]Iodine need not be included in trace mineral mix if 0.25–0.5% *iodized* salt is included in diets.

recommended levels of trace minerals to add to swine diets (14). These suggested levels are somewhat different than those recommended by the National Research Council and discussed in Chapters 3, 4, and 5. But this is to be expected since swine scientists tend to add a safety level to take care of many factors which can increase nutrient needs and which are discussed in Chapter 2.

The suggested levels recommended by the University of Kentucky are excellent and can serve as a starting point or guideline for anyone considering levels of nutrients to use in swine diets.

REFERENCES

1. Cunha, T. J., E. J. Warwick, M. E. Ensminger, and N. K. Hart, *J. Anim. Sci.* **7,** 117 (1948).
2. Cunha, T. J., O. B. Ross, P. H. Phillips, and G. Bohstedt, *J. Anim. Sci.* **3,** 415 (1944).
3. Krider, J. L., B. W. Fairbanks, R. F. Van Poucke, D. E. Becker, and W. C. Carroll, *J. Anim. Sci.* **5,** 256 (1946).
4. Forshaw, R. P., H. M. Maddock, P. G. Homeyer, and D. V. Catron, *J. Anim. Sci.* **12,** 263 (1953).
5. Weaver, L. A., and R. Bogart, *M., Agric. Exp. Stn., Bull.* **461** (1943).
6. Winters, L. M., J. N. Cummings, and H. A. Stewart, *J. Anim. Sci.* **6,** 288 (1947).
7. Bustad, L. K., W. E. Ham, and T. J. Cunha, *Arch. Biochem.* **17,** 247 (1948).
8. Becker, D. E., D. E. Ullrey, and S. W. Terrill, *Arch. Biochem. Biophys.* **48,** 178 (1954).
9. Becker, D. E., D. E. Ullrey, S. W. Terrill, and R. A. Notzold, *Science* **120,** 345 (1954).
10. Johnson, S. R., *Fed. Proc., Fed. Am. Soc. Exp. Biol.* **8,** 387 (1949).
11. Bowland, J. P., R. H. Grummer, P. H. Phillips, and G. Bohstedt, *J. Anim. Sci.* **8,** 199 (1949).
12. Jones, H. W., *Purdue Agric. Exp. Stn., Swine Day Rep.* pp. 9–11 (1970).

13. Holden, P., V. C. Speer, E. J. Stevermer, and D. R. Zimmerman, *Iowa, Agric. Exp. Stn., Rev.* **PM-489** (1974).
14. Hays, V. W., Kentucky, Agricultural Experiment Station, Lexington, personal communication (1976).
15. Nelson, L. F., L. N. Hazel, A. A. Moore, H. M. Maddock, G. C. Ashton, C. C. Culbertson, and D. V. Catron, *Iowa Farm Sci.* **7,** 3 (1953).
16. Terrill, S. W., R. J. Meade, R. O. Nesheim, and D. E. Becker, *Ill., Agric. Exp. Stn.,* **AS-213** (1951).
17. Wallace, H. D., Florida Agricultural Experiment Station, Gainesville, unpublished data (1955).
18. Jensen, A. H., J. E. Launer, S. W. Terrill, and D. E. Becker, *Ill. Agric. Exp. Stn.,* **AS-418** (1955).
19. England, D. C., *J. Anim. Sci.* **38,** 1045 (1974).
20. Galster, K. E., W. T. Ahlschwede, E. R. Peo, Jr., M. Danielson, R. D. Fritschen, and B. D. Mosher, *Nebr., Agric. Exp. Stn., Rep.* **EC-74-210** (1974)

13

Feeding the Growing-Finishing Pig

I. INTRODUCTION

After the pig is weaned, there are many alternative methods of feeding it. They will depend on the home-grown feeds available, whether pasture or confinement feeding will be used, the quality of the pig produced and many other factors. The weight of the pig at weaning will also determine the best program to use. If pigs are weaned at 3–5 weeks of age, a grower diet is used after the starter diet.

II. GROWER DIETS

After the pigs have a good start and weigh 25–35 lb, they can be switched from the starter diet to a grower diet. This diet is not quite as highly fortified and consequently is less costly. At this stage, pigs start eating considerably more feed; thus, costs must be given more consideration. In grower diets, soybean meal may replace the dried skim milk, and ground yellow corn may replace sugar in the diet. Also, the amount of vitamin, mineral, and antibiotic supplementation may be reduced.

If 30–35 lb pigs are not too thrifty in appearance, it might be well to continue the starter diet for awhile longer before switching to the grower diet. Some swine producers use a mixture of starter and grower diet until the pigs look thrifty in appearance.

Table 13.1 gives examples of grower diets recommended by the Kentucky Station (1). This is a much more simplified diet than the starter diets. It is recommended for feeding until the pigs weigh 75 lb. Table 13.2 presents grower diets recommended by the Purdue Station (2). The Michigan Station recommendations on grower diets are shown in Table 13.3 (3). A study of Tables 13.1, 13.2, and 13.3 shows that there are differences in the recommendations made by the three university scientists. This is to be expected since opinion on nutritional

TABLE 13.1
Grower Diets[a]—University of Kentucky (1)

Ingredient	Diet[b]			
	1	2	3	4
Ground yellow corn	1413.5	1436.5	1476.5	1451.5
Soybean meal (44%)	390	416	424	482
Meat and bone meal	50	50	50	—
Dried whey	50	50	—	—
Distillers' dried solubles	50	—	—	—
Calcium carbonate	11	10	11	16
Dicalcium phosphate	9	11	12	24
Iodized salt	10	10	10	10
Trace mineral premix	1.5	1.5	1.5	1.5
Vitamin premix	15	15	15	15
Antibiotics[c]	+	+	+	+
	2000.0	2000.0	2000.0	2000.0
Calculated analysis				
Protein (%)	17.0	17.0	17.0	17.0
Calcium (%)	0.65	0.64	0.65	0.64
Phosphorus (%)	0.55	0.55	0.55	0.55

[a]For pigs weighing 35–75 lb.
[b]Values in pounds per ton.
[c]Add antibiotics to provide 50–100 gm/ton.

requirements and feeds to use varies between scientists. Moreover, the weights of the pigs differ in the three tables. This has an effect on the levels of nutrients to use. The information shown in Tables 13.1, 13.2, and 13.3 can be used as a guide as to grower diets and of what they consist. Most grower diets are fed as complete feeds. Many are fed as pelleted diets. Most swine producers buy a commercial grower diet rather than trying to mix their own.

III. FEEDING PIGS FROM 75 POUNDS TO MARKET WEIGHT

Once the pig has reached 75 lb and is thrifty in appearance, it has passed a nutritionally critical period. After the pigs weigh 75 lb, they will do well on relatively simple diets which are less highly fortified.

The diets for pigs from 75 to 125 lb in weight are usually referred to as developing diets, grower diets or growing-finishing diets. Diets fed to pigs from 125 lb to market weight are usually referred to as finisher-diets or growing-finishing diets.

A. Dry Lot versus Pasture

The trend in the United States is toward confinement feeding of growing-finishing pigs. This is because the returns obtained from the land will be greater if it is used to grow corn, soybeans and other crops rather than pasture for finishing pigs. The feed saved by pasture will, in almost all cases, not pay for the cost of the pasture. Faster gains are obtained in dry lot than on pasture if properly balanced diets are used. High quality pastures, however, will still continue to be used for breeding stock. Herd replacements should also be grown out on pasture as much as possible. Less leg and stiffness problems will be encountered with breeding animals grown and kept on high quality pasture as much as possible. The use of confinement feeding emphasizes the need for well-balanced diets since pasture is no longer available to make up for some of the nutrients which are lacking in poor quality diets. The feeder who is not using well-balanced diets will benefit from pasture feeding since it will cover up many of his mistakes. In many areas of the world, and in some localities in the United States, pigs will continue to be grown and finished on pasture. This will especially be the case in areas with small farms and where the land is rough, steep, and thus cannot be well used for food crop production. Also, in some countries the cost of concentrate feeds and the economics of pork production necessitate maximum use of pasture. High quality pasture can, on the average, replace 10–15% of the concentrates and 25–50% of the protein supplement in the diet. Usually, the protein level in the total diet is reduced 2 percentage points on pasture as compared to dry lot. So, in many situations one needs to carefully consider the alternative of growing-finishing pigs in confinement as compared to pasture.

B. Complete Diets versus Corn and Supplement Free-Choice

Both methods are used on swine farms. There is some difference of opinion as to which is the best method to use. Self-feeding grain and supplement free-choice offers the most simple program. But the efficiency of this method may vary, depending on the palatability of the supplement and that of the grain and pasture. Protein supplements must be formulated so they will not be too palatable or they will encourage overconsumption by the pigs. The addition of alfalfa meal, minerals, and other feeds will lessen the palatability of the protein supplement mixture. Palatability of the corn and pasture will also vary depending on variety, soil fertility, maturity, and many other factors.

Thus, the feed manufacturer needs to formulate supplements which do an efficient job with the pastures and grains available in the area. The swine producer must keep accurate records to determine which method of feeding will be the most economical under his conditions. He must balance the cost of grinding and mixing against the inefficiency of overeating or undereating which may

TABLE 13.2
Grower Diets for Swine 50–75 Pounds–Purdue University (2)

Ingredient	Diet number[a]										
	1	2	3	4	5	6	7	8	9	10	11
Corn, ground yellow	1591	1732	1192	1228	808	—	1596	1598	1352	1565	—
Corn, opaque-2	—	—	—	—	—	—	—	—	—	—	1690
Grain sorghum	—	—	—	—	—	1464	—	—	—	—	—
Oats	—	—	400	—	—	—	—	—	—	—	—
Wheat	—	—	—	—	800	—	—	—	—	—	—
Lysine, synthetic[b]	—	5	—	—	—	—	—	—	—	—	—
Cottonseed meal (expeller)	—	—	—	—	—	—	—	—	—	60	—
Soybean meal (44%)	358	210	358	324	342	487	310	300	—	324	258
Soybeans, full-fat	—	—	—	—	—	—	—	—	597	—	—
Wheat midds	—	—	—	400	—	—	—	—	—	—	—
Meat and bone meal (50%)	—	—	—	—	—	—	60	—	—	—	—
Tankage	—	—	—	—	—	—	—	60	—	—	—

Limestone, ground (38% Ca)	17	16	17	18	17	18	11	13	18	18	16
Dicalcium phosphate	22	25	21	18	21	19	11	17	21	21	24
Salt	5	5	5	5	5	5	5	5	5	5	5
Trace mineral premix	1	1	1	1	1	1	1	1	1	1	1
Vitamin premix	6	6	6	6	6	6	6	6	6	6	6
Selenium premix	1	1	1	1	1	1	1	1	1	1	1
	2000	2000	2000	2000	2000	2000	2000	2000	2000	2000	2000
Calculated analysis											
Protein (%)	14.9	12.4	15.5	15.7	15.8	18.1	15.3	15.4	17.2	15.2	15.0
Lysine (%)	0.70	0.70	0.72	0.73	0.70	0.83	0.71	0.74	0.84	0.70	0.70
Methionine + cystine (%)	0.50	0.50	0.50	0.50	0.50	0.50	0.50	0.50	0.50	0.51	0.51
Tryptophan (%)	0.17	0.13	0.19	0.19	0.19	0.23	0.17	0.17	0.20	0.18	0.19
Calcium (%)	0.60	0.60	0.60	0.60	0.60	0.60	0.60	0.60	0.60	0.60	0.60
Phosphorus (%)	0.50	0.50	0.50	0.50	0.50	0.50	0.50	0.50	0.50	0.50	0.50
Metab. energy (kcal/lb)	1586	1582	1523	1427	1605	1586	1574	1576	1633	1572	1587

[a] Values in pounds per ton.

[b] 98% lysine hydrochloride; 78% L-lysine.

TABLE 13.3
Grower Diets for Swine 30–75 Pounds—Michigan Station (3)

Ingredient (lb/ton)	Diet number												
	1	2	3	4	5	6	7	8	9	10	11	12	13
Corn	1507			1185	1142		1167	1620	1537	1633	1351	1440	1639
Opaque-2 corn		1600											
Sorghum grain			1458										
Oats				340									
Rye					400								
Wheat						1540							
Wheat middlings							400						
Fish meal								120					
Meat and bone meal									138				
Tankage										188			
Dried whey											200		
Soybeans, heated												400	
Lysine, 78% L-													5
Soybean meal (44%)	437	342	488	420	403	406	382	222	306	160	400	125	300
Limestone	20	20	20	20	20	20	25	16	3	3	17	20	18
Dicalcium phosphate	20	22	18	19	19	18	10	6	0	0	16	19	22
Salt	5	5	5	5	5	5	5	5	5	5	5	5	5
Selenium premix	1	1	1	1	1	1	1	1	1	1	1	1	1
MSU VTM premix	10	10	10	10	10	10	10	10	10	10	10	10	10
	2000	2000	2000	2000	2000	2000	2000	2000	2000	2000	2000	2000	2000
Analyses													
DE(kcal/lb)	1562	1567	1500	1500	1551	1566	1510	1578	1577	1570	1562	1665	1557
Protein (%)	16.5	16.4	18.3	16.8	16.4	17.6	16.8	15.7	17.1	16.5	16.3	16.6	14.0
Lysine (%)	0.80	0.80	0.83	0.80	0.80	0.80	0.80	0.80	0.80	0.80	0.80	0.80	0.81
Methionine + cystine (%)	0.55	0.54	0.50	0.54	0.53	0.54	0.53	0.56	0.53	0.51	0.54	0.51	0.50
Tryptophan (%)	0.19	0.20	0.22	0.20	0.19	0.22	0.20	0.17	0.17	0.17	0.18	0.19	0.15
Calcium (%)	0.67	0.68	0.66	0.67	0.67	0.67	0.65	0.65	0.65	0.65	0.65	0.67	0.65
Phosphorus (%)	0.50	0.50	0.50	0.50	0.50	0.50	0.50	0.50	0.56	0.54	0.50	0.50	0.50

occur if the protein supplement is self-fed. Moreover, he has the advantage of having the grain already on his farm for self-feeding. This can be changed, however, if arrangements can be made for him to supply his own grain to be mixed or deliver grain to be credited to future complete mixed diets. If these arrangements can be made, many swine producers favor custom-mixing or bulk deliveries of complete mixed diets.

Some experiments show no difference between complete diets and free-choice diets, whereas others do. The differences obtained are due to the conditions under which the experiments are conducted. So, conditions on the farm will determine which system of feeding will be the most economical. Under some conditions, complete diets will be best, whereas free-choice feeding of grain and protein supplements will be superior in others. In most cases, the farmer will use some complete diets and will also free-choice feed others, depending on the stage of the life cycle of the pig and how its overall requirements can best be met. Many feeders are using complete ground and mixed diets for young pigs—when nutritional needs are most critical. There is also an increasing trend toward the use of mixed diets for older pigs. This is done as a means of better controlling the nutrient and antibiotic (or other feed additives) intake rather than trusting the pigs to consume the correct amount when they are self-fed corn and supplement free-choice. As bulk facilities become more available and as automatic feed handling and feeding systems are developed, there will be more complete feeds used.

C. Pelleted Feeds

The trend is toward the greater use of pelleted feeds. Their use is most beneficial with the young pig. Most prestarter and starter feeds are pelleted. Some of the grower diets are also pelleted. For older animals, the cost of pelleting will probably outweigh the advantages of using a pelleted feed. Pelleting is more helpful with bulky or high fiber feeds. Pelleting, for example, increases rate of gain about 14% with barley diets as compared to about 4% with corn diets. The use of pelleted feeds will increase as engineering developments decrease the cost of pelleting. The economics of pelleting must always be considered to ensure that the advantages derived from pelleting are greater than the cost involved.

D. Housing for Confinement

The swine industry is rapidly becoming more mechanized and this trend will continue in the future. There is an increase in sophisticated housing which will permit the ease of feeding, handling, and management of animals, efficiency of operation and maximum comfort for the pigs. Extremes of heat and cold, which decrease efficiency of production, are being minimized or even eliminated in some housing. The newborn pig is highly sensitive to cold, and the bigger pig becomes more susceptible to heat as it approaches market weight. Waste disposal is becoming more sophisticated and odors around swine establishments are being minimized and eliminated in some cases. Confinement feeding, which facilitates mechanization, will stimulate more studies to determine the feasibility of liquid versus dry feeding, limited versus full-feeding, floor versus conventional feeding, slotted versus conventional floors and many other questions which still are not fully answered, such as space requirements and number of pigs per pen. Eventually, the pig will be exposed to less changes in environmental conditions than the humans taking care of them. Caution should be exercised, however, to prevent overinvestment in facilities and mechanization until sufficient studies are available to indicate their economic feasibility.

E. Subclinical Disease Level and Stress Factors

Much more attention should be paid to this area in the future. It is assumed, for example, that the higher the subclinical disease level and stress factors on the farm, the greater the response which is obtained from antibiotic supplementation. It seems logical that these same factors may also affect nutritional needs. Most of the standards on nutrient needs or requirements, however, have been established with University and Experiment Station trials which are almost always conducted with good animals, good quality feeds as well as good management and sanitation. The application of these requirements, though, may be altered on the average farm where poorer quality animals, feeds, management, and sanitation exist. This means that nutrient requirement standards are only a guide, and modification may be needed depending on the type of operation or program being followed. It is hoped that studies involving subclinical disease levels and other stress factors will be increasingly studied in the future. Studies of this type will result in information badly needed on the average farm. More attention must be paid to field trials that are properly conducted. Results of these may apply better to the average farm than those of some university trials. There is a need to supplement university trials with field trials as well.

IV. DIETS FOR PIGS FROM 75 TO 125 POUNDS

There are many diets which can be fed during this period. Examples from three universities will be presented to indicate the variation in diets which can be used. Table 13.4 shows the diets recommended by the University of Kentucky (1). Table 13.5 gives thirteen diets recommended by the Michigan Station (3) and

Table 13.6 gives eleven diets recommended by Purdue University (2). Other feeds can also be used in place of some of those recommended. Chapter 11 discusses other feeds, their relative feeding values, and limitations on their use. With that information, one can modify the diets shown in Tables 13.4, 13.5, and 13.6 and use other feeds in them. The same can be done with diets recommended in Tables 13.7, 13.8, and 13.9 for finishing swine.

V. DIETS FOR PIGS FROM 125 POUNDS TO MARKET WEIGHT

The finishing period from 125 lb to market weight requires the lowest level of protein and nutrient supplementation. Tables 13.7, 13.8, and 13.9 give recommended diets for finishing pigs as recommended by the Kentucky, Michigan and Purdue Stations, respectively. But, other feeds and other diets can also be used.

TABLE 13.4
Developer Diets—University of Kentucky (1)[a]

	Diet[b]		
Ingredient	1	2	3
Ground yellow corn	1584	1606	1577
Solvent soybean meal (44%)	276	302	364
Meat and bone meal	50	50	—
Distillers' dried solubles	50	—	—
Calcium carbonate	12	12	17
Dicalcium phosphate	7	9	21
Iodized salt	10	10	10
Trace mineral premix	1	1	1
Vitamin premix	10	10	10
Antibiotics[c]	+	+	+
	2000	2000	2000
Calculated analysis			
Protein (%)	15.0	15.0	14.9
Calcium (%)	0.61	0.62	0.61
Phosphorus (%)	0.50	0.50	0.50

[a]For pigs weighing 75–125 lb.
[b]Values in pounds per ton.
[c]Add antibiotics to provide 20–50 gm/ton.

TABLE 13.5
Grower Diets for Swine 75–125 Pounds—Michigan State (3)

Ingredient (lb/ton)	Diet number												
	1	2	3	4	5	6	7	8	9	10	11	12	13
Corn	1626			1126	1260		1285	1742	1654	1710	1472	1537	1789
Opaque-2 corn		1730											
Sorghum grain			1600										
Oats				522									
Rye					400								
Wheat						1655							
Wheat middlings							400						
Fish meal								120					
Meat and bone meal									120				
Tankage										120			
Dried whey											200		
Soybeans, heated												410	
Lysine, 78%L-													6
Soybean meal (44%)	320	216	350	300	288	285	265	104	206	140	280		150
Limestone	16	14	14	16	16	16	22	10	4	7	14	17	14
Dicalcium phosphate	22	24	20	20	20	18	12	8	0	7	18	20	25
Salt	5	5	5	5	5	5	5	5	5	5	5	5	5
Selenium premix	1	1	1	1	1	1	1	1	1	1	1	1	1
MSU VTM premix	10	10	10	10	10	10	10	10	10	10	10	10	10
	2000	2000	2000	2000	2000	2000	2000	2000	2000	2000	2000	2000	2000
Analyses													
DE (kcal/lb)	1568	1574	1502	1466	1557	1572	1519	1585	1583	1575	1568	1647	1570
Protein (%)	14.3	14.3	15.8	14.8	14.2	15.5	14.6	13.6	14.9	14.3	14.0	14.4	11.2
Lysine (%)	0.65	0.65	0.65	0.65	0.65	0.65	0.65	0.65	0.65	0.65	0.65	0.65	0.67
Methionine + cystine (%)	0.49	0.49	0.43	0.48	0.47	0.48	0.48	0.51	0.48	0.47	0.48	0.46	0.41
Tryptophan (%)	0.16	0.17	0.18	0.17	0.16	0.19	0.16	0.13	0.14	0.14	0.15	0.16	0.11
Calcium (%)	0.60	0.60	0.60	0.60	0.60	0.60	0.60	0.60	0.60	0.60	0.60	0.60	0.60
Phosphorus (%)	0.50	0.50	0.50	0.50	0.50	0.50	0.50	0.50	0.51	0.50	0.50	0.50	0.50

TABLE 13.6
Grower Diets for Swine 75–125 Pounds—Purdue University (2)

Ingredient	Diet number[a]										
	1	2	3	4	5	6	7	8	9	10	11
Corn, ground yellow	1623	1752	1240	1283	841	—	1638	1661	1529	1598	—
Corn, opaque-2	—	—	—	—	—	—	—	—	—	—	1723
Grain sorghum	—	—	—	—	—	1595	—	—	—	—	—
Oats	—	—	400	—	—	—	—	—	—	—	—
Wheat	—	—	—	—	800	—	—	—	—	—	—
Lysine, synthetic[b]	—	4.4	—	—	—	—	—	—	—	—	—
Cottonseed meal (expeller)	—	—	—	—	—	—	—	—	—	60	—
Soybean meal (44%)	326	192	311	271	311	357	269	237	—	292	225
Soybeans, full-fat	—	—	—	—	—	—	—	—	420	—	—
Wheat midds	—	—	—	400	—	—	—	—	—	—	—
Mean and bone meal (50%)	—	—	—	—	—	—	60	—	—	—	—
Tankage	—	—	—	—	—	—	—	60	—	—	—

continued

TABLE 13.6 *(continued)*

	Diet number[a]										
Ingredient	1	2	3	4	5	6	7	8	9	10	11
Limestone, ground (38% Ca)	17	16	17	18	17	18	11	14	18	18	17
Dicalcium phosphate	22	24	20	16	21	18	10	16	21	20	23
Salt	5	5	5	5	5	5	5	5	5	5	5
Trace mineral premix	1	1	1	1	1	1	1	1	1	1	1
Vitamin premix	6	6	6	6	6	6	6	6	6	6	6
Selenium premix	1	1	1	1	1	1	1	1	1	1	1
	2000	2000	2000	2000	2000	2000	2000	2000	2000	2000	2000
Calculated analysis											
Protein (%)	14.3	12.1	14.7	14.8	15.2	15.8	14.6	14.3	14.7	14.7	14.4
Lysine (%)	0.66	0.66	0.66	0.66	0.66	0.66	0.66	0.66	0.66	0.66	0.66
Methionine + cystine (%)	0.49	0.42	0.48	0.47	0.48	0.43	0.48	0.47	0.43	0.50	0.49
Tryptophan (%)	0.16	0.12	0.18	0.18	0.18	0.19	0.15	0.15	0.15	0.17	0.18
Calcium (%)	0.60	0.60	0.60	0.60	0.60	0.60	0.60	0.60	0.60	0.60	0.60
Phosphorus (%)	0.50	0.50	0.50	0.50	0.50	0.50	0.50	0.50	0.50	0.50	0.50
Metab. energy (kcal/lb)	1587	1584	1525	1538	1606	1588	1575	1577	1613	1572	1588

[a]Values in pounds per ton.
[b]98% lysine hydrochloride, 78% L-lysine.

TABLE 13.7
Finisher Diets—University of Kentucky (1)[a]

Ingredient	Diet[b] 1	2	3
Ground yellow corn	1691	1712	1684
Solvent soybean meal (44%)	168	196	256
Meat and bone meal	50	50	—
Distillers' dried solubles	50	—	—
Calcium carbonate	11	10	16
Dicalcium phosphate	9	11	23
Iodized salt	10	10	10
Trace mineral premix	1	1	1
Vitamin premix	10	10	10
Antibiotics[c]	+	+	+
	2000	2000	2000
Calculated analysis			
Protein (%)	13.1	13.1	13.0
Calcium (%)	0.60	0.59	0.60
Phosphorus (%)	0.50	0.50	0.50

[a]For pigs weighing 125 lb to market weight.
[b]Values in pounds per ton.
[c]Add antibiotics to provide 20–50 gm/ton.

VI. FEEDING HERD REPLACEMENT ANIMALS DURING GROWTH

Some unidentified factor (or factors) present in high quality alfalfa meal (and other feeds such as pasture, distillers' grains with solubles, fish meal, fish solubles, and others), is definitely needed for reproduction and lactation. This factor (or factors) is stored for a long period of time. Such storage is so important that the diet a pig receives during growth will definitely influence the ability of the animal to conceive, reproduce, and lactate many months later (as many as 6–8 months later). Thus, prospective herd replacement gilts should be fed differently and better than pigs being finished for market.

A farmer who feeds his herd replacement gilts and market hogs together may run into difficulties in his breeding program. When they reach market weight, the herd replacement gilts are sorted out. That procedure is all right if excellent diets

TABLE 13.8
Finisher Diets for Swine 125–220 Pounds–Michigan State (3)

	Diet number[a]												
Ingredient	1	2	3	4	5	6	7	8	9	10	11	12	13
Corn	1666			1044	1204		1325	1782	1677	1706	1512	1586	1821
Opaque-2 corn		1771											
Sorghum grain			1638										
Oats				648									
Rye					500								
Wheat						1703							
Wheat middlings							400						
Fish meal								120					
Meat and bone meal									60				
Tankage										60			
Dried whey											200		
Soybeans, heated												360	
Lysine, 78%L-													5
Soybean meal (44%)	280	175	310	254	242	245	225	64	225	192	240		120
Limestone	16	14	14	16	16	16	22	10	12	10	14	16	14
Dicalcium phosphate	22	24	22	22	22	20	12	8	10	16	18	22	25
Salt	5	5	5	5	5	5	5	5	5	5	5	5	5
Selenium premix	1	1	1	1	1	1	1	1	1	1	1	1	1
MSU VTM premix	10	10	10	10	10	10	10	10	10	10	10	10	10
	2000	2000	2000	2000	2000	2000	2000	2000	2000	2000	2000	2000	2000
Analyses													
DE (kcal/lb)	1570	1576	1503	1444	1556	1575	1520	1587	1578	1574	1571	1640	1573
Protein (%)	13.6	13.7	15.2	13.9	13.6	14.9	13.9	12.9	13.9	13.6	13.2	13.7	10.7
Lysine (%)	0.60	0.60	0.60	0.60	0.60	0.60	0.60	0.60	0.60	0.60	0.60	0.60	0.60
Methionine + cystine (%)	0.47	0.47	0.41	0.46	0.45	0.46	0.46	0.49	0.47	0.46	0.46	0.42	0.40

Tryptophan (%)	0.15	0.16	0.17	0.16	0.15	0.18	0.15	0.12	0.14	0.14	0.14	0.14	0.10
Calcium (%)	0.60	0.60	0.60	0.60	0.60	0.60	0.60	0.60	0.60	0.60	0.60	0.60	0.60
Phosphorus (%)	0.50	0.50	0.50	0.50	0.50	0.50	0.50	0.50	0.50	0.50	0.50	0.50	0.50

Free-Choice Protein Supplement

Ingredients	Supplement A	Recommended vitamin addition/ton Suppl. A	
Soybean meal (44%)	1295	Vitamin A (IU) millions	20
Meat and bone meal	350	Vitamin D (IU) millions	2.7
Alfalfa meal	200	Vitamin E (IU) thousands	30.0
Limestone	60	Riboflavin (gm)	15.0
Dicalcium phosphate	40	Pantothenic acid (gm)	50.0
Trace mineral salt (Hi Zinc)	50	Nicotinic acid (gm)	80.0
Selenium premix	5	Choline (gm)	500.0
	—	Vitamin B_{12} (mg)	70.0
	2000	Antibiotic (gm)	100.0
Analyses			
Crude protein (%)	39.0		
Calcium (%)	3.77		
Phosphorus (%)	1.59		

[a]Values in pounds per ton.

TABLE 13.9
Finisher Diets for Swine 125–220 Pounds—Normal Protein Levels—Purdue University (2)

Ingredient	Diet number[a]										
	1	2	3	4	5	6	7	8	9	10	11
Corn, ground yellow	1654	1811	1271	1314	871	—	1668	1693	1569	1628	—
Corn, opaque-2	—	—	—	—	—	—	—	—	—	—	1755
Grain sorghum	—	—	—	—	—	1624	—	—	—	—	—
Oats	—	—	400	—	—	—	—	—	—	—	—
Wheat	—	—	—	—	800	—	—	—	—	—	—
Lysine, synthetic[b]	—	5.4	—	—	—	—	—	—	—	—	—
Cottonseed meal (expeller)	—	—	—	—	—	—	—	—	—	60	—
Soybean meal (44%)	295	131	279	240	279	327	238	206	—	261	192
Soybeans, full-fat	—	—	—	—	—	—	—	—	380	—	—
Wheat midds	—	—	—	400	—	—	—	—	—	—	—
Meat and bone meal (50%)	—	—	—	—	—	—	60	—	—	—	—
Tankage	—	—	—	—	—	—	—	60	—	—	—

Limestone, ground (38% Ca)	17	16	17	18	17	18	11	13	18	18	17
Dicalcium phosphate	22	25	21	16	21	19	11	16	21	21	24
Salt	5	5	5	5	5	5	5	5	5	5	5
Trace mineral premix	1	1	1	1	1	1	1	1	1	1	1
Vitamin premix	6	6	6	6	6	6	6	6	6	6	6
Selenium premix	1	1	1	1	1	1	1	1	1	1	1
	2000	2000	2000	2000	2000	2000	2000	2000	2000	2000	2000
Calculated analysis											
Protein (%)	13.8	11.0	14.0	14.3	14.7	15.3	14.1	13.8	14.0	14.1	13.9
Lysine (%)	0.62	0.62	0.62	0.62	0.62	0.62	0.62	0.62	0.62	0.62	0.62
Methionine + cystine (%)	0.47	0.39	0.46	0.46	0.47	0.42	0.47	0.45	0.42	0.48	0.48
Tryptophan (%)	0.16	0.11	0.17	0.17	0.17	0.18	0.14	0.15	0.14	0.16	0.17
Calcium (%)	0.60	0.60	0.60	0.60	0.60	0.60	0.60	0.60	0.60	0.60	0.60
Phosphorus (%)	0.50	0.50	0.50	0.50	0.50	0.50	0.50	0.50	0.50	0.50	0.50
Metab. energy (kcal/lb)	1587	1584	1525	1538	1607	1588	1575	1577	1610	1573	1588

[a]Values in pounds per ton.
[b]98% lysine hydrochloride, 78% L-lysine.

are fed during growth. Many farmers, however, do not feed highly nutritious diets to pigs to be sold as market hogs. As a result, when they sort out the replacement animals they may select some unsuitable for breeding purposes.

The best diets—adequate nutritionally—should be fed to prospective breeding animals. In other words, one should sort out their prospective herd replacement gilts soon after weaning and feed them excellent diets during the growing period. This will pay big dividends later, because the gilts will settle quickly and will farrow larger litters and wean a higher precentage of their pigs.

If gilts are farrowing small litters and losing most of their pigs before weaning due to nutritional problems, the diet fed during gestation and lactation should not be blamed entirely; the diet fed during the growth period may have been inadequate and thus responsible for the poor results. Breeding animals should always be fed good, well-balanced diets before and after they are bred or have farrowed a litter. If poor diets are fed too long during growth, gilts may become so depleted that it is too late to build them up later on with a good diet and so they are not able to raise large, healthy litters of pigs.

The addition of 100–200 lb of dehydrated alfalfa (in place of corn) to the first diets in Tables 13.4 and 13.7 would result in excellent diets for raising herd replacements. The important thing is to include 5–10% of high quality alfalfa in the diet during the growing period from 75 lb to market weight. The use of either 2.5–5.0% of distillers grains with solubles, 2.5% fish solubles, 2–5% of high quality fish meal or access to dirt and/or pasture will also help. The objective in using one or more sources of unidentified factors is to insure the development of a normal reproductive tract by the gilt during her growth period.

The first diet shown in Table 14.4 and recommended by the University of Kentucky is an excellent example of a diet to feed herd replacement gilts and boars from 75 lb to market weight (or breeding time). It contains fish solubles, distillers solubles and alfalfa meal. If possible, the diet should be modified by replacing 300 lb of the corn with 300 lb of oats. Oats are also valuable for reproduction. I would also increase the alfalfa level from 100 to 200 lb in the diet.

The first diet in Table 14.4, with these suggested changes, should be ideal for growing prospective herd replacement gilts and boars. It provides a number of good feed sources of unidentified factors. The diet is also properly fortified with vitamins, minerals, and protein. Moreover, it has some bulk and should be a good diet for growth without getting the animals too fat. This kind of a diet should result in the development of a normal reproductive tract in both the gilts and boars. This is necessary in order to obtain high conception rates as well as large litters at birth and at weaning time.

REFERENCES

1. Hays, V. W., Kentucky, Agricultural Experiment Station, Lexington, personal communication (1976).
2. Fister, J. R., V. B. Mayrose, T. R. Cline, H. W. Jones, M. P. Plumlee, D. M. Forsyth, and W. H. Blake, *Purdue Agric. Exp. Stn., Swine Day Rep.* pp. 73–87 (1975).
3. Miller, E. C., E. R. Miller, and D. E. Ullrey, *Mich., Agric. Ext. Bull.* **537,** 1–32 (1975).

14

Feeding the Breeding Herd

I. INTRODUCTION

To make profit, a swine producer must develop a feeding and management program which results in large, heavy litters at birth and at weaning time. At present, it is estimated that 20–25% of all pigs farrowed die before they are weaned. Nutritional deficiencies account for a good part of those losses. Thus, proper feeding of the breeding herd can eliminate, to a large extent, the handicap of these ''small pig losses'' which the farmer has to contend with yearly.

The feeding program has a great deal to do with the number, weight, and strength of the pigs in each litter. Their survival rate and weaning weights are also dependent on diet adequacy. A poor feeding program will raise havoc with raising and producing a market pig.

II. FEEDING PROSPECTIVE BREEDING ANIMALS

Prospective breeding gilts need to be fed excellent diets during the time they are growing. The diet fed during growth, from weaning to breeding age, has been shown to affect the results obtained many months later during conception, gestation, and lactation (1–4). This means that a producer should sort out his prospective herd replacement gilts soon after weaning and feed them a well-balanced diet during growth. This will result in the gilts' developing a normal reproductive tract which is necessary for production of large, thrifty litters.

Alfalfa meal contains some unidentified factor or factors of considerable value for pigs during growth and influences their ability to conceive, reproduce, and lactate many months later (1–3). A level of 5% alfalfa is not high enough; at least 10% alfalfa is needed to supply this factor or factors with corn-soybean diets. High quality pasture would most likely supply this same factor or factors.

This means that prospective replacement animals should be fed on good pasture or have higher levels of alfalfa meal in the diet during growth. Experimental information is not available on the effect of growth diet on their breeding ability, but it would probably be a good idea to treat prospective herd boars just like the gilts. Therefore, the first step in feeding the breeding herd is to grow replacement animals properly so they will have normal reproductive organs when they enter the breeding herd. This will eliminate sterile or poor-breeding boars and gilts.

III. LEVEL OF FEED DURING BREEDING PERIOD

An excellent diet, adequate in energy, vitamins, minerals, and protein should be fed during the breeding season. This diet can be the same as the one used during the gestation period. The practice of "flushing" or having the gilts gaining weight just before breeding is practiced by some breeders. The beneficial effect of flushing, beginning about 21 days before breeding, has been substantiated by experimental studies. Flushing increases the number of eggs produced by the gilt at breeding time, which in turn increases litter size.

The rate of gain desired would depend on the condition and size of the sow. If they are in thrifty condition, neither too fat nor too thin, sows should be gaining in the neighborhood of about a pound per day prior to breeding. Sows and boars should be in thrifty condition during the breeding season. It is usually necessary to decrease gradually, before the breeding season, the condition of fat sows and boars so they will breed well. On the other hand, animals that are too thin should increase in flesh to get into thrifty condition by breeding time. Gilts that respond to flushing are those that are too thin to begin with. The extra feeding of thin gilts and boars evidently stimulates the endocrine and reproductive system to greater activity.

IV. LEVEL OF FEED DURING GESTATION

An excellent, well-balanced diet is very important during gestation. Gilts have greater requirements than mature sows because their diet will have to take care of their growth as well as that of the developing fetus. Thus, gilts need more feed per 100 lb of body weight.

It is recommended that gilts and sows average 4.4 lb of feed daily during the entire gestation period. Most swine producers will feed 5–6 lb during the breeding period, and then decrease to 3–4 lb during the major part of gestation. Then they increase feed intake to 5 lb or more during the 3- or 4-week period prior to farrowing. This feeding schedule should vary depending on the condition of the gilts or sows. If this level of feeding results in the animals becoming "over-fat," it should be decreased. If, conversely, this level of feeding results in the gilts and sows becoming too thin, it should be increased accordingly. The gilts and sows should be kept in a thrifty condition, not too fat nor too thin, but somewhere in between.

The developing young achieve over half of their growth during the last one-third of the gestation period (Table 14.1). However, one should not wait until then to feed a well-balanced diet. The thriftiness of the litter at birth and their survival rate will depend on the diet fed during the whole gestation period; it will also depend on what the gilts were fed during their growing period (1,2,4). The developing fetus achieves most of its growth during the last third of gestation, but its fate is determined before and during the entire gestation period.

V. GAIN DURING GESTATION

A gilt should gain enough to compensate for the weight of the litter and afterbirth, as well as for a normal increase in her growing body. This means that gilts should gain from 80 to 110 lb during gestation. Sows need to gain less, since

TABLE 14.1
Computed Weight and Composition of a Litter of Eight Pigs during Gestation (5)[a]

Week of gestation	Total fresh weight (gm)[b]	Crude protein (gm)	Ash (gm)	Calcium (gm)	Phosphorus (gm)	Iron (mg)[b]
1	99	1.5	0.06	0.0002	0.002	0.82
2	366	8.5	0.6	0.005	0.028	4.2
3	787	23	2	0.036	0.12	11
4	1,354	47	5	0.14	0.36	22
5	2,062	83	10	0.40	0.82	37
6	2,909	130	18	0.96	1.61	57
7	3,891	191	30	2.0	2.8	82
8	5,005	265	45	3.8	4.7	113
9	6,251	356	66	6.6	7.2	149
10	7,625	462	92	10.9	10.6	191
11	9,127	585	125	17.1	15.1	239
12	10,755	726	165	26	21	294
13	12,507	886	213	38	28	355
14	14,385	1,065	269	54	37	423
15	16,384	1,263	335	74	47	499
16	18,504	1,483	411	101	60	581

[a]These data were obtained from Poland China gilts bred to the boar of the same breed. They were all fed the same diet in amounts to produce an average daily gain of 1–1.25 lb.

[b]One pound contains 454 gm or 454,000 mg.

they are not growing and only have to maintain their body weight. Sows should gain 60–85 lb during gestation. This gain would allow the sow to take care of weight losses during parturition and lactation and still weigh approximately the same when she weans her pigs as when she was bred. These weight gains can be decreased somewhat if the pigs are weaned early. Usually, the heaviest milking sows are the biggest weight losers.

These weight gains for gilts and sows will vary, depending on their original condition. The kind of diet used may also modify the weight gains needed. Swine producers can use these recommended gains for gilts and sows as a guide. Table 14.2 gives information on the weight losses which occurred with gilts and sows during parturition and the suckling period in a 15-year-study period at Purdue University (6).

During gestation, the gilt or sow needs to build up body reserves to be used during lactation. The weight loss during lactation will vary with the litter size, the condition of the animal, the adequacy of the diet fed, the milking ability of the sow and age when the pigs are weaned. Thus, weight losses during lactation will vary somewhat from the figures shown in Table 14.2; however, the figures given there can be used as a guide.

VI. FEEDING BEFORE AND AFTER FARROWING

The diet fed should be reduced and made more bulky and laxative a few days before farrowing. This will prevent constipation and reduce or prevent a feverish condition in the sow. Wheat bran is a good laxative feed and can be substituted for about half of the regular diet. On the day the sow farrows, she should be given no feed for about 10–12 hours. She should be given plenty of water, however. It should be warmed in cold weather to remove the chill. If the sow appears hungry, she might be given a pound or so of feed.

The day after she farrows, the sow can be fed 2 or 3 lb of feed. The diet can be increased gradually, until the sow is on full-feed 5–7 days after farrowing. After this, the diet may be self-fed or hand-fed. It is very important not to force the sow too fast during the first week after farrowing. Otherwise, she may develop milk fever, her udder may become caked, or scouring may develop in the small pigs because of overeating.

While sows are being brought to full-feed, their udders and the droppings of the young pigs should be observed. Whether they are being brought to full-feed too quickly should be evident from these observations. After the danger of overfeeding is past, the sow should be fed an excellent, well-balanced diet to stimulate good milk production. There is no one best program to recommend at farrowing time. A swine producer should adjust and modify the feeding schedule according to the results obtained under his conditions.

TABLE 14.2
Weights, Gains, and Weight Losses in Gilts and Mature Sows from Breeding to Weaning (6)

	No. of sows, 445	No. of gilts, 248
Average weight when bred	404.05	260.41
Average weight before farrowing	492.96	354.63
Average gain during gestation	88.91	94.22
Average gestation days	113.36	113.55
Average daily gain during gestation	0.78	0.83
Average weight after farrowing	455.20	327.14
Average weight loss at farrowing	37.76	27.49
Average weight at weaning	423.39	299.00
Average loss during suckling[a]	−31.82	−28.15
Total loss at farrowing and during suckling	−69.58	−55.64

[a]Occasionally there was a year when the sows or gilts actually gained during the suckling period. These data were obtained over a 15-year period.

Wheat bran and oats are good bulky, laxative feeds. Some producers use beet pulp, citrus pulp, or linseed meal as a laxative feed.

Some swine producers use high levels of antibiotics shortly before and after farrowing. They feel this helps prevent infections and complications which sometimes occur during and after farrowing. Other feed additives are also used.

VII. FEEDING DURING LACTATION

The feed requirements of the sow during lactation are considerably greater than during gestation. This is because the increase of nutrients required by the sow for milk production is greater than for producing young (Table 14.1). For example, California workers (7) summarized the information obtained by various investigators on milk production by sows. During an 8-week lactation period the average milk production per sow was 413.2 lb of milk with a range of 388.2–638.4 lb. The amount of milk produced daily by sows averaged 6.8 lb with a range of 3.4–11.6 lb daily. A study at the Missouri Station (8) showed that sows produced from 236.2 to 306.2 lb of milk during a 6-week lactation period. They found a significant difference between sows of different breeds and crosses in the amount of milk produced.

There evidently is considerable variation in the amount of milk produced by sows. This is dependent on the diet fed, the inherent ability to produce milk, the number of pigs in the litter, the condition of the sow, and other factors. The vigor, condition, thriftiness, and weight of pigs at weaning will depend largely on their milk supply. Thus, it is important that sows be fed adequate diets in sufficient amount to provide all the essential nutrients needed for good milk flow.

Gilts will usually average 11 lb of feed daily during lactation and sows 12 lb. But this level of feed intake can vary between 8 to 15 lb per day. Level of feed consumption will depend on litter size, milking ability of the sow, size of pigs, quality of the diet, condition of the sow, and other factors. The age at which the pigs are weaned will also affect feed intake. If the gilts or sows have small litters of six pigs or less, they should be limited in feed intake. Sows with eight or more pigs should be fed as much as the sow will consume (so long as they do not get fat) as a means of stimulating as much milk production as possible. In areas of scarce and expensive feed supply, such as in Europe, each sow is fed individually daily (both during gestation and lactation) according to condition and number of pigs being suckled. Some of this is done in the United States but not to the same degree as in many foreign countries.

VIII. METHOD OF FEEDING

Many farmers are interested in self-feeding sows because it saves time and labor. It also gives the timid sow a chance to get her share of the diet. For example, Minnesota workers (9) observed that pigs have a definite social order. This behavior pattern is emphasized especially when the feed supply is restricted and the pigs are fed in fairly large groups. The more aggressive pigs get more than their share of the feed, and the timid, less aggressive ones get less. This behavior pattern is not a problem when hogs are self-fed in groups with plenty of feeder space, since even the greedy ones do not eat all the time. The Minnesota group successfully self-fed a diet containing 15% ground corncobs to start and gradually increased this to 35%.

Hand-feeding is still used considerably, especially in countries where feed is scarce and expensive (such as in Europe). It allows control in rate of gain and in the proper amount of supplementation of the sow or gilt. The same end can be accomplished by self-feeding a diet made bulky by the addition of large amounts of high fiber feeds such as oats, alfalfa, wheat bran, and corn and cob meal. A well-balanced supplement is included in these bulky diets to make them adequate in protein, minerals, vitamins, and other nutrients.

Care must be taken to follow closely the condition of sows and gilts on self-fed diets. If they start to become too fat, the amount of bulk in the diet should be increased a little. Likewise, if they do not look thrifty enough, the bulk in the diet should be reduced. The hazard in self-feeding diets to breeding animals is that they may overeat and become too fat.

The Minnesota (10,11) and Purdue (12) Stations, as well as others, have successfully self-fed diets to sows and gilts. Both methods of feeding (hand-fed and self-fed) have resulted in satisfactory and about equal litters. Thus, if the diet is bulky enough so that females do not become too fat, diets can be self-fed successfully to gilts and sows.

The Purdue Station (12) tried ground corncobs as a means of adding bulk to a diet for self-feeding sows. Ear corn contains about 20% cobs by weight. Cobs are bulky and fibrous and thus are a bulky feed. Purdue obtained excellent results with a gestation diet containing 35% ground corncobs, 43.5% ground corn, 5% blackstrap molasses, 5% soybean meal, 5% meat and bone scraps, and 5% alfalfa meal, plus minerals. The cobs were ground through a 1/2 inch hammermill screen initially. But sorting of the large particles by the pigs resulted in wastage and in the consumption of an insufficient quantity of the cob. Grinding the cobs through a 1/4 inch hammermill screen, however, greatly reduced the amount of wastage and produced favorable results. The addition of 5% of blackstrap molasses in the diet increased consumption of the ground cob portion of the diet and reduced wastage.

Interval feeding is used by some swine producers. This is a method to limit-feed gilts or sows during gestation. With this system, they are allowed to eat every third day by turning them into a self-feeder for 2–8 hours. This system should reduce the labor requirement compared to daily hand-feeding. The daily feed consumption would average about 4.4 lb (13.2 lb consumed every third day). The average feed intake can be regulated somewhat by limiting the time allowed at the self-feeder. Reproductive performance in several trials with interval feeding has been essentially the same as with the normal hand, limit-feeding method (13).

More self-feeding of lactating sows is practiced than with sows during gestation. This is because feed intake needs to be restricted during gestation to keep the sow from becoming too fat. During lactation, however, even hand-fed sows are virtually on full-feed. Self-feeding lactating sows is one way to have feed available for the sows all the time and thus insure maximum milk production. Sows with small litters, however, should not be self-fed, since the extra feed is not needed for milk production. These sows should be separated and fed restricted diets in line with the number of pigs suckling them.

IX. DIETS TO USE DURING GESTATION

There is considerable variation in the diets recommended during gestation. This is to be expected since many different diets can be used. The objective is to feed a well-balanced diet which results in large, thrifty litters at birth and at

TABLE 14.3
Feed Costs Are Lower with Bigger Litters (14)

	Pounds of feed/pig		
Number of pigs weaned	At weaning	From weaning to 225 lb	Total feed/pig at 225 lb
2	722	653	1375
4	368.5	653	1021.5
6	250.6	653	903.6
8	191.7	653	844.7
10	156.4	653	809.4

weaning. A look at Table 14.3 shows that feed costs decrease considerably as more pigs are weaned per litter. Swine producers should set as a goal the weaning of ten pigs per litter (Fig. 14.1). A few top swine producers are accomplishing this, so it can be done and should be strived for.

Table 14.4 from the Kentucky Station lists some diets suggested for use during the gestation period (15). Table 14.5 and Table 14.6 list recommended gestation

Fig. 14.1. Sow suckling ten pigs. Ten pigs at weaning should be the goal of swine producers. (Courtesy of T. J. Cunha and University of Florida.)

TABLE 14.4
Replacement Gilt and Boar Developer and Gestation Diets Recommended by University of Kentucky[a] (15)

	Diet[b]			
Ingredient	1	2	3	4
Ground yellow corn	1190	1250	1527	1541
Oats	—	300	—	—
Wheat middlings	300	—	—	—
Meat and bone meal	—	100	100	—
Soybean meal (44%)	232	204	226	374
Fish solubles	50	—	—	—
Distillers' solubles	50	—	—	—
Alfalfa meal	100	100	100	—
Calcium carbonate	23	6	6	21
Dicalcium phosphate	23	8	9	32
Iodized salt	10	10	10	10
Trace mineral premix	2	2	2	2
Vitamin premix	20	20	20	20
	2000	2000	2000	2000
Calculated analysis				
Protein (%)	15.0	15.1	15.0	
Calcium (%)	0.81	0.81	0.81	
Phosphorus (%)	0.61	0.60	0.60	

[a]Diets for hand feeding developing herd replacements, boars, and sows during gestation. Hand feed 4–6 lb/day. Do not let sows get too fat.

[b]Values in pounds per ton.

diets from the Purdue (6) and Michigan Stations (17), respectively. The diets recommended by these three universities can be used as a guide as to the type of gestation diets to use. Other feeds not included in these diets can also be used. Considerable information on the relative feeding value and limitations in the use of other feeds is given in Chapter 11.

X. DIETS TO USE DURING LACTATION

Tables 14.7, 14.8, and 14.9 from the Kentucky, Purdue and Michigan Stations, respectively, list diets recommended for use during lactation. Other diets and other feeds can also be used in the lactation diets. A review of Chapter 11 provides information on the value and limitations of other feeds in lactation diets.

TABLE 14.5
Sow Gestation Diets—Purdue University (16)[a]

Ingredient	Diet number										
	1	2	3	4	5	6	7	8	9	10	11
Corn, ground yellow	1721	1874	1338	1379	1378	1746	1360	1565	1785	1396	—
Corn, opaque-2	—	—	—	—	—	—	—	—	—	—	1832
Oats	—	—	400	—	—	—	200	—	—	—	—
Lysine, synthetic[b]	—	5.25	—	—	—	—	—	—	—	—	—
Alfalfa meal (17%)	—	—	—	—	—	—	—	200	—	200	—
Soybean meal (44%)	203	43	187	147	150	108	169	165	54	139	96
Soybeans, full-fat	—	—	—	—	—	—	—	—	—	200	—
Wheat bran	—	—	—	—	400	—	200	—	—	—	—
Wheat midds	—	—	—	400	—	—	—	—	—	—	—
Meat and bone meal (50%)	—	—	—	—	—	100	—	—	—	—	—
Tankage	—	—	—	—	—	—	—	—	100	—	—

Limestone ground (38% Ca)	30	29	30	32	30	19	32	23	24	27	30
Dicalcium phosphate	24	27	23	20	20	5	17	25	15	16	20
Salt	5	5	5	5	5	5	5	5	5	5	5
Trace mineral premix	1	1	1	1	1	1	1	1	1	1	1
Vitamin premix	10	10	10	10	10	10	10	10	10	10	10
Choline chloride[c]	6	6	6	6	6	6	6	6	6	6	6
Selenium premix	1	1	1	1	1	1	1	1	1	1	1
	2000	2000	2000	2000	2000	2000	2000	2000	2000	2000	2000
Calculated analysis											
Protein (%)	12.1	9.4	12.4	12.4	12.3	12.6	12.3	12.3	12.1	12.3	12.2
Lysine (%)	0.50	0.50	0.50	0.50	0.50	0.50	0.50	0.50	0.50	0.50	0.50
Methionine + cystine (%)	0.43	0.35	0.41	0.41	0.40	0.41	0.41	0.42	0.40	0.41	0.43
Tryptophan (%)	0.13	0.08	0.14	0.14	0.13	0.11	0.14	0.15	0.11	0.15	0.14
Calcium (%)	0.85	0.85	0.85	0.85	0.84	0.85	0.85	0.85	0.85	0.85	0.85
Phosphorus (%)	0.50	0.50	0.50	0.55	0.62	0.50	0.53	0.50	0.50	0.50	0.50
Metab. energy (kcal/lb)	1573	1571	1511	1415	1482	1554	1497	1475	1557	1429	1574

[a]Values in pounds per ton.
[b]98% lysine hydrochloride, 78% L-lysine.
[c]Choline chloride premix containing 25% choline.

TABLE 14.6
Sow Gestation Diets—Michigan Station (17)[a]

Ingredient	Diet number											
	1	2	3	4	5	6	7	8	9	10	11	12
Corn	1735			1352		1566	1394	1574	1755	1802	1678	1887
Opaque-2 corn		1843										
Sorghum grain			1705									
Oats				400								
Wheat					1781							
Wheat bran						200						
Wheat middlings							400					
Alfalfa meal								200				
Meat and bone meal									100			
Tankage										100		
Soybeans, heated											260	
Lysine, 78% L-												5
Soybean meal (44%)	202	93	235	186	157	177	147	170	110	55		43
Limestone	23	22	22	23	25	27	30	16	14	16	23	22
Dicalcium phosphate	24	26	22	23	21	14	13	24	5	11	23	27
Salt	5	5	5	5	5	5	5	5	5	5	5	5
Selenium premix	1	1	1	1	1	1	1	1	1	1	1	1
MSU VTM premix	10	10	10	10	10	10	10	10	10	10	10	10
	2000	2000	2000	2000	2000	2000	2000	2000	2000	2000	2000	2000
Analyses												
DE (kcal/lb)	1565	1570	1560	1495	1570	1520	1515	1470	1550	1560	1625	1575
Protein (%)	12.2	12.2	13.8	12.5	13.3	12.4	12.5	12.4	12.7	12.1	12.3	9.3
Lysine (%)	0.50	0.50	0.50	0.50	0.50	0.50	0.50	0.50	0.50	0.50	0.50	0.50
Methionine + cystine (%)	0.43	0.38	0.37	0.42	0.42	0.44	0.42	0.42	0.42	0.40	0.41	0.36
Tryptophan (%)	0.12	0.14	0.15	0.13	0.16	0.13	0.13	0.15	0.11	0.11	0.12	0.08
Calcium (%)	0.75	0.75	0.75	0.75	0.75	0.75	0.75	0.75	0.75	0.75	0.75	0.75
Phosphorus (%)	0.50	0.50	0.50	0.50	0.50	0.50	0.50	0.50	0.50	0.50	0.50	0.50

[a]Values in pounds per ton.

TABLE 14.7
Lactation Diets Recommended by University of Kentucky (15)

Ingredient	Diet[c]			
	1	2	3	4
Ground yellow corn	1287	1368	1340	1270
Ground oats	100	—	200	300
Wheat bran[a]	100	200	100	—
Alfalfa meal	100	100	—	—
Meat and bone meal	50	100	100	—
Soybean meal (44%)	202	190	212	346
Fish solubles	50	—	—	—
Distillers' dried solubles	50	—	—	—
Calcium carbonate	14	10	12	21
Dicalcium phosphate	15	—	4	31
Iodized salt	10	10	10	10
Trace mineral premix	2	2	2	2
Vitamin premix	20	20	20	20
Antibiotic[b]	+	+	+	+
	2000	2000	2000	2000
Calculated analysis				
Protein (%)	15.0	15.0	15.0	15.0
Calcium (%)	0.80	0.80	0.82	0.80
Phosphorus (%)	0.62	0.60	0.61	0.60

[a]Dried beet pulp may be substituted for wheat bran.
[b]50–100 gm of antibiotics/ton.
[c]Values in pounds per ton.

XI. FEEDING VALUE OF SILAGE FOR SOWS

Many swine producers have been feeding sows more total energy than they actually need during pregestation and gestation. As a result the sows get too fat. A fat sow usually kills more pigs by laying or stepping on them. As a result, many swine producers have been feeding silage diets as a means of decreasing the amount of total energy fed their sows. Florida data (18) showed that limited-fed sows usually farrowed in a few hours, whereas full-fed sows, which were overfat, took approximately 12 or more hours to farrow.

Information available on the feeding value of silage for sows is still limited. Some work has been conducted at the Indiana, Iowa, Illinois, and other experiment stations. So this discussion on silage feeding will be based on limited information.

TABLE 14.8
Sow Lactation Diets—Purdue University (16)

Ingredient	Diet number[a]										
	1	2	3	4	5	6	7	8	9	10	11
Corn, ground yellow	1574	1713	1190	1230	1226	1598	1210	1417	1637	1247	—
Corn, opaque-2	—	—	—	—	—	—	—	—	—	—	1668
Oats	—	—	400	—	—	—	200	—	—	—	—
Lysine, synthetic[b]	—	5	—	—	—	—	—	—	—	—	—
Alfalfa meal (17%)	—	—	—	—	—	—	—	200	—	200	—
Soybean meal (44%)	359	212	344	304	308	264	325	322	211	296	262
Soybeans, full-fat	—	—	—	—	—	—	—	—	—	—	—
Wheat bran	—	—	—	—	400	—	200	—	—	200	—
Wheat midds	—	—	—	400	—	—	—	—	—	—	—
Meat and bone meal (50%)	—	—	—	—	—	100	—	—	—	—	—
Tankage	—	—	—	—	—	—	—	—	100	—	—

Limestone, ground (38% Ca)	30	30	30	30	30	20	33	23	24	28	30
Dicalcium phosphate	21	24	20	20	20	2	16	22	12	13	24
Salt	5	5	5	5	5	5	5	5	5	5	5
Trace mineral premix	1	1	1	1	1	1	1	1	1	1	1
Vitamin premix	10	10	10	10	10	10	10	10	10	10	10
Selenium premix	1	1	1	1	1	1	1	1	1	1	1
	2000	2000	2000	2000	2000	2000	2000	2000	2000	2000	2000
Calculated analysis											
Protein (%)	14.8	12.3	15.2	15.2	15.0	15.4	15.1	15.0	14.8	15.1	15.0
Lysine (%)	0.70	0.70	0.70	0.70	0.70	0.70	0.70	0.70	0.70	0.70	0.77
Methionine + cystine (%)	0.50	0.43	0.49	0.49	0.48	0.49	0.48	0.49	0.47	0.48	0.48
Tryptophan (%)	0.17	1.13	0.19	0.18	0.18	0.16	0.18	0.20	0.16	0.20	0.21
Calcium (%)	0.85	0.85	0.85	0.85	0.85	0.85	0.85	0.85	0.85	0.85	0.85
Phosphorus (%)	0.50	0.50	0.50	0.58	0.65	0.50	0.53	0.50	0.50	0.50	0.50
Metab. energy (kcal/lb)	1572	1569	1510	1413	1480	1552	1495	1473	1556	1428	1439

[a]Values in pounds per ton.

[b]98% lysine hydrochloride, 78% L-lysine.

TABLE 14.9
Sow Lactation Diets—Michigan State (17)[a]

Ingredient	Diet number											
	1	2	3	4	5	6	7	8	9	10	11	12
Corn	1579			1200		1409	1244	1424	1605	1661	1478	1711
Opaque-2 corn		1678										
Sorghum grain			1550									
Oats				400								
Wheat					1612							
Wheat bran						200						
Wheat middlings							400					
Alfalfa meal								200				
Meat and bone meal									120			
Tankage										120		
Soybeans, heated											462	
Lysine, 78% L-												5
Soybean meal (44%)	360	260	390	340	327	334	300	322	248	182		222
Limestone	24	24	23	24	24	29	30	16	12	15	25	23
Dicalcium phosphate	21	22	21	20	21	12	10	22	0	6	19	23
Salt	5	5	5	5	5	5	5	5	5	5	5	5
Selenium premix	1	1	1	1	1	1	1	1	1	1	1	1
MSU VTM premix	10	10	10	10	10	10	10	10	10	10	10	10
	2000	2000	2000	2000	2000	2000	2000	2000	2000	2000	2000	2000
Analyses												
DE (kcal/lb)	1560	1565	1505	1500	1560	1505	1505	1465	1540	1550	1660	1560
Protein (%)	15.0	15.0	16.5	15.3	16.2	15.2	15.3	15.2	15.6	15.1	15.1	12.5
Lysine (%)	0.70	0.70	0.70	0.70	0.70	0.70	0.70	0.70	0.70	0.70	0.70	0.72
Methionine + cystine (%)	0.51	0.46	0.44	0.50	0.49	0.52	0.49	0.50	0.49	0.47	0.46	0.44
Tryptophan (%)	0.17	0.18	0.19	0.18	0.20	0.18	0.18	0.19	0.15	0.15	0.17	0.13
Calcium (%)	0.75	0.75	0.75	0.75	0.75	0.75	0.75	0.75	0.75	0.75	0.75	0.75
Phosphorus (%)	0.50	0.50	0.50	0.50	0.50	0.50	0.50	0.50	0.52	0.50	0.50	0.50

[a]Values in pouinds per ton.

Some of the advantages and problems involved in feeding silage to sows are as follows.

Advantages to feeding silage to sows

1. Feeding silage may reduce the cost of pigs at farrowing.
2. It keeps sows from getting too fat. This is especially true for sows that wean early, since these sows will not lose as much weight during lactation.
3. High quality silage, properly made, is a close substitute for pasture from a nutritional standpoint.
4. Silage is usually available on farms with beef and dairy cattle. This makes it available for swine at a reasonable cost.
5. It will supply nutritional factors during the winter which might not otherwise be supplied unless the farmer is feeding well-balanced diets.
6. It makes possible the use of greater amounts of grass and legume forage in swine diets.

Problems involved in feeding silage

1. Feeding silage alone, without proper supplementation, will result in poor litters.
2. Feeding silage to average or small herds is hard to justify unless it is already available on the farm for use with other livestock.
3. Handling and feeding silage daily are difficult unless adequate equipment and facilities are available.
4. Silage fed to sows should be of high quality so as to be palatable. Silage made from corn nearing maturity, or matured, is unpalatable. Corn silage is palatable when the kernels are dented and most of the leaves are still green when ensiled. The use of corn or molasses as a preservative will increase the palatability of grass or legume silage.
5. Baby pigs may have digestive troubles if they eat silage. This can be prevented by separating the pigs and sows at feeding time.
6. Moldy silage should never be fed to sows or gilts. Sows evidently have a large digestive system which is not utilized to capacity. They will consume quite a large amount of silage. Gilts will eat 8–12 lb (average 10) and sows 10–15 lb (average 12) of corn silage daily (19). Gilts will eat 5–10 lb (average 7) and sows 7–12 lb (average 10) of grass silage per head per day (19). These figures will vary somewhat depending on the size of the animal, the quality and palatability of the silage, and the remainder of the diet given the animal. They can, however, be used as guides for daily silage consumption.

Tables 14.10 and 14.11 list recommendations of the Iowa Station (19) for supplements and level of supplementation for sows fed corn silage or grass silage during pregestation or gestation. These supplements and recommended levels might be modified somewhat as more information is obtained on feeding silage to sows. They can be used, however, as a guide for silage supplementation.

TABLE 14.10
Level of Supplementation of Silage for Sows during Pregestation and Gestation Recommended by Iowa Station (19)

	Corn Silage							
	Silage (full-fed) made from corn yielding							
	Up to 30 bu/acre			30–60 bu/acre			Over 60 bu/acre	
	Grain and supplement/head/day (lb)							
Grain (shelled corn equivalent)	2 (Ranging from 1 to 3)			1.5 (Ranging from 1 to 2)			1 (Ranging from 0 to 1)	
35% balanced supplement	1.5 (Ranging from 1 to 2)			1.5 (Ranging from 1 to 2)			1.5 (Ranging from 1 to 2)	
	Percent protein in total diet when feeding 1.5 lb 35% balanced supplement with variable amounts of grain							
	(Pounds shelled corn)			(Pounds shelled corn)			(Pounds shelled corn)	
	1	2	3	1	1.5	2	0	1
	(Percent protein)			(Percent protein)			(Percent protein)	
Sows	12.0	13.3	14.4	12.7	13.4	14.0	13.4	14.2
Gilts	13.3	14.8	17.2	14.0	14.9	15.5	14.6	15.6

Grass Silage

	Methods used to make grass silage			
	Wilted silage (preserved with molasses or corn)		Nonwilted silage (preserved with molasses or corn)	
	Grain and supplement/head/day (lb)			
Grain (shelled corn equivalent)	2 (Ranging from 1 to 3)		2.5 (Ranging from 2 to 3)	
35% balanced supplement	0.5 (Ranging from 0.5 to 1)		0.5 (Ranging from 0.5 to 1)	
	Percent protein in total diets when feeding 0.5 lb 35% balanced supplement with variable amounts of grain			
	(Pounds shelled corn)		(Pounds shelled corn)	
	1.5	2.0	2.0	3.0
	(Percent protein)		(Percent protein)	
Sows	17.0	16.2	15.3	14.0
Gilts	16.6	15.7	15.4	14.0

TABLE 14.11
Suggested Supplements Recommended by Iowa Station for Silage Fed to Sows (19)

	Pounds per head daily		
	1.5 (1 to 2)[a]	0.5 (0.5 to 1)[a]	1.0 (0.75 to 1.5)[a]
Ingredients	No. 1 Corn silage supplement[b]	No. 2 Grass silage supplement[c]	No. 3 General silage supplement[d]
50% meat and bone scraps	500	700	600
44% soybean oil meal	865	800	820
17% dehydrated alfalfa meal	300	—	200
25% dried distillers' solubles	100	—	—
Steamed bonemeal	180	400	300
Iodized salt	40	80	60
Trace mineral premix	15	25	20
	2000	2005	2000[e]
Vitamin-antibiotic premix			
Vitamin D_2	5 million IU	10 million IU	8 million IU
Riboflavin	9 gm	—	6 gm
Calcium pantothenate	30 gm	—	20 gm
Niacin	50 gm	—	30 gm
Choline chloride	50 gm	—	25 gm
Vitamin B_{12}	60 mg	120 mg	90 mg
Antibiotic(s)	60 gm	120 gm	90 gm

[a]Range of supplement per head daily.

[b]Corn silage needs supplementation primarily with protein, minerals, and B vitamins.

[c]Grass silage needs supplementation mainly with phosphorus and the right kind of protein to balance the amino acids.

[d]Supplements No. 1 and No. 2 are for corn or grass silage. The general silage supplement is suggested when it is desirable to use only one supplement for both corn or grass silage.

[e]Values in pounds.

The Purdue Station (20,21) obtained good results using corn silage or grass silage for sows. The silage-fed sows averaged 0.3–1.3 more pigs per litter at 8 weeks than those fed the control diet without silage. Feed costs were also lowered with silage feeding. They also showed that ground corn and minerals alone are not a satisfactory supplement for grass silage for sows. Additional protein and vitamins are also needed. A high quality mixed-protein supplement, including vitamins and minerals, is needed to balance the deficiencies of corn silage for sows.

There is virtually no information available on feeding silage to sows during lactation. Some swine authorities feel it can be used during lactation, but in much smaller amounts than during gestation, since it is bulky and would limit total energy intake of the sow. If too much silage is fed, it would restrict total feed intake and thus would limit milk production. It might be used in lactation diets as a substitute, or a partial substitute, for pasture or alfalfa meal ordinarily included in the diet.

Until experimental information is obtained, there is no definite recommendation which can be made on levels of silage to use in lactating diets for sows. Those who wish to try, however, should start with low levels and gradually increase them without reducing milk production adversely. In some cases, it might be economical to feed a certain amount of silage, even though it may reduce weaning weights to a certain extent. Silage would certainly have a place in sow diets after they have weaned their pigs. It could serve as a large part of the diet, depending on the weight gains desired. It could especially serve as a large part of the diet where maintenance and only small gains are desired.

XII. PROBLEMS IN FEEDING SOWS IN CONFINEMENT

A. Introduction

A great deal remains to be learned about managing sows in complete confinement (Fig. 14.2), especially if sows are to farrow four or more litters. A few farmers are doing a satisfactory job in complete confinement, but they are the exception, not the general rule. A top swine producer using high quality pasture can wean ten pigs per litter. Therefore, this should be the goal of swine producers in confinement. Until swine producers can wean ten pigs per litter in complete confinement, there are problems to be solved. Most of the top producers are now weaning 8–8.6 pigs per litter in confinement away from pasture and dirt. They are also experiencing reduced conception rate, infantile reproductive tracts, more services per conception, increased sterility, decreased litter size, poor feet and legs, as well as other problems. Reproduction appears satisfactory when sows are

Fig. 14.2. An example of a complete confinement swine operation. Note large number of pigs in a small area. (Courtesy of T. J. Cunha and University of Florida.)

first moved from pasture into confinement, but reproductive problems usually start to occur after the first or second litter.

Complete confinement of sows in sophisticated and highly mechanized sow units will be the trend in the future. Sows will eventually do as well there as on pasture. But before this occurs, there is much to learn on nutrition, housing, management, breeding, air circulation, waste disposal, and other factors needed for a successful and profitable swine operation. Moreover, considerable emphasis will need to be placed on breeding and selecting animals which will do well in complete confinement. Increasing productivity of the sow is one of the real challenges ahead and one which can greatly increase profitability. If it takes six pigs to break even in a sow operation, every extra pig weaned will add a great deal to the profit obtained.

B. Problems in Complete Confinement

1. Gilts which are kept in individual pens, or are tethered in confinement (Fig. 14.3), show less incidence of estrus and greater failure to mate. Some of them have infantile reproductive tracts. Sexual development with these gilts may also be delayed. Housing gilts in groups (5–7 gilts) rather than individually helps in this regard. Having boars nearby also helps. This means that some animals fail to exhibit normal estrus and mating behavior in complete confinement. Therefore, confinement systems require a more experienced and higher level of management skill than does a pasture system.

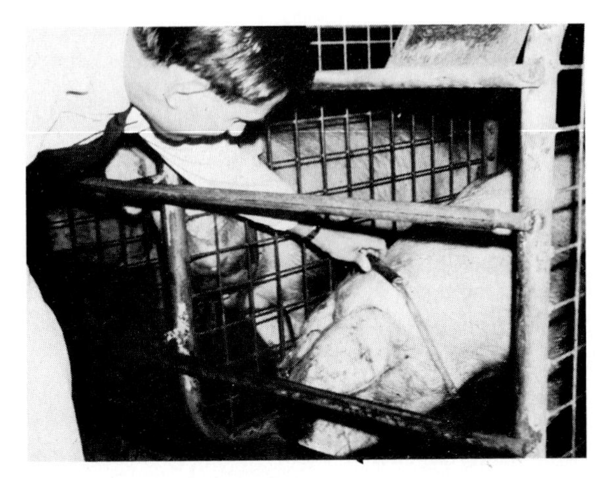

Fig. 14.3. An example of a tethered sow. A neck chain or strap is used to keep them tethered. (Courtesy of T. J. Cunha and University of Florida.)

2. Lameness and other feet and leg problems are increased in complete confinement. Pen conditions, space allocation, exercise, mineral level and other factors may be involved. Some think that breeding and selection against poor feet and legs can eliminate much of this problem. Others have helped it by having a sandbox (Fig. 14.4) in the pen or by having part of the pen with a sand foundation (which also helps keep the sows cool).

Fig. 14.4. An example of a sow with a sandbox which occupies about one-fourth of the pen. This helps with feet and legs at the University of Hawaii. (Courtesy of T. J. Cunha and University of Florida.)

3. Complete confinement houses need to be kept cool during the hot weather. Otherwise, the high temperature will delay estrus, decrease conception rate and increase embryo deaths. Therefore, proper air circulation or zone air conditioning to keep the sows cool is needed for good reproduction from sows bred during the hot months. Studies have shown that one or two less pigs per litter are born with sows kept under high temperature conditions during the breeding season.

4. Some boars show reduced libido when confined or when they have to breed sows on slotted floors or in tight quarters. In the most successful complete confinement systems, the boars are located near the sows in the same barn and the sows are bred twice during a 24-hour period. This increases conception rate and litter size.

5. Endocrine imbalances may occur. It is possible that total confinement may affect the production of certain hormones which are involved with some of the reproduction problems being encountered.

6. More problems are occurring with metritis, mastitis, and agalactia (MMA). Damp floors and other pen conditions may be involved. It is also possible that nutrition or endocrine balance may be involved in these problems.

7. Toxic or harmful levels of obnoxious gases such as ammonia, carbon dioxide and hydrogen sulfide are found in many confinement houses. Respiratory diseases are more prevalent in confinement and respiratory infections can spread quickly when ventilation is poor. Therefore, proper ventilation is very important in confinement housing.

8. Floor surface is very important. It should not be too slippery which predisposes pigs to injury from slipping. Nor should it be too rough because this causes tenderness or abrasions on the feet and legs which in turn causes infections to occur via the bruises. These infections usually cause enlarged legs or joints rather than poor bone structure.

9. Age and social order must be considered in handling sows in groups. This will help minimize injury due to competitive fighting for feed or to establish social rank in the pen. Frequent regrouping of different sows will accentuate this problem. There are indications that regrouping sows shortly after breeding may increase embryo mortality. It is recommended that gilts selected for use in continuous confinement have a gentle disposition. This will minimize the problem of fighting which is a serious one in certain swine operations.

10. Much needs to be learned on designing housing and a total management and feeding program that pleases the animal. Animals have likes and dislikes just as humans do. But not much attention has been given to how the animal would like to be housed and managed. Rather, "least cost" has been used as the main criteria in housing and managing them. Until animal comfort is maximized, their performance will be suboptimal.

C. Suggestions to Increase Litter Size in Confinement

1. Raise herd replacement animals during their growth period (from 75 lb to breeding time) on pasture and/or dirt in order to ensure the development of a more normal reproductive tract that would result in larger litter size. This procedure would also lessen feet and leg problems. Breed the gilts on pasture or dirt for the first time and then bring them into confinement. Once gilts are bred for the first time outside, it is much easier to get them to breed inside in complete confinement for the second litter, and, thus, conception rate greatly increases.

2. If possible, the sows should be taken out to clean dirt lots (if pasture is not available) for at least the middle 2 months of the gestation period. This helps feet and leg problems and gives sows exercise. It is not known if exercise is needed, but it should not hurt the sows. There are some who think exercise is important. This procedure should be followed if sows are kept for 4–8 litters or more.

3. If gilts are to be grown and bred in confinement, then the boars should be kept close to them. The odor, sound and sight of the boar seems to help in getting them to come into estrus earlier, show it more clearly, conceive at a higher rate and decrease the number of gilts with infantile reproductive tracts. More needs to be learned about the "boar influence" and how it can be used to improve reproductive performance in the gilt or sow.

4. Add a source of unidentified factors such as high quality fish meal, fish solubles, distillers' solubles, alfalfa meal, and others to the diet. This is not needed as much if the sows have access to high quality pasture and dirt during a good part of the gestation period in between litters. The need for unidentified factors shows up more under stress, under intensified confined farm conditions, especially in the higher producing herds (which evidently have a higher requirement for them).

5. Use a high quality, properly fortified diet with vitamins, minerals, and protein (amino acids). Pasture covers up many "sins" (or omissions) in feeding. Therefore, as one moves away from pasture, a properly balanced diet becomes more important, especially if one is interested in optimum long-term performance. In Europe and Australia where confinement feeding has been practiced for a long time, it is felt that feeding green chopped forage is very beneficial to sows in confinement. This beneficial effect is obtained with good, practical diets properly fortified with vitamins, protein and minerals.

6. A confined sow uses less energy and thus needs less total feed intake daily. As a result, the diet fed needs to have a higher level of vitamins, minerals and protein per pound in order to supply the sow's total daily needs. More attention should be paid to the amount of total daily nutrient intake of the sow as her total feed intake is decreased. Otherwise, a supposedly well-balanced diet may become deficient in certain nutrients. Sows are feed less during hot weather than in

cold weather. This means warm weather diets need a higher level of nutrients per pound of feed.

REFERENCES

1. Cunha, T. J., O. B. Ross, P. H. Phillips, and G. Bohstedt, *J. Anim. Sci.* **3,** 415 (1944).
2. Cunha, T. J., E. J. Warwick, M. E. Ensminger, and N. K. Hart, *J. Anim. Sci.* **7,** 117 (1948).
3. Krider, J. L., B. W. Fairbanks, R. F. Van Poucke, D. E. Becker, and W. E. Carroll, *J. Anim. Sci.* **5,** 256 (1946).
4. Vestal, C. M., W. M. Beeson, F. N. Andrews, L. M. Hutchings, and L. P. Doyle, *Purdue, Agric. Exp. Stn., Mimeo* No. 34 (1948).
5. Mitchell, H. H., W. E. Carroll, T. S. Hamilton, and G. E. Hunt, *Ill., Agric. Exp. Stn., Bull.* **375** (1931).
6. Vestal, C. M., *Purdue, Agric. Exp. Stn., Bull.* **413** (1938).
7. Hughes, E. H., and G. H. Hart, *J. Nutr.* **9,** 311 (1935).
8. Allen, A. D., and J. F. Lasley, *J. Anim. Sci.* **19,** 150 (1960).
9. Hanson, L. E., E. F. Ferrin, and W. J. Aunan, *Minn., Agric. Exp. Stn.,* **H-131** (1955).
10. Hanson, L. E., L. H. Holt, and E. F. Ferrin, *Minn., Agric. Exp. Stn.,* **H-130** (1955).
11. Peters, W. H., and E. F. Ferrin, *Minn., Agric. Exp. Stn., Mimeo* No. H-88 (1946).
12. Conrad, J. H., and W. M. Beeson, *Purdue, Agric. Exp. Stn., A. H. Mimeo* No. 152 (1955).
13. Holden, P., V. C. Speer, E. J. Stevermer, and D. R. Zimmerman, *Iowa, Agric. Exp. Stn., Rev.* PM-489 (1974).
14. Hoefer, J. A., H. F. Moxley, and R. E. Rust, *Mich., Agric. Ext. Bull.* **335** (1955).
15. Hays, V. W. Kentucky, Agricultural Experiment Station, Lexington, personal communication (1976).
16. Foster, J. R., V. B. Mayrose, T. R. Cline, H. W. Jones, M. P. Plumlee, D. M. Forsythe, and W. H. Blake, *Purdue, Agric. Exp. Stn., Swine Day Rep.* pp. 20–26 (1975).
17. Miller, E. C., E. R. Miller, and D. E. Ullrey, *Mich., Agric. Ext. Bull.* **537** (1975).
18. Cunha, T. J., and S. J. Folks, Florida, Agricultural Experiment Station, Gainesville, unpublished data (1951).
19. Catron, D. V., G. Ashton, V. Speer, C. C. Culbertson, and E. L. Quaife, *Iowa, Agric. Exp. Stn., A. H., Mimeo* No. 680 (1955).
20. Conrad, J. H., and W. M. Beeson, *Purdue, Agric. Exp. Stn., A. H. Mimeo* No. 133 (1954).
21. Conrad, J. H., and W. M. Beeson, *Purdue, Agric. Exp. Stn., A. H. Mimeo* No. 151 (1955).

Appendix

Swine Management Recommendations*

I. SOW AND LITTER

A. Housing and Shelter

1. A farrowing house temperature of 55° to 65°F with adequate ventilation is recommended.
2. Heat lamps placed in a corner accessible only to pigs are recommended when farrowing house temperature falls below 65°F. If a 250-watt heat lamp is used, suspend it approximately 24 inches above bedding. Condition pigs to doing without the lamp by raising or turning it off during warmer periods.
3. Minimum size for farrowing pens in a central farrowing house or individual farrowing houses is 6 by 8 feet for gilts, and 8 by 8 feet for sows. Guard rails 8 inches above bedding and 8 inches from wall are recommended.
4. Farrowing stalls or crates should have widths of 20 inches for gilts and 24 inches for sows, and minimum lengths of 6 feet for gilts and 7 feet for sows. Space beneath bottom board should be one-half stall width. Recommended minimum width on each side of stall or crate for pigs up to 2 weeks of age is 18 inches.
5. Bed farrowing pen or individual farrowing house lightly with chopped or short straw or hay, shavings, ground corncobs, bagasse, peanut hulls, cottonseed hulls, oat hulls, or other suitable bedding material. Use more bedding in unheated houses during cold weather, provided it is short or fine material that will not interfere with movement of pigs.
6. Recommended shade area for summer is 50 square feet per gilt and litter, and 60 square feet per sow and litter.

---

*These were developed by the Nutrition Council of the American Feed Manufacturers Association in cooperation with many scientists in universities, USDA, industry and other agencies and are reproduced from the "AFMA Management Guide." They were provided through the courtesy of Lee H. Boyd, Director, Feed Control and Nutrition of AFMA.

B. Feed and Water

1. A minimum of 1 linear foot† of self-feeder space or 1 self-feeder hole per sow and litter is recommended for self-feeding either in dry lot or on pasture. Provide young pigs with additional creep-feeding space.

2. For hand-feeding in troughs either in dry lot or on pasture, a minimum of 1½ linear feet of feeding space is recommended per sow and litter. Again, young pigs should have additional creep-feeding space.

3. With automatic cup watering, provide at least 1 cup, not less than 6 inches in diameter or equivalent, for each 4 sows and their litters. (Consider an automatic waterer with 2 openings to be 2 cups.) For hand-watering, provide at least 2 linear feet of trough space per sow and litter. Additional watering space may be required during warm weather.

4. Begin creep-feeding the first week. Replace uneaten creep-feed with fresh feed each morning until pigs begin to eat. Number of pigs per linear foot of feeder space should not exceed 5. Edge of the feeder trough should not be more than 4 inches above the floor. Allow a maximum of 40 pigs per creep. Place creep-feeders close to water supply and near area where animals congregate. In cool weather, feeders should be inside a well-lighted area. In warm weather, they may be outside in a covered, protected area.

C. Weaning

1. With proper nutrition and management, pigs may be successfully weaned at 3 to 4 weeks.
2. When weaning at five weeks or earlier, follow these recommendations:

	Age in weeks				
	5	4	3	2	1
Nursery house temperature (°F)	60	65	70	75	75
Minimum square feet of floor space/pig	3	2½	2½	2	2
Maximum number of pigs/ linear foot of feeder space	4	4	4	5	5
Maximum number of pigs/ linear foot of waterer space	10	10	12	12	12
Maximum number of pigs/group	25	20	10	10	10

3. Early-weaned pigs require warm, dry, draft-free housing. Supplemental heat, such as a heat lamp, and special feeders and waterers are recommended.

D. General Management

1. When possible, adjust size of litters to number of functioning teats or nursing ability of the sow. Transfer pigs from sow to sow no later than 3 to 4 days after farrowing. Masking pig odor makes transfer at a later time possible.

---

†Linear foot: One foot of feeding or watering space. For example, a 6-foot self-feeder open on both sides has 12 linear feet of feeder space. The same principle applies to trough space for sows and litter.

2. Clip needle teeth of pigs at birth or first day for large litters, for pigs to be transferred, or when injuries to pigs or to sows' teats are a problem. Clip only tips of these teeth.

3. To prevent anemia in pigs farrowed in houses, a supplemental source of iron must be provided beginning the first week. Some suggested methods are to make clean soil or sod available, apply iron as solution to sow's udder, give individual iron pills, or inject pigs at 3 to 5 days with an ''approved'' injectable iron preparation. This also may be necessary for pigs farrowed on pasture in unfavorable weather.

4. Sows and litters usually can be run together at 2 weeks, although small groups may be put together at 1 week. Age difference between litters should not be more than 1 week in a central farrowing house or 2 weeks on pasture. Group no more than 4 sows with litters in a central farrowing house or 6 on pastures.

5. On good legume or legume-grass pasture, allow 6 to 8 sows or gilts and their litters per acre.

6. Castrate during first 2 weeks. Pigs weaned at 4 weeks or earlier should not be castrated within 1 week of weaning.

II. GROWING-FINISHING SWINE

A. Feeding Management

1. Number of pigs per self-feeder hole should be:

	Dry Lot	Pasture
Under 25 pounds	2	2
25 to 75 pounds	4	4–5
76 pounds to market	5	5–6

2. Percentage of self-feeder space given to protein supplement will depend on palatability of grain, supplement and pasture, but generally should be:

Weaning to 75 pounds	20–25%
76 to 125 pounds	15–20%
126 pounds to market	10–15%

3. Allot 3 self-feeder holes or 3 linear feet of mineral box space for 100 pigs when feeding salt or mineral mixture free-choice.

4. For hand-feeding in troughs or for hand-watering, length of trough per pig should be:

Weaning to 75 pounds	9 inches
76 to 125 pounds	12 inches
126 pounds to market	15 inches

(A 10-foot trough is considered to provide 10 feet of feeder space for growing-finishing swine whether pigs eat from one or both sides.)

5. For complete confinement from weaning to market, provide 15 square feet of floor space per pig if pigs are fed from troughs and 10 square feet of floor space if fed from self-feeders.
6. Provide 1 automatic watering cup for 25 pigs. (Automatic waterer with 2 openings should be considered 2 cups.)
7. Minimum capacity waterer provided for 10 pigs per day should be 25 gallons in summer and 15 gallons in winter.
8. Drinking water temperature should not fall below 35° to 40° during winter.

B. General Management

1. Shelter area provided should be:

	Summer (shade or housing) square feet	Winter (housing) square feet
Weaning to 75 pounds	7	6
76 to 125 pounds	9	8
126 pounds to market	12	10

With slotted floors, these recommendations do not apply.
2. Ring pigs where rooting becomes a problem.
3. Strict sanitation is recommended for control of roundworm infestation. Where such a program is not effective, worm soon after weaning and repeat later if needed.
4. Effective mange and lice treatment is recommended at weaning and whenever needed thereafter.
5. On good legume or legume-grass pasture, allow 20 growing-fattening pigs per acre on full feed and 10 to 15 per acre on limited feeding.
6. Pigs of widely varying weights should not be run together. Range in weight should not exceed 20% above or below average.

III. GROWING AND FINISHING IN CONTROLLED ENVIRONMENT

A. Floor Space

1. Optimum floor space for slotted and partially slotted floors:

	Square feet, per pig	
Pig weights (pounds)	Winter	Summer
25–50	4	4
50–100	5	5
100–150	6	6
150–200 or market	8	9

2. Optimum floor space for solid floors:

	Square feet, per pig	
Pig weights (pounds)	Winter	Summer
25–50	4	5
50–100	6	7
100–150	8	9
150–200 or market	10	12

B. Pigs Per Pen

1. Number of pigs per pen can vary from 20 to 40 depending on weight, ventilation and other factors, with no practical difference resulting from type of floor. NOTE: A higher pig population usually results in cleaner pens or slotted or partially slotted floors.

C. Flooring

1. Concrete has proven most satisfactory for both slotted and solid flooring. It should be smooth or troweled, sloping $\frac{1}{2}$ to $\frac{3}{4}$ inch per foot. Certain types of satisfactory metal slats are being developed. Wood is not recommended because of durability and sanitation problems.

2. Width of slats and slots. From a practical, investment standpoint, 4- to 6-inch slats spaced 1 inch apart will work for pigs weighting 25 to 200 pounds. Place partitions either directly over slots or an inch or so above slats.

D. Pen Shape

1. Solid and partially slotted floors: rectangular with length approximately twice width.
2. Slotted floors: pens should be square or nearly square.

E. Feeding Management

1. Self-feeding is the preferred and most acceptable method of feeding on any types of floor.
2. Number of pigs per feeder hole:

Pig weight (pounds)	Pigs per hole
25–50	4
50–100	4
100–200	5

3. Place feeders away from dunging and watering area.

4. Generally, increased frequency of feeding increases feed intake up to a point. Feed 4 to 6 times daily the amount pigs will readily consume, or about 90 to 95% full feed.

5. Provide 1 automatic waterer for each 25 pigs or $1\frac{1}{2}$ or $2\frac{1}{2}$ gallons daily per pig depending on season and temperature. Maintain water temperature at optimum of 50 to 55°F. Place waterers over slotted areas where possible.

F. Optimum Temperature

Pig weight (pounds)	Suggested temperature (°F)
25–50	70°
50–100	70°
100–150	65°
150–200	65°

G. Optimum Humidity

1. 50 to 75% humidity is satisfactory with lower levels desirable.
2. When temperature and humidity exceed 80, problems can be expected.

H. Ventilation

1. Create some air movement at all times regardless of outside temperature. Mortality has occurred when ventilation failed or was inadequate. Have at least 1 fan operating any time temperature is above 35°F.

2. Provide circulation fans to move air in corners and other areas if ventilation fans do not create uniform temperature and humidity in all areas.

3. Recommended fan capacities at $\frac{1}{8}$-inch static pressure.‡

		Ventilation rate (CFM)		
		Winter		
Age (weeks)	Animal weight (pounds)	Minimum (continuous)	Normal	Summer
Sow and litter				
0–6	2–20	20	80	210
Growing pigs				
6–9	20–40	1¼	10	36
9–13	40–100	1½	12	48
13–18	100–150	2½	15	72
18–23	150–200	3¼	18	100
Gilt, sow, or boar				
20–32	200–250	2	20	120
32–52	250–300	3	25	180
52+	300–500+	4	30	250

‡From Midwest Plan Service, Swine Handbook MWPS-8.

Other recommendations:

a. Provide one square foot of inlet or vent area for each 500 CFM fan capacity.

b. Install inlets which allow incoming fresh, uncontaminated air to be tempered during cold weather. Provide attic ventilation during hot weather.

c. Provide adjustable controls over inlets and outlets to vary the air flow.

d. Do not be misled by using the fan diameter to rate ventilation capacity.

e. Sprinkler systems are frequently used in the corn belt areas. Do not use foggers as they increase humidity with little effect as a cooling agent.

f. Evaporative cooling is frequently used in the Southwest and certain Pacific areas. It is usually satisfactory in dry or arid areas.

g. Mechanical refrigeration is usually practical as a snout cooler for sows and gilts in confinement farrowing systems.

I. Waste Management

1. Nonaerated, underfloor holding tanks are satisfactory provided ventilation and/or removal programs are adequate.
2. Follow state waste disposal regulations when choosing a waste handling method.
3. Pit size should be large enough to hold 3 to 4 months' waste accumulation.
4. Figure waste at 10% of body weight daily, which means 25 to 30 cubic feet of pit per pig should be adequate for 3 to 4 months.
5. With aerobic oxidation, provide enough flow to prevent setttling out and to provide 1 milligram per liter of dissolved oxygen throughout the mixture. To do this use:

a. 1 foot of rotor per 25 hogs.

b. 250 cubic feet of liquid per foot of rotor.

c. A 28-inch rotor at 100 rpm with 6-inch immersion.

J. Noxious Gas Control (See Table A.1)

1. Common noxious gases from waste decomposition are hydrogen sulfide, ammonia, carbon dioxide, and methane. Other noxious gases may be present in small amounts.
2. Cleaning solid floors daily maintains a low ammonia concentration.
3. Waste level in pits should not rise higher than 12 inches from slats.
4. Have sufficient water in pits to keep solids submerged.
5. In pressurized ventilated buildings, vents along side of pits below slats permit gases to escape.
6. Have an emergency generator for ventilation in case power supply fails.
7. Maintain pH of waste as near 7 as possible to prevent organic gas formation which leads to bad odors and noxious gases.
8. If agitation is intermittent, do not start agitation without extra ventilation.
9. Do not enter a freshly cleaned pit without first exhausting carbon dioxide and other gases.

K. Cannibalism

1. Dock pigs' tails at birth.
2. Watch for and remove tail biters.
3. Avoid overcrowding, drafts, cold wet pens and empty feeders.
4. Breed for docile animals or obtain pigs from ancestors with a good behavior record.

TABLE A.1
Properties of Noxious Gases and Their Physiological Effects[a]

Gas	Specific gravity[b]	Odor	Concentration[c] (ppm)	Exposure period[d]	Physiological effects[e]
Ammonia	0.6	Sharp, pungent	4	—	Irritant
(NH_3)			400	—	Irritation of throat
(Colorless)			700	—	Irritation of eyes
			1,700	—	Coughing and frothing
			3,000	30 min	Asphyxiating
			5,000	40 min	Could be fatal
Carbon	1.5	None			Asphyxiant
Dioxide			20,000	—	Safe
(CO_2)			30,000	—	Increased breathing
(Colorless)			40,000	—	Drowsiness, headaches
			60,000	30 min	Heavy, asphyxiating breathing
			300,000	30 min	Could be fatal
Hydrogen	1.2	Rotten eggs			Poison
Sulfide		smell,	100	hours	Irritation of eyes and nose
(H_2S)		nauseating			
(Colorless)			200	60 min	Headaches, dizziness
			500	30 min	Nausea, excitement, insomnia
			1,000	—	Unconsciousness, death
Methane	0.5	None			Asphyxiant
(CH_4)			50,000	—	Headache, nontoxic
(Colorless)					

[a]From Ohio State University Research Foundation Bulletin, "Origin, Identification, Concentration and Control of Noxious Gases in Animal Confinement Production Units." by E. Taiganides and R. K. White, Grant No. SW-00015.

[b]Specific Gravity: The ratio of the weight of pure gas to standard atmospheric air. If number is less than 1 the gas is lighter than air; if greater than 1 it is heavier than air.

[c]Concentration, in parts of the pure gas in million parts of atmospheric air; to change concentration to % by volume, divide the listed numbers by 10,000.

[d]Exposure Period: The time during which effects of noxious gas are felt by an adult human and an animal (especially pig) of about 150 pounds weight.

[e]Physiological Effects: Those found to occur in adult humans; similar effects would be felt by animals weighing 150 pounds; lighter animals will be affected sooner and at lower levels; heavier animals at later times and higher concentrations.

L. Parasites

1. Follow a definite program for control of parasites. Worm and spray growing pigs soon after weaning and repeat as needed thereafter.

M. Feet and Leg Problems

1. Select for strong feet and legs to avoid problems.
2. Follow flooring and housing recommendations.
3. Use litter on solid floors.

N. Loading and Shipping

1. Use a long, narrow, curved or L-shaped alley 20 to 24 inches wide to chute.
2. Use bedding or sand on chute and in truck.
3. Avoid sharp edges on equipment and loose material that could cause injury.
4. Use only hog slapper as necessary. Do not pound and bruise.
5. Fill the truck or partition as necessary, and avoid mixing different lots of hogs.
6. During hot weather, load and ship in the coolest part of the day.
7. Use patience and good sense.

IV. SWINE BREEDING AND GESTATION

A. Feeding Management

1. Hand-feeding sows and gilts is recommended during gestation to more closely check their condition and to attain greater utilization of pasture and other desirable roughages. However, some specially adapted bulk rations can be successfully self-fed. Keep sows and gilts from getting too fat.

2. For self-feeding gilts and sows during gestation, number per linear foot of feeder space, or self-feeder hole, should be as follows: Pasture 3 to 4; dry lot 2 to 30.

3. When hand-feeding in troughs during gestation, or for hand watering, space required is $1\frac{1}{2}$ to 2 linear feet.

4. When feeding alfalfa hay in a rack, provide 1 linear foot of rack space for 4 sows.

5. Let bred sows and gilts glean corn left in the field, provided there is not an excessive amount of corn on the ground and supplement is available.

6. Provide 1 automatic watering cup for each 12 gilts or sows. Additional watering space may be required during warmer weather. (An automatic waterer with 2 openings should be considered 2 cups.)

B. General Management

1. Separate breeding gilts from market herd at 4 to 5 months of age or at 150 to 175 pounds—whichever occurs first. They should have at least 12 well-developed teats.

2. Gilts should be at least 8 months old and weigh near 250 pounds before breeding. Vaccinate gilts and sows before each breeding, in accordance with veterinarian's recommendations.

3. Worm and spray sows and gilts before breeding. Then follow sanitary measures to prevent reinfection.

4. "Flushing" (increasing feed intake) during breeding season is recommended. Increase feed 7 to 10 days before breeding starts and maintain until all sows or gilts are bred. Do not exceed 28 days of increased feed per individual animal.

5. With hand or individual mating, 2 services per sow or gilt are recommended. First mating of gilts should be on the first day of estrus and first mating of sows on the second day of estrus. The second service should follow the first by 24 hours. (Note: When only one mating can be made during estrus period, both gilts and sows should be served on second day of estrus.)

6. When weaning under 3 weeks of age, breed sows on second heat period after weaning. It is generally satisfactory to breed sows on the first heat period following weaning at 3 or more weeks.

7. Keep gilts and sows separated during gestation.

8. Mange and lice treatment is recommended during gestation.

9. Boars should be 8 months old before being used in the breeding herd.

10. Whenever practical, have boars serve several sows or gilts outside the breeding herd prior to serving those in the breeding herd. Accustom boars to environment before breeding season.

11. Do not run boars of different ages, junior and mature, together. However, boars of same age or size can be run together during off-breeding season.

12. Recommend size of exercise lot for holding a boar is $\frac{1}{4}$ acre.

13. Maximum number of services per board should be:

	Per day	Per week	Per month
Mature boar	3	12	40
Junior boar	2	8	25

Mature boar considered to be 15 months or older, Junior boar under 15 months.

14. The use of a breeding crate is recommended when breeding gilts to old boars. It is often desirable to use a breeding crate when mating old sows to young boars.

15. Hand or individual mating is recommended over field-mating. If field-mating is practiced, 2 methods are recommended. One method is to split the sow or gilt herd so as to have 1 boar per group. Another method is to alternate boars in the sow or gilt herd. This method requires 2 boars or sets of boars to be used on alternate days.

16. Postpone breeding if boars and sows are experiencing or recently had a fever from any cause.

17. On good legume or legume-grass pasture, allow 10 to 12 gilts or 8 to 10 sows per acre.

18. Keep boars and sows free of draft in winter and as cool and comfortable as possible in summer. Square feet of housing or shade per animal should be as follows:

	Square feet per animal	
	Winter (housing)	Summer (shade or housing)
Gilt, or Junior boar	15	17
Sow, or Mature boar	18	20

Index

C

E

F

H

M

O

P

R

S

W

X

Y

Z